Perspectives on Noncommutative Geometry

Fields Institute Communications

The Fields Institute for Research in Mathematical Sciences

Perspectives on Noncommutative Geometry

Masoud Khalkhali
Guoliang Yu
Editors

American Mathematical Society
Providence, Rhode Island

The Fields Institute for Research in Mathematical Sciences
Toronto, Ontario

FIELDS

The Fields Institute for Research in Mathematical Sciences

The Fields Institute is a center for mathematical research, located in Toronto, Canada. Our mission is to provide a supportive and stimulating environment for mathematics research, innovation and education. The Institute is supported by the Ontario Ministry of Training, Colleges and Universities, the Natural Sciences and Engineering Research Council of Canada, and seven Ontario universities (Carleton, McMaster, Ottawa, Toronto, Waterloo, Western Ontario, and York). In addition there are several affiliated universities in both Canada and the United States, and five Corporate Affiliate Members (Algorithmics, General Motors, QWeMA Group Inc., R2 Financial Technologies Inc., and Sigma Analysis and Management).

2000 *Mathematics Subject Classification.* Primary 58B34;
Secondary 19D55, 16T05, 18G30.

Library of Congress Cataloging-in-Publication Data

Perspectives on noncommutative geometry / Masoud Khalkhali, Guoliang Yu, editors.
p. cm. — (Fields Institute Communications)
Includes bibliographical references.
ISBN 978-0-8218-4849-4 (alk. paper)
1. KK-theory. 2. Hopf algebras. 3. Algebra, Homological. I. Khalkhali, Masoud, 1956– II. Yu, Guoliang, 1963 Aug. 11–

QA612.33.P47 2011
512′.55—dc23

2011032554

∞ The paper used in this book is acid-free and falls within the guidelines
established to ensure permanence and durability.
This publication was prepared by the Fields Institute.
`http://www.fields.utoronto.ca`
Visit the AMS home page at `http://www.ams.org/`

10 9 8 7 6 5 4 3 2 1   16 15 14 13 12 11

Contents

Preface

During May 27–31, 2008, the *Noncommutative Geometry Workshop* was held at the Fields Institute as part of the thematic program on operator algebras. This was the second major conference on the subject organized at the Fields Institute, the first being in 1995. It was interesting to appreciate how much the subject has progressed between these two events. The present volume consists primarily of articles by speakers in this workshop.

Roughly speaking, noncommutative geometry concerns itself with the study of noncommutative spaces. These 'spaces' are usually represented by a noncommutative algebra that replaces the coordinate algebra in the commutative case. Examples include highly singular spaces such as the space of leaves of a foliation, the unitary dual of a noncompact group, and more generally, 'bad quotients' of classical spaces.

Initiated and pioneered by Alain Connes since 1980, in its initial stage noncommutative geometry was mostly inspired by global analysis, topology, operator algebras, and quantum physics, as they show up in areas such as index theory, foliation theory, and quantum statistical mechanics. Its main applications were to settle some long standing conjectures such as the Novikov conjecture, and the Baum-Connes conjecture in topology and analysis. The main tools here are cyclic cohomology, K-theory and K-homology, and KK-theory. Next came the impact of spectral geometry and the way the spectrum of a geometric operator like the Laplacian informs us about the geometry and topology of a manifold, as in the celebrated Weyl's law. This is now subsumed and vastly generalized through Connes' notion of spectral triples, which is a centerpiece of noncommutative Riemannian geometry and applications of noncommutative geometry to particle physics. Finally, in recent years we have witnessed the impact of number theory, algebraic geometry, and the theory of motives and quantum field theory on noncommutative geometry, and a strong interaction between these areas is gradually emerging. All these aspects of the field were reflected and touched upon in lectures by the invited speakers at the workshop.

During the workshop, Alain Connes delivered his Fields Institute Distinguished Lectures Series. He gave a series of three lectures on the frontiers of research in the subject, with titles *The spectral characterization of manifolds, A CKM invariant in Riemannian geometry*, and *About the field with one element*. The reader can refer to the web page[1] of the Fields Institute for abstracts of these lectures.

We would like to thank Carl Riehm and Debbie Iscoe of Fields Institute Publications for their patience and professional assistance. We also thank all the contributors for their contributions. It is a pleasure to thank George Elliott for supporting the original idea of a workshop on noncommutative geometry during the Fields Institute thematic program on operator algebras. The workshop was financially

[1] http://www.fields.utoronto.ca/programs/scientific/07-08/noncommutative/

supported by the Fields Institute and the NSF, and we would like to thank both institutions for their support. It is a pleasure to thank Arthur Greenspoon for checking the entire manuscript and suggesting many improvements. Finally we would like to warmly thank Matilde Marcolli who coorganized this workshop with us.

Masoud Khalkhali, University of Western Ontario, Canada
Guoliang Yu, Vanderbilt University, USA

Fields Institute Communications
Volume **61**, 2011

Local Index Theorem for Projective Families

Moulay-Tahar Benameur
UMR 7122 du CNRS
Université de Metz
Ile du Saulcy, Metz, France
benameur@univ-metz.fr

Alexander Gorokhovsky
Department of Mathematics
University of Colorado at Boulder
Boulder, CO 80305-0395 USA
Alexander.Gorokhovsky@colorado.edu

Abstract. We give a superconnection proof of the Mathai-Melrose-Singer index theorem for the family of twisted Dirac operators [28, 29].

1 Introduction

Let $\pi : M \to B$ be a smooth fibration. Given a class $\delta \in H^3(B, \mathbb{Z})$, Mathai, Melrose and Singer defined an algebra of families of pseudodifferential operators twisted by this class (also called projective). In order to give this definition, in [28] it is assumed that δ is a torsion class. In [29] the corresponding assumption is that $\delta = \alpha \cup \beta$, $\alpha \in H^1(B, \mathbb{Z})$, $\beta \in H^2(B, \mathbb{Z})$ and $\pi^*\beta = 0$.

There is a notion of ellipticity for such a twisted pseudodifferential family and for such an elliptic family one can then define its index as an element of the K-theory of the algebra of smoothing operators. Mathai, Melrose and Singer then prove in [28, 29] an index theorem for such elliptic families, thus giving an extension of the Atiyah-Singer family index theorem [1] to the twisted case. In this paper we give a superconnection proof of the cohomological version of this index theorem for a projective family of Dirac operators. We assume that we are given a gerbe $\mathcal{L}$ on B with a class $\delta = [\mathcal{L}] \in H^3(B, \mathbb{Z})$. Our conditions on this class are somewhat weaker than in [28, 29]; namely we assume that $\pi^*\delta$ is a torsion element in $H^3(M, \mathbb{Z})$. Assume that we are given a $\mathbb{Z}_2$-graded $\mathcal{L}$ -twisted Clifford module $\mathcal{E}$ on M as defined in Section 4.1. We note that such Clifford modules always exist under our assumptions on the class of the gerbe $\mathcal{L}$. We define the algebra $\Psi_{\mathcal{L}}(M|B; \mathcal{E})$ of projective

2010 *Mathematics Subject Classification.* Primary 58J20; Secondary 19K56, 19L50, 46L80, 53C08.

Key words and phrases. Gerbe, Twisted K-theory, Family index, Superconnection.
The second author was partially supported by the NSF grant DMS-0900968.

families of pseudodifferential operators on $\mathcal{E}$. One can then define the index of (the positive part of) the twisted Dirac operator $\operatorname{ind} D^+ \in K_0(\Psi_{\mathcal{L}}^{-\infty}(M|B;\mathcal{E}))$. Here, by $\Psi_{\mathcal{L}}^{-\infty}(M|B;\mathcal{E})$ we denote the algebra of smoothing $\mathcal{L}$-twisted pseudodifferential operators on $\mathcal{E}$.

In order to define the Chern character of the index we use cyclic homology and the map constructed in [30]. This Chern character takes values in the twisted cohomology of B defined as follows. Let $\Omega \in \Omega^3(B)$ be a form representing the class $[\mathcal{L}]$ and let u be a formal variable of degree -2. Twisted cohomology $H^\bullet_{\mathcal{L}}(B)$ is then the homology of the complex $\Omega^*(B)[u]$ with the differential $d_\Omega = ud + u^2\Omega \wedge \cdot$. Note that the form Ω and thus the complex depend on the choice of connection on $\mathcal{L}$. Nevertheless the homologies of all the complexes thus obtained are canonically isomorphic. Following Mathai and Stevenson [30] one defines a morphism of complexes $\Phi_{\nabla^{\mathcal{H}}} \colon CC^-_\bullet(\Psi_{\mathcal{L}}^{-\infty}(M|B;\mathcal{E})) \to (\Omega^*(B)[u], d_\Omega)$. Here $CC^-_\bullet$ denotes the negative cyclic complex. By composing $\Phi_{\nabla^{\mathcal{H}}}$ with the Chern character $\operatorname{ch} \colon K_0(\Psi_{\mathcal{L}}^{-\infty}(M|B;\mathcal{E})) \to HC^-_0(\Psi_{\mathcal{L}}^{-\infty}(M|B;\mathcal{E}))$ we obtain the class $[\Phi_{\nabla^{\mathcal{H}}}(\operatorname{ch}(\operatorname{ind} D^+)) \in H^\bullet_{\mathcal{L}}(B)]$. The main result of the paper is the proof of the following theorem expressing the Chern character of the index in terms of characteristic classes:

Theorem 1.1 *Let D be a projective family of Dirac operators on a horizontally $\mathcal{L}$-twisted Clifford module $\mathcal{E}$ on a fibration $\pi : M \to B$. Then the following equality holds in $H^\bullet_{\mathcal{L}}(B)$:*

$$\left[\Phi_{\nabla^{\mathcal{H}}}(\operatorname{ch}(\operatorname{ind} D^+))\right] = \left[u^{-\frac{k}{2}} \int_{M|B} \widehat{A}\left(\frac{u}{2\pi i} R^{M|B}\right) \operatorname{Ch}_{\mathcal{L}}(\mathcal{E}/\mathcal{S})\right],$$

where $k = \dim M - \dim B$ is the dimension of the fibers.

Here, as usual, $\widehat{A}$ is the power series defined by $\widehat{A}(x) = \det^{1/2}\left(\frac{x/2}{\sinh x/2}\right)$ and $\operatorname{Ch}_{\mathcal{L}}(\mathcal{E}/\mathcal{S})$ is the twisted relative Chern character form of $\mathcal{E}$; see Proposition 2.13 and equation (4.1).

The characteristic classes appearing in the right hand side are defined in the Section 4.1.

Our proof uses the Bismut superconnection formalism [5]. We extend the notion of superconnection to the twisted context and show in Theorem 4.6 that the Chern character of the superconnection adapted to D computes the Chern character of $\operatorname{ind} D^+$. Then in Theorem 4.8, using the results of [5] and [6] we compute the limit of the Chern character of the rescaled superconnection obtaining the expression in the right hand side of the index formula. Together these results imply Theorem 1.1.

The paper is organized as follows. In Section 2 we give a brief review of cyclic homology, gerbes and connections on them. We also discuss twisted bundles and their characteristic forms. In Section 3 we give the definitions of the algebra of twisted families of pseudodifferential operators. Finally, in Section 4 we define the projective family of Dirac operators on a twisted Clifford module and give the proofs of the main theorems of the paper.

Some work in related direction recently appeared in [13, 12, 14]. Related questions of deformation theory are considered in [9, 10]. This paper is a byproduct of a joint project with E. Leichtnam. The authors would like to thank him and C. Blanchet, M. Karoubi, M. Lesch, V. Mathai, R. Nest and B. Wang for helpful discussions. The authors worked on this paper while visiting CIRM, Luminy as well as the Hausdorff Institute and Max Planck Institute for Mathematics in Bonn. Part of this work was done while the second author was visiting the Laboratoire de Mathématiques et Applications of Metz and he is grateful for their hospitality.

2 Preliminaries

2.1 Cyclic homology. The general reference for this material is the book [27].

Let A be a complex unital algebra. Set $C_k(A) = A \otimes (A/\mathbb{C}1)^{\otimes k}$. Let u be a formal variable of degree -2. The space of negative cyclic chains of degree $l \in \mathbb{Z}$ is defined by

$$CC_l^-(A) = (C_\bullet(A)[[u]])_l = \prod_{-2n+k=l,\ n\geq 0} u^n C_k(A).$$

The boundary is given by $b + uB$, where b and B are the Hochschild and Connes boundaries of the cyclic complex. The homology of this complex is denoted $HC_\bullet^-(A)$. When the algebra A is $\mathbb{Z}_2$ graded the relevant signs are incorporated. If A is not necessarily unital denote by A^+ its unitalisation and set $CC_l^-(A) = CC_l^-(A^+)/CC_l^-(\mathbb{C})$. If I is an ideal in a unital algebra A the relative cyclic complex is defined by $CC_\bullet^-(A, I) = \operatorname{Ker}(CC_\bullet^-(A) \to CC_\bullet^-(A/I))$. One has a natural morphism of complexes $\iota\colon CC_\bullet^-(I) \to CC_\bullet^-(A, I)$ induced by the homomorphism $I^+ \to A$.

Recall that for an algebra A we have the Chern character in cyclic homology $\operatorname{ch}\colon K_0(A) \to HC_0^-(A)$. It is defined by the following formula. Let $P, Q \in M_n(A^+)$ be two idempotents in $n \times n$ matrices of the algebra A^+, representing a class $[P - Q] \in K_0(A)$. Then

$$\begin{aligned}\operatorname{Ch}([P-Q]) &= \operatorname{tr}(P-Q) + \sum_{n=1}^{\infty} (-u)^n \frac{(2n)!}{n!} \\ &\quad \times \operatorname{tr}\left(\left(P - \frac{1}{2}\right) \otimes P^{\otimes(2n)} - \left(Q - \frac{1}{2}\right) \otimes Q^{\otimes(2n)}\right) \qquad (2.1)\end{aligned}$$

We will use the notation $\operatorname{Ch}([P-Q])$ for the cyclic cycle defined above and $\operatorname{ch}([P-Q])$ for its class in cyclic homology $HC_0^-(A)$.

We will also need to use the entire cyclic complex. For our purpose the algebraic version from [17], IV.7.α Remark 7 b. will be sufficient. First recall that one has the periodic cyclic complex $(CC_\bullet^{per}(A), b + uB)$, where $CC_\bullet^{per}(A) = C_\bullet(A)[u^{-1}, u]]$. Assume we are given a periodic chain $\alpha = \sum_{k\geq 0} \alpha_k u^k \in CC_m^{per}(A)$, $\alpha_k \in C_{2k+m}(A)$. Then α is called entire if there exist a finite dimensional subspace $V \subset A$, $1 \in V$ and $C > 0$ (depending on α) such that $\alpha_k \in V \otimes (V/\mathbb{C}1)^{\otimes k}$ and $\|\alpha_k\| \leq C^k k!$. Here the norms on $V \otimes (V/\mathbb{C}1)^{\otimes k}$ are induced by an arbitrary norm on V. We denote the entire cyclic complex of A by $CC_\bullet^{entire}(A)$.

Note that the chain $\operatorname{Ch}([P-Q])$ defined in (2.1) is an element in $CC_0^{entire}(A)$.

2.2 Notion of a gerbe. We give here only the general overview, referring the reader to [11] and [24] for the details. The differential geometry of not necessarily abelian gerbes is described in [8]. We will describe gerbes in terms of their descent data.

Let M be a smooth manifold. Given an open cover $(U_\alpha)_{\alpha\in\Lambda}$ of M, we set as usual

$$U_{\alpha_1\cdots\alpha_k} = \bigcap_{1\leq j\leq k} U_{\alpha_j}.$$

Definition 2.1 A descent datum for a gerbe $\mathcal{L}$ on M is the collection $(U_\alpha, \mathcal{L}_{\alpha\beta}, \mu_{\alpha\beta\gamma})$, where $(U_\alpha)_{\alpha\in\Lambda}$ is an open cover of M, $(\mathcal{L}_{\alpha\beta} \to U_{\alpha\beta})_{\alpha,\beta\in\Lambda}$ is a collection of line bundles and $\mu_{\alpha\beta\gamma} : \mathcal{L}_{\alpha\beta} \otimes \mathcal{L}_{\beta\gamma} \to \mathcal{L}_{\alpha\gamma}$ are line bundle isomorphisms over each triple intersection $U_{\alpha\beta\gamma}$ such that over each quadruple intersection $U_{\alpha\beta\gamma\delta}$, the following diagram commutes

$$\begin{array}{ccc}
\mathcal{L}_{\alpha\beta}\otimes\mathcal{L}_{\beta\gamma}\otimes\mathcal{L}_{\gamma\delta} & \xrightarrow{\mu_{\alpha\beta\gamma}\otimes\mathrm{id}} & \mathcal{L}_{\alpha\gamma}\otimes\mathcal{L}_{\gamma\delta} \\
{\scriptstyle \mathrm{id}\otimes\mu_{\beta\gamma\delta}}\downarrow & & \downarrow{\scriptstyle \mu_{\alpha\gamma\delta}} \\
\mathcal{L}_{\alpha\beta}\otimes\mathcal{L}_{\beta\delta} & \xrightarrow{\mu_{\alpha\beta\delta}} & \mathcal{L}_{\alpha\delta}
\end{array}$$

Notice that we don't assume in this definition that the open sets $(U_\alpha)_{\alpha\in\Lambda}$ are contractible.

Given two descent data $(U_\alpha, \mathcal{L}_{\alpha\beta}, \mu_{\alpha\beta\gamma})$ and $(U_\alpha, \mathcal{L}'_{\alpha\beta}, \mu'_{\alpha\beta\gamma})$ on the same open cover $\{U_\alpha\}$, an isomorphism between them is given by line bundles S_α on U_α and isomorphisms of line bundles $\lambda_{\alpha\beta}\colon S_\alpha^{-1}\otimes\mathcal{L}_{\alpha\beta}\otimes S_\beta \to \mathcal{L}'_{\alpha\beta}$ over $U_{\alpha\beta}$ so that the diagram

$$\begin{array}{ccc}
S_\alpha^{-1}\otimes\mathcal{L}_{\alpha\beta}\otimes S_\beta\otimes S_\beta^{-1}\otimes\mathcal{L}_{\beta\gamma}\otimes S_\gamma & \xrightarrow{\mathrm{id}\otimes\mu_{\alpha\beta\gamma}\otimes\mathrm{id}} & S_\alpha^{-1}\otimes\mathcal{L}_{\alpha\gamma}\otimes S_\gamma \\
{\scriptstyle \lambda_{\alpha\beta}\otimes\lambda_{\beta\gamma}}\downarrow & & \downarrow{\scriptstyle \lambda_{\alpha\gamma}} \\
\mathcal{L}'_{\alpha\beta}\otimes\mathcal{L}'_{\beta\gamma} & \xrightarrow{\mu'_{\alpha\beta\gamma}} & \mathcal{L}'_{\alpha\gamma}
\end{array}$$

commutes.

Given two isomorphisms $(S_\alpha, \lambda_{\alpha\beta})$ and $(S'_\alpha, \lambda'_{\alpha\beta})$ between $(U_\alpha, \mathcal{L}_{\alpha\beta}, \mu_{\alpha\beta\gamma})$ and $(U_\alpha, \mathcal{L}'_{\alpha\beta}, \mu'_{\alpha\beta\gamma})$, a two-morphism between them is a collection of line bundle isomorphisms $\nu_\alpha\colon S_\alpha \to S'_\alpha$ such that

$$\lambda'_{\alpha\beta}\circ(\nu_\alpha^{-1}\otimes\mathrm{id}\otimes\nu_\beta) = \lambda_{\alpha\beta}$$

where we denote by ν_α^{-1} the isomorphism $S_\alpha^{-1}\to(S'_\alpha)^{-1}$ induced by ν_α.

Let $(V_i, \varrho)_{i\in I}$ be a refinement of the open cover of $(U_\alpha)_{\alpha\in\Lambda}$ of M. So $(V_i)_{i\in I}$ is an open cover of M and

$$\varrho : I \to \Lambda \quad \text{such that} \quad V_i \subset U_{\varrho(i)}.$$

Then, restriction to the refinement of $(U_\alpha, \mathcal{L}_{\alpha\beta}, \mu_{\alpha\beta\gamma})$ is the descent datum $\mathcal{L}' = (V_i, \mathcal{L}'_{ij}, \mu'_{ijk})$ given by

$$\mathcal{L}'_{ij} := \mathcal{L}_{\varrho(i)\varrho(j)}|_{V_{ij}} \text{ and } \mu'_{ijk} := \mu_{\varrho(i)\varrho(j)\varrho(k)}|_{V_{ijk}}.$$

Similarly one defines restriction of the isomorphisms and 2-morphisms to a refinement. We do not distinguish between a descent datum, isomorphisms of descent data, etc., and their restrictions to a refinement. Thus for instance, the isomorphism between two descent data $(U_\alpha, \mathcal{L}_{\alpha\beta}, \mu_{\alpha\beta\gamma})$ and $(U'_\alpha, \mathcal{L}'_{\alpha\beta}, \mu'_{\alpha\beta\gamma})$ is an isomorphism between their restrictions to some common refinement of $\{U_\alpha\}$ and $\{U'_\alpha\}$.

An equivalence class of Dixmier-Douady gerbes on M is an equivalence class of the descent data on M. More precisely, a gerbe is a maximal collection of descent data $\mathcal{D}_i$, $i\in A$ together with the isomorphisms $s_{ij}\colon \mathcal{D}_j\to\mathcal{D}_i$ for each $i, j\in A$ and

2-morphisms $\nu_{ijk}\colon s_{ij}s_{jk} \to s_{ik}$ satisfying the natural associativity condition. We refer the reader to the book [11] for the details.

If the cover U_α is good [7] all the bundles $\mathcal{L}_{\alpha\beta}$ are trivializable. After a choice of such trivialization the collection $(\mu_{\alpha\beta\gamma})$ can be viewed as a Čech 2-cochain with coefficents in the sheaf $\underline{\mathbb{C}^*}$ of smooth functions with values in the nonzero complex numbers $\mathbb{C}^*$. The compatibility condition over $U_{\alpha\beta\gamma\delta}$ tells us that μ is a 2-cocycle and hence defines a cohomology class $[\mu] \in H^2(M; \underline{\mathbb{C}^*}) \cong H^3(M, \mathbb{Z})$. This class is a well defined invariant of the gerbe called the Dixmier-Douady class. We denote this class by $[\mathcal{L}]$. Every class in $H^3(M, \mathbb{Z})$ is a class of a gerbe defined by this class uniquely up to isomorphism (see [11]).

Given a smooth map $f : M' \to M$ between smooth manifolds M' and M, we can pull back any decent datum for a gerbe on M to a descent datum on M'. The pullbacks of isomorphic descent data are isomorphic and thus we obtain a well-defined pullback of a gerbe. Clearly the Dixmier-Douady class of the pullback is the pullback of the Dixmier-Douady class.

A unitary descent datum is $(U_\alpha, \mathcal{L}_{\alpha\beta}, \mu_{\alpha\beta\gamma})$ together with a choice of metric on each $\mathcal{L}_{\alpha\beta}$ such that each $\mu_{\alpha\beta\gamma}$ is an isometry. A notion of unitary equivalence of two unitary descent data on the same open cover U_α is obtained from the notion of equivalence above by requiring that each line bundle S_α is Hermitian and each $\lambda_{\alpha\beta}$ is an isometry. The definition of 2-morphisms is modified by requiring each ν_α to be an isometry. It is clear that the restriction of a unitary descent datum to a refinement is again unitary. Then a unitary gerbe is an equivalence class of unitary descent data in the sense described above.

2.3 Connections on gerbes.

Lemma 2.2 *Let $(U_\alpha, \mathcal{L}_{\alpha\beta}, \mu_{\alpha\beta\gamma})$ be a descent datum on M. There exists a collection $(\nabla_{\alpha\beta})$ of connections on $(\mathcal{L}_{\alpha\beta})$ such that for any $(\alpha, \beta, \gamma) \in \Lambda^3$ with $U_{\alpha\beta\gamma} \neq \emptyset$:*

$$\mu^*_{\alpha\beta\gamma}\nabla_{\alpha\gamma} = \nabla_{\alpha\beta} \otimes \mathrm{id} + \mathrm{id} \otimes \nabla_{\beta\gamma}.$$

If the descent datum is unitary each $\nabla_{\alpha\beta}$ can be chosen Hermitian.

Proof Fix for any α, β with $U_{\alpha\beta} \neq \emptyset$ a connection $\nabla'_{\alpha\beta}$ on $\mathcal{L}_{\alpha\beta}$. We set for $U_{\alpha\beta\gamma} \neq \emptyset$,

$$A_{\alpha\beta\gamma} := \mu^*_{\alpha\beta\gamma}\nabla_{\alpha\gamma} - \nabla_{\alpha\beta} \otimes \mathrm{id} - \mathrm{id} \otimes \nabla_{\beta\gamma}.$$

Then, using the identification

$$\mathrm{End}(\mathcal{L}_{\alpha\beta} \otimes \mathcal{L}_{\beta\gamma}) \simeq U_{\alpha\beta\gamma} \times \mathbb{C},$$

we see that $A_{\alpha\beta\gamma}$ is identified with a differential 1-form on the open set $U_{\alpha\beta\gamma}$. We also have for $U_{\alpha\beta\gamma\delta} \neq \emptyset$

$$A_{\beta\gamma\delta} + A_{\alpha\beta\delta} = A_{\alpha\gamma\delta} + A_{\alpha\beta\gamma}.$$

Therefore, there exists $A' = (A'_{\alpha\beta})$ such that

$$A_{\alpha\beta\gamma} = A'_{\beta\gamma} - A'_{\alpha\gamma} + A'_{\alpha\beta}.$$

The collection of connections $(\nabla_{\alpha\beta} = \nabla'_{\alpha\beta} + A'_{\alpha\beta})$ is then a connection on the gerbe $\mathcal{L}$. □

Lemma 2.3 *Let $(\nabla_{\alpha\beta})$ be as above, and denote by $\omega_{\alpha\beta} = \nabla^2_{\alpha\beta}$ the curvatures of the connections $\nabla_{\alpha\beta}$. Then there exists a collection of differential 2-forms $\omega_\alpha \in \Omega^2(U_\alpha)$ such that*

$$\omega_{\alpha\beta} = \omega_\alpha - \omega_\beta, \qquad \text{for } U_{\alpha\beta} \neq \emptyset.$$

Proof The collection $\omega_{\alpha\beta}$ satisfies

$$\omega_{\alpha\gamma} = \omega_{\alpha\beta} + \omega_{\beta\gamma}.$$

This shows that $(\omega_{\alpha\beta})$ is a Čech cocycle and hence by the acyclicity of the Čech complex of forms, there exist (ω_α) such that

$$\omega_{\alpha\beta} = \omega_\alpha - \omega_\beta.$$

□

We will say that the collection $(\nabla_{\alpha\beta}, \omega_\alpha)$ is a connection on the descent datum $(U_\alpha, \mathcal{L}_{\alpha\beta}, \mu_{\alpha\beta})$.

A connection on descent datum yields a connection on its restriction to a refinement in an obvious manner; we will identify a connection with its restriction. An isomorphism between a descent datum $(U_\alpha, \mathcal{L}_{\alpha\beta}, \mu_{\alpha\beta})$ with connection $(\nabla_{\alpha\beta}, \omega_\alpha)$ and descent datum $(U_\alpha, \mathcal{L}'_{\alpha\beta}, \mu'_{\alpha\beta})$ with connection $(\nabla'_{\alpha\beta}, \omega'_\alpha)$ is given by $(S_\alpha, \lambda_{\alpha\beta}, \nabla_\alpha)$, where $(S_\alpha, \lambda_{\alpha\beta})$ is an isomorphsm between the descent data (without connections) and each ∇_α is a connection on S_α satsfying the following conditions. Let $\pi_\alpha = \nabla^2_\alpha$ be the curvatures of these connections and let ∇^{-1}_α be the dual connection on S^{-1}_α. Then we require the following identities (using the isomorphism $\mathcal{L}_{\alpha\beta} = S_\alpha \otimes (S^{-1}_\alpha \otimes \mathcal{L}_{\alpha\beta} \otimes S_\beta) \otimes S^{-1}_\beta$):

$$\nabla_{\alpha\beta} = (\lambda_{\alpha\beta})^* (\nabla_\alpha \otimes \mathrm{id} \otimes \mathrm{id} + \mathrm{id} \otimes \nabla'_{\alpha\beta} \otimes \mathrm{id} + \mathrm{id} \otimes \mathrm{id} \otimes \nabla^{-1}_\beta) \tag{2.2}$$

and

$$\omega_\alpha = \omega'_\alpha + \pi_\alpha. \tag{2.3}$$

If $s = (S_\alpha, \lambda_{\alpha\beta}, \nabla_\alpha)$ and $s' = (S'_\alpha, \lambda'_{\alpha\beta}, \nabla'_\alpha)$ are two isomorphisms between the descent data with connections as above, the 2-morphisms between s and s' are the same as the 2-morphisms between $(S_\alpha, \lambda_{\alpha\beta})$ and $(S'_\alpha, \lambda'_{\alpha\beta})$.

Assume $\mathcal{L}$ is a gerbe given by a collection of descent data $\mathcal{D}_i$, $i \in A$, isomorphisms $s_{ij} \colon \mathcal{D}_j \to \mathcal{D}_i$ for each $i, j \in A$ and 2-morphisms $\nu_{ijk} \colon s_{ij} s_{jk} \to s_{ik}$. A connection on a gerbe $\mathcal{L}$ is a lift of each $\mathcal{D}_i$ to a descent datum with connection $\widetilde{\mathcal{D}_i}$ and each s_{ij} to an isomorphsm $\widetilde{s_{ij}} \colon \widetilde{\mathcal{D}_j} \to \widetilde{\mathcal{D}_i}$. An easy argument using Lemma 2.2 shows that on a given gerbe there always exists a connection.

Lemma 2.4 *Let $\mathcal{L}$ be a gerbe represented by a descent datum $(U_\alpha, \mathcal{L}_{\alpha\beta}, \mu_{\alpha\beta\gamma})$. Choose a connection on $\mathcal{L}$ represented by a connection $(\nabla_{\alpha\beta}, \omega_\alpha)$ on the descent datum.*

1. *There exists a well-defined closed form $\varOmega \in \Omega^3(M)$ – the curvature 3-form of the connection – such that $\varOmega|_{U_\alpha} = \frac{d\omega_\alpha}{2\pi i}$.*
2. *Let $(\nabla'_{\alpha\beta}, \omega'_\alpha)$ be another connection on $\mathcal{L}$ and let $\varOmega'$ be the corresponding curvature 3-form. Then there exists a canonical $\eta \in \Omega^2(M)/d\Omega^1(M)$ such that $\varOmega' = \varOmega + d\eta$.*
3. *Let $(\nabla_{\alpha\beta}, \omega_\alpha)$ $(\nabla'_{\alpha\beta}, \omega'_\alpha)$, $(\nabla''_{\alpha\beta}, \omega''_\alpha)$ be 3 connections on $\mathcal{L}$ with the corresponding curvature 3-forms $\varOmega$, $\varOmega'$, $\varOmega''$. Let $\eta, \eta', \eta'' \in \Omega^2(M)/d\Omega^1(M)$ be the canonical elements constructed above such that $\varOmega' - \varOmega = d\eta$, $\varOmega'' - \varOmega' = d\eta'$, $\varOmega'' - \varOmega = d\eta''$. Then $\eta'' = \eta + \eta'$.*

The 3-form Ω is a de Rham representative of the Dixmier-Douady class of the gerbe.

Proof We use the notation of the previous Lemma. For the first part notice that since $d\omega_{\alpha\beta} = 0$ for any α, β we see that

$$d\omega_\alpha|_{U_{\alpha\beta}} = d\omega_\beta|_{U_{\alpha\beta}}$$

which shows the existence of Ω; it is clearly closed. Ω does not change if one restricts the descent datum to a refinement. Equation (2.3) implies that if one uses a different descent datum for the same gerbe one obtains the same 3-curvature form.

We proceed to the proof of the second part. Let $(\nabla_{\alpha\beta}, \omega_\alpha)$ and $(\nabla'_{\alpha\beta}, \omega'_\alpha)$ be two connections on the descent datum $(U_\alpha, \mathcal{L}_{\alpha\beta}, \mu_{\alpha\beta\gamma})$. Set $\delta_{\alpha\beta} = \nabla'_{\alpha\beta} - \nabla_{\alpha\beta} \in \Omega^1(U_{\alpha\beta})$. Choose $\delta_\alpha \in \Omega^1(U_\alpha)$ such that $\delta_{\alpha\beta} = \delta_\alpha - \delta_\beta$ on $U_{\alpha\beta}$. Then $\omega'_{\alpha\beta} - \omega_{\alpha\beta} = d\delta_{\alpha\beta} = d\delta_\alpha - d\delta_\beta$. Hence $\omega'_\alpha - \omega_\alpha - d\delta_\alpha = \omega'_\beta - \omega_\beta - d\delta_\beta$ on $U_{\alpha\beta}$. Therefore there exists $\eta \in \Omega^2(M)$ such that $\eta|_{U_\alpha} = \frac{1}{2\pi i}(\omega'_\alpha - \omega_\alpha - d\delta_\alpha)$. Then $\Omega' - \Omega = d\eta$. The formula for η given above depends on the choice of δ_α. If $\bar{\delta}_\alpha$ is a different such choice, then $\bar{\delta}_\alpha - \delta_\alpha = \epsilon|_{U_\alpha}$ for some $\epsilon \in \Omega^1(M)$. Then $\bar{\eta} = \eta - d\epsilon$. Hence the class of $\eta \in \Omega^2(M)/d\Omega^1(M)$ does not depend on the choice made. It is easy to see that η does not depend on the choice of the particular descent datum used.

The verification of the third statement is straightforward and is left to the reader. □

2.4 Twisted cohomology. For a smooth manifold M let $\Omega \in \Omega^3(M)$ be a closed 3-form. Denote by u a formal variable of degree -2. The twisted de Rham complex is defined as the complex $\Omega^*(M)[u]$ with the differential $d_\Omega = ud + u^2\Omega\wedge\cdot$. Note that if $\Omega' = \Omega + d\eta$ is cohomologous to Ω then the complexes $(\Omega^*(M)[u], d_\Omega)$ and $(\Omega^*(M)[u], d_{\Omega'})$ are isomorphic via the isomorphism

$$I_\eta : \xi \mapsto e^{-u\eta} \wedge \xi \tag{2.4}$$

Lemma 2.5 *The map induced by I_η on cohomology depends only on the class of η in $\Omega^2(M)/d\Omega^1(M)$.*

Proof Indeed, for $\epsilon \in \Omega^1(M)$ define $h_\epsilon : (\Omega^*(M)[u])_\bullet \to (\Omega^*(M)[u])_{\bullet-1}$ by

$$h_\epsilon = u\epsilon \left(\sum_k \frac{(u\eta)^k}{(k+1)!} \right) \wedge \cdot.$$

Then

$$I_{\eta+d\epsilon} - I_\eta = h_\epsilon \circ d_\Omega + d_{\Omega'} \circ h_\epsilon$$

and therefore the maps I_η and $I_{\eta+d\epsilon}$ are chain homotopic. □

The identity $d_{\Omega+\Omega'}(\xi\wedge\eta) = d_\Omega\xi\wedge\eta + (-1)^{|\xi|}\xi\wedge d_{\Omega'}\eta$ implies that the product of forms induces the product $H^\bullet_\Omega(M) \otimes H^\bullet_{\Omega'}(M) \to H^\bullet_{\Omega+\Omega'}(M)$ and in particular endows $H^\bullet_\Omega(M)$ with the structure of $H^\bullet(M)[u]$-module.

We will also need to consider the following situation. Let $\pi : M \to B$ be an oriented fibration with compact fibers. Then we have an integration along the fibers map $\int_{M|B} : \Omega^*(M) \to \Omega^{*-k}(B)$, $k = \dim M - \dim B$. Let $\Omega \in \Omega^3(B)$ be a closed form. Then $\int_{M|B} d\eta = d\int_{M|B} \eta$, $\int_{M|B} \pi^*\Omega \wedge \eta = \Omega \wedge \int_{M|B} \eta$ and therefore

$$\int_{M|B} d_{\pi^*\Omega}\eta = d_\Omega \int_{M|B} \eta.$$

Hence we obtain a chain map $\int_{M|B}\colon (\Omega^*(M), d_{\pi^*\Omega})_\bullet \to (\Omega^*(B), d_\Omega)_{\bullet-k}$.

Let $\mathcal{L}$ be a gerbe. The choice of connection defines a closed 3-form Ω. We can therefore consider the complex $(\Omega^*(M)[u], d_\Omega)$. A different choice of connection leads to a different complex $(\Omega^*(M)[u], d_{\Omega'})$. Lemma 2.4 however implies that there exists a canonical isomorphism I of the homologies of these complexes. We denote this homology by $H^\bullet_{\mathcal{L}}(B)$

2.5 Twisted bundles.

Definition 2.6 A descent datum for a twisted vector bundle $\mathcal{E}$ consists of a descent datum $(U_\alpha, \mathcal{L}_{\alpha\beta}, \mu_{\alpha\beta\gamma})$ for a gerbe $\mathcal{L}$ together with a collection $(\mathcal{E}_\alpha \to U_\alpha)_{\alpha\in\Lambda}$ of vector bundles and a collection of vector bundle isomorphisms $\varphi_{\alpha\beta} : \mathcal{E}_\alpha \otimes \mathcal{L}_{\alpha\beta} \cong \mathcal{E}_\beta$ such that for every α, β, γ the following diagram commutes:

$$\begin{array}{ccc}
\mathcal{E}_\alpha \otimes \mathcal{L}_{\alpha\beta} \otimes \mathcal{L}_{\beta\gamma} & \xrightarrow{\mathrm{id}\otimes\mu_{\alpha\beta\gamma}} & \mathcal{E}_\alpha \otimes \mathcal{L}_{\alpha\gamma} \\
{\scriptstyle \varphi_{\alpha\beta}\otimes\mathrm{id}}\downarrow & & \downarrow{\scriptstyle \varphi_{\alpha\gamma}} \\
\mathcal{E}_\beta \otimes \mathcal{L}_{\beta\gamma} & \xrightarrow{\varphi_{\beta\gamma}} & \mathcal{E}_\gamma
\end{array}$$

Restriction of the descent datum for $\mathcal{E}$ to a refinement is given by the restriction of the descent datum for $\mathcal{L}$ together with restriction of the vector bundles $\mathcal{E}_\alpha$.

Definition 2.7 An (iso)morphism between two descent data $(U_\alpha, \mathcal{L}_{\alpha\beta}, \mu_{\alpha\beta\gamma}, \mathcal{E}_\alpha, \varphi_{\alpha\beta})$ and $(U'_\alpha, \mathcal{L}'_{\alpha\beta}, \mu'_{\alpha\beta\gamma}, \mathcal{E}'_\alpha, \varphi'_{\alpha\beta})$ is given by the collection $(\rho_\alpha, S_\alpha, \lambda_{\alpha\beta})$, where $(S_\alpha, \lambda_{\alpha\beta})$ is an isomorphism between $(U_\alpha, \mathcal{L}_{\alpha\beta}, \mu_{\alpha\beta\gamma})$ and $(U'_\alpha, \mathcal{L}'_{\alpha\beta}, \mu'_{\alpha\beta\gamma})$ and $\rho_\alpha\colon \mathcal{E}_\alpha \otimes S_\alpha \to \mathcal{E}'_\alpha$ is a collection of (iso)morphisms such that the diagram

$$\begin{array}{ccc}
\mathcal{E}_\alpha \otimes S_\alpha \otimes S_\alpha^{-1} \otimes \mathcal{L}_{\alpha\beta} \otimes S_\beta & \xrightarrow{\varphi_{\alpha\beta}\otimes\mathrm{id}} & \mathcal{E}_\beta \otimes S_\beta \\
{\scriptstyle \rho_\alpha\otimes\lambda_{\alpha\beta}}\downarrow & & \downarrow{\scriptstyle \rho_\beta} \\
\mathcal{E}'_\alpha \otimes \mathcal{L}'_{\alpha\beta} & \xrightarrow{\varphi'_{\alpha\beta\gamma}} & \mathcal{E}'_\beta
\end{array}$$

commutes.

An isomorphism between two descent data on two different covers is defined as an isomorphism between their restrictions on a common refinement. A 2-morphism between two isomorphisms is the 2-morphism between the corresponding isomorphisms of the gerbe descent data.

A twisted bundle is then defined as an equivalence class of descent data of twisted vector bundles. "Forgetting" the bundle data we obtain from the descent datum for a twisted vector bundle a descent datum for a gerbe, and the same applies to morphisms and 2-morphisms. We say that $\mathcal{E}$ is an $\mathcal{L}$-twisted vector bundle if "forgetting" the bundle data one obtains the equivalence class of the gerbe descent data defining $\mathcal{L}$.

Assume now that the gerbe $\mathcal{L}$ is unitary. A Hermitian descent datum for $\mathcal{E}$ consists of a unitary descent datum $(U_\alpha, \mathcal{L}_{\alpha\beta}, \mu_{\alpha\beta\gamma})$ for $\mathcal{L}$ and a collection h_α of metrics on $\mathcal{E}_\alpha$ such that the maps $\varphi_{\alpha\beta}$ are isometries. One obtains a notion of isomorphism of Hermitian descent data by requiring the ρ_α to be isometries. A Hermitian twisted bundle is then an equivalence class of Hermitian descent data.

Given a gerbe $(U_\alpha, \mathcal{L}_{\alpha\beta}, \mu_{\alpha\beta\gamma})$ on M, it is well known that (finite dimensional) twisted vector bundles exist if and only if the gerbe is torsion (see, e.g., [19, 25] and references therein).

Lemma 2.8 *Let $\mathcal{L}$ be a gerbe on M and $\mathcal{E}$ an $\mathcal{L}$-twisted bundle. Let $(U_\alpha, \mathcal{L}_{\alpha\beta}, \mu_{\alpha\beta\gamma})$ be a descent datum for $\mathcal{L}$. Then there exists a descent datum for $\mathcal{E}$ isomorphic to the one of the form $(U_\alpha, \mathcal{L}_{\alpha\beta}, \mu_{\alpha\beta\gamma}, \mathcal{E}_\alpha, \varphi_{\alpha\beta})$.*

Proof Let $(V_i, L_{ij}, m_{ijk}, E_i, r_{ij}), i, j, k \in I$ be a descent datum for $\mathcal{E}$. Without loss of generality we may assume that $\{V_i\}$ is refinement of $\{U_\alpha\}$ given by the map $\varrho : I \to \Lambda$ and that there exists an isomorphism (S_i, λ_{ij}) between (V_i, L_{ij}, m_{ijk}) and the restriction of $(U_\alpha, \mathcal{L}_{\alpha\beta}, \mu_{\alpha\beta\gamma})$ to $\{V_i\}$.

We define the vector bundle $\mathcal{E}_\alpha$ as follows. On every non-empty open set $U_\alpha \cap V_i$, we set

$$\mathcal{E}_\alpha^{(i)} := E_i \otimes S_i \otimes \mathcal{L}_{\varrho(i)\alpha}.$$

Notice that $V_i \cap U_\alpha \subset U_{\varrho(i)\alpha}$ so that this definition makes sense as a vector bundle over $V_i \cap U_\alpha$. Moreover, if $(i, j) \in I^2$ is such that $V_{ij} \cap U_\alpha \neq \emptyset$, then we have a bundle isomorphism

$$\psi_\alpha^{(ij)} : \mathcal{E}_\alpha^{(j)}|_{V_{ij} \cap U_\alpha} \longrightarrow \mathcal{E}_\alpha^{(i)}|_{V_{ij} \cap U_\alpha}.$$

defined by the composition

$$\begin{aligned} E_j \otimes S_j \otimes \mathcal{L}_{\varrho(j)\alpha} &\overset{\varphi_{ij}^{-1} \otimes \mathrm{id}}{\longrightarrow} E_i \otimes L_{ij} \otimes S_j \otimes \mathcal{L}_{\varrho(j)\alpha} \longrightarrow E_i \otimes S_i \otimes S_i^{-1} \otimes L_{ij} \otimes S_j \otimes \mathcal{L}_{\varrho(j)\alpha} \\ &\overset{\mathrm{id} \otimes \lambda_{ij} \otimes \mathrm{id}}{\longrightarrow} E_i \otimes S_i \otimes \mathcal{L}_{\varrho(i)\varrho(j)} \otimes \mathcal{L}_{\varrho(j)\alpha} \overset{\mathrm{id} \otimes \mu_{\varrho(i)\varrho(j)\alpha}}{\longrightarrow} E_i \otimes S_i \otimes \mathcal{L}_{\varrho(i)\alpha} \end{aligned}$$

It is easy to see that

$$\psi_\alpha^{(ik)} = \psi_\alpha^{(ij)} \circ \psi_\alpha^{(jk)}.$$

So, we can glue the bundles $(\mathcal{E}_\alpha^{(i)})_i$ for $V_i \cap U_\alpha \neq \emptyset$ together and form a vector bundle $\mathcal{E}_\alpha$ over U_α. More precisely

$$\mathcal{E}_\alpha := \left(\amalg_{V_i \cap U_\alpha \neq \emptyset} \mathcal{E}_\alpha^{(i)} \right) / \{\psi_\alpha^{(ij)}\}.$$

Next we introduce for $i \in I$ with $V_i \cap U_{\alpha\beta} \neq \emptyset$,

$$\varphi_{\alpha\beta}^{(i)} := \mathrm{id} \otimes \mu_{\varrho(i)\alpha\beta} : \mathcal{E}_\alpha^{(i)} \otimes \mathcal{L}_{\alpha\beta} \to \mathcal{E}_\beta^{(i)}.$$

When $V_{ij} \cap U_{\alpha\beta} \neq \emptyset$, a straightforward inspection shows again that the following diagram commutes;

$$\begin{array}{ccc} \mathcal{E}_\alpha^{(i)} \otimes \mathcal{L}_{\alpha\beta} & \xrightarrow{\varphi_{\alpha\beta}^{(i)}} & \mathcal{E}_\beta^{(i)} \\ {\scriptstyle \psi_\alpha^{(ji)} \otimes \mathrm{id}} \downarrow & & \downarrow {\scriptstyle \psi_\beta^{(ji)}} \\ \mathcal{E}_\alpha^{(j)} \otimes \mathcal{L}_{\alpha\beta} & \xrightarrow{\varphi_{\alpha\beta}^{(j)}} & \mathcal{E}_\beta^{(j)} \end{array}$$

Therefore, the isomorphisms $\varphi_{\alpha\beta}^{(i)}$ induce an isomorphism

$$\varphi_{\alpha\beta} : \mathcal{E}_\alpha \otimes \mathcal{L}_{\alpha\beta} \to \mathcal{E}_\beta.$$

We leave it to the reader to show that $(U_\alpha, \mathcal{L}_{\alpha\beta}, \mu_{\alpha\beta\gamma}, \mathcal{E}_\alpha, \varphi_{\alpha\beta})$ is a descent datum for a twisted vector bundle isomorphic to $(V_i, L_{ij}, m_{ijk}, E_i, r_{ij})$. □

We now fix a gerbe $\mathcal{L}$ with a descent datum $(U_\alpha, \mathcal{L}_{\alpha\beta}, \mu_{\alpha\beta\gamma})$ on the smooth manifold M together with a twisted vector bundle $\mathcal{E}$ represented by $(\mathcal{E}_\alpha, \varphi_{\alpha\beta})$. We denote by $\mathcal{A}_\alpha$ the collection of bundles of algebras

$$\mathcal{A}_\alpha := \mathrm{End}(\mathcal{E}_\alpha), \quad \alpha \in \Lambda.$$

For any $U_{\alpha\beta} \neq \emptyset$, we have a canonical vector bundle isomorphism over $U_{\alpha\beta}$

$$, \rho_{\alpha\beta} : \mathrm{End}(\mathcal{E}_\alpha \otimes \mathcal{L}_{\alpha\beta}) \longrightarrow \mathcal{A}_\alpha,$$

extending the canonical isomorphism $\mathrm{End}(\mathcal{L}_{\alpha\beta}) \simeq U_{\alpha\beta} \times \mathbb{C}$. Therefore, the bundle isomorphism $\varphi_{\alpha\beta}$ together with the identification $\rho_{\alpha\beta}$ induce the isomorphism of algebra bundles over $U_{\alpha\beta}$ given by

$$\mathcal{A}_\beta \overset{\varphi^*_{\alpha\beta}}{\longrightarrow} \mathrm{End}(\mathcal{E}_\alpha \otimes \mathcal{L}_{\alpha\beta}) \overset{\rho^*_{\alpha\beta}}{\longrightarrow} \mathcal{A}_\alpha.$$

We denote this isomorphism by $\varphi^*_{\alpha\beta}$. It is then easy to check that

$$\varphi^*_{\alpha\beta} \circ \varphi^*_{\beta\gamma} = \varphi^*_{\alpha\gamma}, \qquad \text{over } U_{\alpha\beta\gamma}.$$

Therefore, the collection $(\mathcal{A}_\alpha)$ defines a bundle $\mathcal{A}$ of algebras over M, which we denote $\mathrm{End}(\mathcal{E})$. We leave to the reader an easy check that the isomorphism class of $\mathrm{End}(\mathcal{E})$ depends only on the isomorphism class of $\mathcal{E}$.

Note that for every α we have the trace $\mathrm{tr}_\alpha\colon \mathrm{End}(\mathcal{E}_\alpha) \to C^\infty(U_\alpha)$. For $a \in \Gamma(U_{\alpha\beta}; \mathcal{A}_\beta)$ $\mathrm{tr}_\alpha(\varphi^*_{\alpha\beta}(a)) = \mathrm{tr}_\beta(a)$. Therefore we obtain the trace $\mathrm{tr}\colon \mathrm{End}(\mathcal{E}) \to C^\infty(M)$ defined by

$$\mathrm{tr}(a)|_{U_\alpha} = \mathrm{tr}_\alpha(a|_{U_\alpha}) \text{ for } a \in \mathrm{End}(\mathcal{E})$$

If the bundle $\mathcal{E}$ is $\mathbb{Z}_2$-graded then $\mathrm{End}(\mathcal{E})$ is a bundle of $\mathbb{Z}_2$-graded algebras with a supertrace $\mathrm{str}\colon \mathrm{End}(\mathcal{E}) \to C^\infty(M)$.

Definition 2.9 The bundle $\mathcal{A} = \mathrm{End}(\mathcal{E})$ is called the Azumaya bundle associated with the $\mathcal{L}$-twisted bundle $\mathcal{E}$.

If $\mathcal{E}$ is a Hermitian twisted bundle, each of the bundles $\mathrm{End}(\mathcal{E}_\alpha)$ is a bundle of *-algebras, with the *-operation given by taking the adjoint endomorphism. This induces a structure of a bundle of *-algebras on $\mathrm{End}(\mathcal{E})$.

Notice that, more generally, if $\mathcal{E}$ and $\mathcal{E}'$ are $\mathcal{L}$-twisted bundles, we have a well defined bundle $\mathrm{Hom}(\mathcal{E}, \mathcal{E}')$, defined similarly by $\mathrm{Hom}(\mathcal{E}, \mathcal{E}')|_{U_\alpha} = \mathrm{Hom}(\mathcal{E}|_{U_\alpha}, \mathcal{E}'|_{U_\alpha})$.

Let $\mathcal{L}$ be a gerbe on M and $\mathcal{E}$ an $\mathcal{L}$-twisted bundle on M. Let $(U_\alpha, \mathcal{L}_{\alpha\beta}, \mu_{\alpha\beta\gamma}, \mathcal{E}_\alpha, \varphi_{\alpha\beta})$ be a descent datum for $\mathcal{E}$.

Definition 2.10 A connection on $(U_\alpha, \mathcal{L}_{\alpha\beta}, \mu_{\alpha\beta\gamma}, \mathcal{E}_\alpha, \varphi_{\alpha\beta})$ is a collection $(\nabla_\alpha, \nabla_{\alpha\beta}, \omega_\alpha)$ where $(\nabla_{\alpha\beta}, \omega_\alpha)$ is a connection on $(U_\alpha, \mathcal{L}_{\alpha\beta}, \mu_{\alpha\beta\gamma})$ and each ∇_α is a connection on $\mathcal{E}_\alpha$ such that the identities

$$\varphi^*_{\alpha\beta} \nabla_\beta = \nabla_\alpha \otimes \mathrm{id} + \mathrm{id} \otimes \nabla_{\alpha\beta} \tag{2.5}$$

hold for $U_{\alpha\beta} \neq \emptyset$.

Lemma 2.11 *Let $(U_\alpha, \mathcal{L}_{\alpha\beta}, \mu_{\alpha\beta\gamma}, \mathcal{E}_\alpha, \varphi_{\alpha\beta})$ be a descent datum for an $\mathcal{L}$-twisted bundle $\mathcal{E}$. Then every connection $(\nabla_{\alpha\beta}, \omega_\alpha)$ on the descent datum $(U_\alpha, \mathcal{L}_{\alpha\beta}, \mu_{\alpha\beta\gamma})$ for $\mathcal{L}$ can be extended to a connection for the descent datum for $\mathcal{E}$.*

Proof We start with a collection (∇'_α) of connections on $(\mathcal{E}_\alpha)$. Then we set for $U_{\alpha\beta} \neq \emptyset$,

$$A_{\alpha\beta} = \varphi^*_{\alpha\beta}\nabla'_\beta - \nabla'_\alpha \otimes \mathrm{id} - \mathrm{id} \otimes \nabla_{\alpha\beta}.$$

So, $A_{\alpha\beta}$ is a differential 1-form on $U_{\alpha\beta}$ with coefficients in the bundle $\mathrm{End}(\mathcal{E}_\alpha \otimes \mathcal{L}_{\alpha\beta})$. This latter being canonically isomorphic to $\mathrm{End}(\mathcal{E}_\alpha)$, we see that

$$A_{\alpha\beta} \in \Omega^1(U_{\alpha\beta}, \mathrm{End}(\mathcal{E}_\alpha)).$$

Using the functoriality conditions on the isomorphisms $(\varphi_{\alpha\beta})$, it is then easy to check that

$$A_{\alpha\gamma} = A_{\alpha\beta} + \varphi^*_{\alpha\beta} A_{\beta\gamma},$$

where $\varphi_{\alpha\beta}$ is viewed here as the isomorphism over $U_{\alpha\beta}$ between $\mathrm{End}(\mathcal{E}_\beta)$ and $\mathrm{End}(\mathcal{E}_\alpha)$. Since the sheaf of sections of the bundle $\mathrm{End}(\mathcal{E})$ is soft, there exists a collection $A_\alpha \in \Omega^1(U_\alpha; \mathrm{End}(\mathcal{E}_\alpha))$ such that

$$A_{\alpha\beta} = A_\alpha - \varphi^*_{\alpha\beta} A_\beta.$$

The collection $\nabla_\alpha = \nabla'_\alpha + A_\alpha$ is then a connection on the $\mathcal{L}$-twisted vector bundle $\mathcal{E}$ compatible with the curving $\nabla_{\alpha\beta}$. □

An isomorphism between two descent data $(U_\alpha, \mathcal{L}_{\alpha\beta}, \mu_{\alpha\beta\gamma}, \mathcal{E}_\alpha, \varphi_{\alpha\beta})$ and $(U'_\alpha, \mathcal{L}'_{\alpha\beta}, \mu'_{\alpha\beta\gamma}, \mathcal{E}'_\alpha, \varphi'_{\alpha\beta})$ for $\mathcal{E}$ with connections $(\nabla_\alpha, \nabla_{\alpha\beta}, \omega_\alpha)$ and $(\nabla'_\alpha, \nabla'_{\alpha\beta}, \omega'_\alpha)$ respectively is given by the collection $s = (\rho_\alpha, S_\alpha, \lambda_{\alpha\beta}, \nabla^S_\alpha)$ where $(\rho_\alpha, S_\alpha, \lambda_{\alpha\beta})$ is a morphism between the descent data without connections, ∇^S_α are connections on S_α such that $\left(S_\alpha, \lambda_{\alpha\beta}, \nabla^S_\alpha\right)$ is a morphism of the corresponding gerbe descent data with connections (i.e., Equatons (2.2), (2.3) are satisfied) and the equality

$$\rho^*_\alpha \nabla'_\alpha = \nabla_\alpha \otimes \mathrm{id} + \mathrm{id} \otimes \nabla^S_\alpha$$

holds (in the notations of (2.2) and (2.3)).

A connection on a twisted bundle is then a choice of connection on each descent datum of this twisted bundle and lifting of isomorphisms of descent data to isomorphisms of descent data with connections. Every connection on a gerbe $\mathcal{L}$ can be extended to a connection on any $\mathcal{L}$-twisted bundle.

Proposition 2.12 *Let $\mathcal{E}$ be an $\mathcal{L}$-twisted bundle with connection. Choose a descent datum $(U_\alpha, \mathcal{L}_{\alpha\beta}, \mu_{\alpha\beta\gamma}, \mathcal{E}_\alpha, \varphi_{\alpha\beta})$ with a connection $(\nabla_\alpha, \nabla_{\alpha\beta}, \omega_\alpha)$ representing $\mathcal{E}$. Then the collection $(\theta_\alpha + \omega_\alpha)$, where $\theta_\alpha = \nabla^2_\alpha$ is the curvature of ∇_α, defines a global differential 2-form θ on M with coefficients in the Azumaya bundle $\mathcal{A} = \mathrm{End}(\mathcal{E})$. This form is independent of the choice of the representing descent datum.*

Proof We have for any $\alpha \in \Lambda$, $\theta_\alpha \in \Omega^2(U_\alpha, \mathcal{A}_\alpha)$ where $\mathcal{A}_\alpha = \mathrm{End}(\mathcal{E}_\alpha))$ is the Azumaya bundle associated with the twisted bundle $\mathcal{E}$. Equation (2.5) implies that

$$\varphi^*_{\alpha\beta}\theta_\beta = \theta_\alpha + \omega_{\alpha\beta} \quad \in \quad \Omega^2(U_{\alpha\beta}, \mathrm{End}(\mathcal{E}_\alpha)).$$

Therefore, the collection $(\theta_\alpha + \omega_\alpha)$ of elements of $\Omega^2(U_\alpha, \mathcal{A}_\alpha)$ satisfies the relations

$$\varphi^*_{\alpha\beta}(\theta_\beta + \omega_\beta) = \theta_\alpha + \omega_\alpha.$$

It is easy to see that the form θ is independent of the equivalence class of the connection and is functorial with respect to the isomorphism of descent data. □

In the notation above let $\nabla\colon \mathcal{A} \to \Omega^1(M,\mathcal{A})$ be the connection defined for a fixed descent datum by

$$(\nabla\xi)|_{U_\alpha} = [\nabla_\alpha, \xi]. \tag{2.6}$$

It is easy to see that ∇ is well defined and by derivation with respect to the product on $\mathcal{A}$.

Note that

$$\nabla^2 = [\theta, \cdot] \text{ and } \nabla\theta = 2\pi i\Omega \tag{2.7}$$

where Ω is the 3-curvature form of the connection on $\mathcal{L}$, see Lemma 2.4.

Proposition 2.13 1. *Let $\mathcal{E}$ be an $\mathcal{L}$-twisted bundle and ∇ a connection on $\mathcal{E}$. Set $\mathrm{Ch}_{\mathcal{L}}(\nabla) = \operatorname{tr} e^{-\frac{u\theta}{2\pi i}} \in \Omega^*(M)[u]$. Then $d_\Omega \mathrm{Ch}_{\mathcal{L}}(\nabla) = 0$*

2. *The class of $\mathrm{Ch}_{\mathcal{L}}(\nabla)$ in $H_\Omega(M)$ is independent of the choice of connection. Namely, assume we are given a different connection ∇' on $\mathcal{E}$ (and therefore on $\mathcal{L}$) and let Ω' be the associated 3-curvature form. Then $I([\mathrm{Ch}_{\mathcal{L}}(\nabla)]) = [\mathrm{Ch}_{\mathcal{L}}(\nabla')]$, where I is the canonical isomorphism of the cohomology of $(\Omega^*(M)[u]$,*
$d_\Omega)$ with the cohomology of $(\Omega^*(M)[u], d_{\Omega'})$.

We denote the class of $\mathrm{Ch}_{\mathcal{L}}(\nabla)$ by $\mathrm{Ch}_{\mathcal{L}}(\mathcal{E})$.

Proof Since Ω is central, $\nabla(e^{-\frac{u\theta}{2\pi i}}) = -u\Omega e^{-\frac{u\theta}{2\pi i}}$. Hence

$$d \operatorname{tr} e^{-\frac{u\theta}{2\pi i}} = \operatorname{tr} \nabla(e^{-\frac{u\theta}{2\pi i}}) = -u\Omega \wedge \mathrm{Ch}_{\mathcal{L}}(\nabla)$$

and $d_\Omega \mathrm{Ch}_{\mathcal{L}}(\nabla) = 0$, which proves the first statement.

Fix now a descent datum $(U_\alpha, \mathcal{L}_{\alpha\beta}, \mu_{\alpha\beta\gamma}, \mathcal{E}_\alpha, \varphi_{\alpha\beta})$ for $\mathcal{E}$ together with a connection $\nabla = (\nabla_\alpha, \nabla_{\alpha\beta}, \omega_\alpha)$ and let $\nabla' = (\nabla'_\alpha, \nabla'_{\alpha\beta}, \omega'_\alpha)$ be another connection on $\mathcal{E}$. Set $\delta_{\alpha\beta} = \nabla'_{\alpha\beta} - \nabla_{\alpha\beta}$ and let δ_α be such that $\delta_\alpha - \delta_\beta = \delta_{\alpha\beta}$. Recall that the isomorphism I is induced by the map of complexes

$$\xi \longmapsto e^{-u\eta} \wedge \xi$$

where $\eta \in \Omega^2(M)$ is defined by $\eta|_{U_\alpha} = \frac{1}{2\pi i}(\omega'_\alpha - \omega_\alpha - d\delta_\alpha)$.

Consider now the manifold $\widetilde{M} = \mathbb{R} \times M$. Denote by $\pi\colon \widetilde{M} \to M$ the projection on the second factor and by $t\colon \widetilde{M} \to \mathbb{R}$ the projection on the first factor. Consider on $\widetilde{M}$ the gerbe $\widetilde{\mathcal{L}} = \pi^*\mathcal{L}$ given by the descent datum $(\pi^{-1}U_\alpha, \pi^*\mathcal{L}_{\alpha\beta}, \pi^*\mu_{\alpha\beta\gamma})$. Then $\pi^*\mathcal{E}$, $\pi^*\phi_{\alpha\beta}$ describes an $\widetilde{\mathcal{L}}$-twisted bundle $\widetilde{\mathcal{E}}$. Moreover,

$$\widetilde{\nabla}_{\alpha\beta} := (1-t)\pi^*\nabla_{\alpha\beta} + t\pi^*\nabla'_{\alpha\beta} \text{ and } \widetilde{\omega}_\alpha = (1-t)\pi^*\omega_\alpha + t\pi^*\omega'_\alpha + dt \wedge \delta_\alpha$$

define respectively, a connective structure and curving of $\widetilde{L}$. Similarly, $\widetilde{\nabla}_\alpha = (1-t)\pi^*\nabla_\alpha + t\pi^*\nabla'_\alpha$ defines a compatible connection on $\widetilde{\mathcal{E}}$. The corresponding 3-form is given by

$$\widetilde{\Omega} = (1-t)\pi^*\Omega + t\pi^*\Omega' + \frac{1}{2\pi i} dt \wedge (\omega'_\alpha - \omega_\alpha - d\delta_\alpha) = (1-t)\pi^*\Omega + t\pi^*\Omega' + dt \wedge \pi^*\eta.$$

Now, the map $\xi \mapsto e^{ut\eta} \wedge \xi$ is an isomorphism of complexes $\left(\Omega^*(\widetilde{M})[u], d_{\widetilde{\Omega}}\right)$ and $\left(\Omega^*(\widetilde{M})[u], d_{\pi^*\Omega}\right)$. Set $\widetilde{\mathcal{A}} = \pi^*\mathcal{A}$ and let $\widetilde{\theta} \in \Omega^2(\widetilde{M}, \widetilde{\mathcal{A}})$ be the form defined by $\pi^*\theta_\alpha + \widetilde{\omega}_\alpha$. By the result of the first part of the proposition the differential form

$\operatorname{tr} e^{-\frac{u\tilde{\theta}}{2\pi i}}$ is a cocycle in the complex $\left(\Omega^*(\widetilde{M})[u], d_{\widetilde{\Omega}}\right)$. Hence $\operatorname{tr} e^{ut\eta} \wedge e^{-\frac{u\tilde{\theta}}{2\pi i}}$ is a cocycle in the complex $\left(\Omega^*(\widetilde{M})[u], d_{\pi^*\Omega}\right)$. This implies the relation

$$\operatorname{tr} e^{u\eta} \wedge e^{-\frac{u\theta'}{2\pi i}} - \operatorname{tr} e^{-\frac{u\theta}{2\pi i}} = d_\Omega \int_0^1 \iota_{\frac{\partial}{\partial t}} \operatorname{tr} e^{ut\eta} \wedge e^{-\frac{u\tilde{\theta}}{2\pi i}} dt.$$

Therefore, we finally deduce that the differential forms $\operatorname{tr} e^{u\eta} \wedge e^{-\frac{u\theta'}{2\pi i}}$ and $\operatorname{tr} e^{-\frac{u\theta}{2\pi i}}$ are cohomologous in $(\Omega^*(M)[u], d_\Omega)$, and hence that the differential forms $\operatorname{tr} e^{-\frac{u\theta'}{2\pi i}}$ and $\operatorname{tr} e^{-u\eta} \wedge e^{-\frac{u\theta}{2\pi i}}$ are cohomologous in $(\Omega^*(M)[u], d_{\Omega'})$, which finishes the proof. □

We will also need a notion of superconnection on the twisted bundle. We now briefly indicate the modifications which need to be made to the notion of connection to obtain that of superconnection. Assume that we are given a gerbe $\mathcal{L}$ and a $\mathbb{Z}_2$-graded $\mathcal{L}$-twisted vector bundle $\mathcal{E} = \mathcal{E}^+ \oplus \mathcal{E}^-$. Let $(U_\alpha, \mathcal{L}_{\alpha\beta}, \mu_{\alpha\beta\gamma}, \mathcal{E}_\alpha, \varphi_{\alpha\beta})$ be a descent datum for $\mathcal{E}$.

Definition 2.14 A superconnection $\mathbb{A}$ on the descent datum is $\mathbb{A} = (\mathbb{A}_\alpha, \nabla_{\alpha\beta}, \omega_\alpha)$ where $(\nabla_{\alpha\beta}, \omega_\alpha)$ is a connection on the descent datum for $\mathcal{L}$ and each $\mathbb{A}_\alpha$ is a superconnection on $\mathcal{E}_\alpha$ satisfying the relations

$$\varphi^*_{\alpha\beta}\mathbb{A}_\beta = \mathbb{A}_\alpha \otimes \operatorname{id} + \operatorname{id} \otimes \nabla_{\alpha\beta}, \qquad \text{for } U_{\alpha\beta} \neq \emptyset.$$

Each superconnection $\mathbb{A}_\alpha$ can be written as $\mathbb{A}_\alpha = \sum_{k\geq 0} \mathbb{A}_\alpha^{[k]}$ where $\mathbb{A}_\alpha^{[k]} \in \Omega^k(U_\alpha; \operatorname{End}(\mathcal{E}_\alpha)^-)$ for k even, $\mathbb{A}_\alpha^{[k]} \in \Omega^k(U_\alpha; \operatorname{End}(\mathcal{E}_\alpha)^+)$ for k odd, $k \neq 1$, and $\mathbb{A}_\alpha^{[1]}$ is a grading-preserving connection on $\mathcal{E}_\alpha$. It is easy to see that for each $k \neq 1$ there exists a form $\mathbb{A}^{[k]} \in \Omega^k(M; \operatorname{End}(\mathcal{E}))$ such that $\mathbb{A}^{[k]}|U_\alpha = \mathbb{A}_\alpha^{[k]}$. For $k = 1$ $(\mathbb{A}_\alpha^{[1]}, \nabla_{\alpha\beta}, \omega_\alpha)$ defines a connection on the (descent datum of) $\mathcal{E}$.

Let now $u^{1/2}$ be a formal variable of degree -1 such that $(u^{1/2})^2 = u$. Define the rescaled superconnection

$$\mathbb{A}_{u^{-1}} := \sum u^{(k-1)/2}\mathbb{A}^{[k]}.$$

Let Ω be as before the curvature 3-form of the connection on $\mathcal{L}$. Define the curvature of the rescaled superconnection $\theta^{\mathbb{A}_{u^{-1}}}$ by

$$\theta^{\mathbb{A}_{u^{-1}}}|_{U_\alpha} = (\mathbb{A}_\alpha)^2_{u^{-1}} + \omega_\alpha.$$

Then $u\theta^{\mathbb{A}_{u^{-1}}} \in \Omega^{even}(M, \operatorname{End}(\mathcal{E})^+)[u] + u^{1/2}\Omega^{odd}(M, \operatorname{End}(\mathcal{E})^-)[u]$. We therefore have the differential form $\exp\left(-\frac{u\theta^{\mathbb{A}_{u^{-1}}}}{2\pi i}\right)$ which belongs to $\Omega^{even}(M, \operatorname{End}(\mathcal{E})^+)[u] + u^{1/2}\Omega^{odd}(M, \operatorname{End}(\mathcal{E})^-)[u]$ and the differential form $\operatorname{str}\exp\left(-\frac{u\theta^{\mathbb{A}_{u^{-1}}}}{2\pi i}\right)$ which belongs to $\Omega^{even}(M)[u]$. The following is an analogue of Proposition 2.13 for the superconnections with an essentially identical proof.

Proposition 2.15

1. *Set* $\operatorname{Ch}_{\mathcal{L}}(\mathbb{A}) = \operatorname{str}\exp\left(-\frac{u\theta^{\mathbb{A}_{u^{-1}}}}{2\pi i}\right)$. *Then* $d_\Omega \operatorname{Ch}_{\mathcal{L}}(\mathbb{A}) = 0$
2. *The class of* $\operatorname{Ch}_{\mathcal{L}}(\mathbb{A})$ *in* $H_\Omega(M)$ *is independent of the choice of superconnection. Specifically, assume we are given a different superconnection* $\mathbb{A}'$ *on* $\mathcal{E}$ *(and therefore a different connection on* $\mathcal{L}$*) and let* Ω' *be the associated 3-curvature form. Then* $I([\operatorname{Ch}_{\mathcal{L}}(\mathbb{A})]) = [\operatorname{Ch}_{\mathcal{L}}(\mathbb{A}')]$, *where* I *is the canonical isomorphism of cohomology of* $(\Omega^*(M)[u], d_\Omega)$ *with the cohomology of* $(\Omega^*(M)[u], d_{\Omega'})$.

2.6 Horizontally twisted bundles. Let $\pi : M \to B$ be a smooth fibration. Let $\mathcal{L}$ be a gerbe on B.

Definition 2.16 A descent datum for a horizontally $\mathcal{L}$-twisted bundle $\mathcal{E}$ on M consists of the descent datum $(U_\alpha, \mathcal{L}_{\alpha\beta}, \mu_{\alpha\beta\gamma})$ for $\mathcal{L}$ together with a collection $(\mathcal{E}_\alpha \to \pi^{-1}U_\alpha)_{\alpha\in\Lambda}$ of vector bundles and a collection of vector bundle isomorphisms $\varphi_{\alpha\beta}\colon \mathcal{E}_\alpha \otimes \pi^*\mathcal{L}_{\alpha\beta} \cong \mathcal{E}_\beta$ so that

$$(\pi^{-1}U_\alpha, \pi^*\mathcal{L}_{\alpha\beta}, \pi^*\mu_{\alpha\beta\gamma}, \mathcal{E}_\alpha, \varphi_{\alpha\beta})$$

is a descent datum for a twisted vector bundle on M.

An (iso)morphism between two such descent data

$$(U_\alpha, \mathcal{L}_{\alpha\beta}, \mu_{\alpha\beta\gamma}, \mathcal{E}_\alpha, \varphi_{\alpha\beta}) \text{ and } (U'_\alpha, \mathcal{L}'_{\alpha\beta}, \mu'_{\alpha\beta\gamma}, \mathcal{E}'_\alpha, \varphi'_{\alpha\beta})$$

is given by the collection $(\rho_\alpha, S_\alpha, \lambda_{\alpha\beta})$ where $(S_\alpha, \lambda_{\alpha\beta})$ is an isomorphism between $(U_\alpha, \mathcal{L}_{\alpha\beta}, \mu_{\alpha\beta\gamma})$ and $(U'_\alpha, \mathcal{L}'_{\alpha\beta}, \mu'_{\alpha\beta\gamma})$ and $\rho_\alpha\colon \mathcal{E}_\alpha \otimes \pi^*S_\alpha \to \mathcal{E}'_\alpha$ is such that $(\rho_\alpha, \pi^*S_\alpha, \pi^*\lambda_{\alpha\beta})$ is an (iso)morphism between $(\pi^{-1}U_\alpha, \pi^*\mathcal{L}_{\alpha\beta}, \pi^*\mu_{\alpha\beta\gamma}, \mathcal{E}_\alpha, \varphi_{\alpha\beta})$ and $(\pi^{-1}U'_\alpha, \pi^*\mathcal{L}'_{\alpha\beta}, \pi^*\mu'_{\alpha\beta\gamma}, \mathcal{E}'_\alpha, \varphi'_{\alpha\beta})$.

With these definitions one can now define a horizontally twisted bundle as an equivalence class of descent data. Let $Tw(\pi^*\mathcal{L})$ denote the set of isomorphism classes of all $\pi^*\mathcal{L}$-twisted bundles on M and $Tw_h(\mathcal{L})$ denote the set of isomorphism classes of all horizontally $\mathcal{L}$-twisted bundles. Then we have an obvious map $Tw_h(\mathcal{L}) \to Tw(\pi^*\mathcal{L})$. According to Lemma 2.8 this map is surjective. In particular $Tw_h(\mathcal{L}) \neq \emptyset$ if and only if $\pi^*[\mathcal{L}]$ is torsion in $H^3(M, \mathbb{Z})$. It is however not injective. Indeed, if $(U_\alpha, \mathcal{L}_{\alpha\beta}, \mu_{\alpha\beta\gamma}, \mathcal{E}_\alpha, \varphi_{\alpha\beta})$ is a descent datum for a horizontally twisted bundle and S is a line bundle on M then $(U_\alpha, \mathcal{L}_{\alpha\beta}, \mu_{\alpha\beta\gamma}, \mathcal{E}_\alpha \otimes S|_{\pi^{-1}U_\alpha}, \varphi_{\alpha\beta} \otimes \mathrm{id})$ is another such descent datum. These data define the same element of $Tw(\pi^*\mathcal{L})$ but, unless S is a pullback of a line bundle from B, different elements of $Tw_h(\mathcal{L})$.

A connection on the descent datum $(U_\alpha, \mathcal{L}_{\alpha\beta}, \mu_{\alpha\beta\gamma}, \mathcal{E}_\alpha, \varphi_{\alpha\beta})$ is a collection $(\nabla_\alpha, \nabla_{\alpha\beta}, \omega_\alpha)$, where $(\nabla_{\alpha\beta}, \omega_\alpha)$ is a connection on the descent datum for $\mathcal{L}$ and ∇_α is a connection on $\mathcal{E}_\alpha$ such that $(\nabla_\alpha, \pi^*\nabla_{\alpha\beta}, \pi^*\omega_\alpha)$ is a connection on the descent datum $(\pi^{-1}U_\alpha, \pi^*\mathcal{L}_{\alpha\beta}, \pi^*\mu_{\alpha\beta\gamma}, \mathcal{E}_\alpha, \varphi_{\alpha\beta})$. With these definitions one can now define a notion of connection on the horizontally $\mathcal{L}$-twisted bundle in complete analogy with the defnitions for the twisted bundles. If ∇ is such a connection and Ω is the curvature 3-form of the gerbe $\mathcal{L}$, one defines $\mathrm{Ch}_{\mathcal{L}}(\nabla)$ – a closed form in $(\Omega^*(M), d_{\pi^*\Omega})_\bullet$. The analogues of Propositions 2.12 and 2.13 hold in this context with the same proofs.

3 Projective families and the analytic index

3.1 Families of pseudodifferential operators. Here we collect several facts about (untwisted) families of pseudodifferential operators.

Let $\pi\colon X \to Y$ be a smooth fibration and E a vector bundle on X.

We denote by $\Psi^m_{\mathcal{L}}(X|Y; E)$ the space of classical fiberwise pseudodifferential operators of order $\leq m$ on π, acting on the sections of the vector bundle E. As usual, we set

$$\Psi(X|Y; E) := \bigcup_{m\in\mathbb{Z}} \Psi^m(X|Y; E)$$

and

$$\Psi^{-\infty}(X|Y; E) := \bigcap_{m\in\mathbb{Z}} \Psi^m(X|Y; E).$$

Recall that composition endows each $\Psi(X|Y;E)$ with the structure of a filtered algebra and that $\Psi^{-\infty}(X|Y;E)$ is an ideal in this algebra. $\Psi(X|Y;E)$ is also a module over $C^\infty(Y)$ and the composition is $C^\infty(Y)$-linear.

We have the following elementary general result:

Lemma 3.1 *Assume $\pi\colon X \to Y$ is a smooth fibration, E a vector bundle on X, L a line bundle on Y. Define a map $\chi_L\colon \Psi(X|Y;E) \to \Psi(X|Y;E\otimes\pi^*L)$ by*

$$\chi_L(D)(e\otimes\pi^*(l)) = D(e)\otimes\pi^*l$$

for $D\in\Psi_{\mathcal{L}}(X|Y;\mathcal{E})$, $e\in\Gamma_c(E)$, $l\in\Gamma(L)$. Then χ_L is a well-defined isomorphism of algebras and $C^\infty(Y)$-modules.

$$\chi_{L_1\otimes L_2} = \chi_{L_2}\circ\chi_{L_1}$$

If we have two vector bundles E, E' on X we denote by $\Psi(X|Y;E,E')$ the set of fiberwise pseudodifferential operators $\Gamma_c(E)\to\Gamma(E')$. For L a line bundle on Y we again have the isomorphism of $C^\infty(Y)$-modules $\chi_L\colon \Psi(X|Y;E,E') \to \Psi(X|Y;E\otimes\pi^*L, E'\otimes\pi^*L)$ defined by the same formula.

We have the vertical cotangent bundle $T^*(X|Y) = T^*X/(\operatorname{Ker}\pi_*)^\perp$. $\mathring{T}^*(X|Y)$ denotes (the total space of) this bundle with the zero section removed, and $p\colon \mathring{T}^*(X|Y)\to X$ is the natural projection. Recall that for $P\in\Psi^m(X|Y;E,E')$ the principal symbol $\sigma_m(P)$ is an m-homogeneous smooth section over $\mathring{T}^*(X|Y)$ of the vector bundle $p^*\operatorname{Hom}(E,E')$. Then, identifying the canonically isomorphic bundles $\operatorname{Hom}(E,E')$ and $\operatorname{Hom}(E\otimes\pi^*L, E'\otimes\pi^*L)$ we have $\sigma_m(P) = \sigma_m(\chi_L(P))$.

3.2 Projective families. Let $\pi: M\to B$ be a smooth fibration with compact fibers. Let $\mathcal{L}$ be a gerbe on B such that $\pi^*[\mathcal{L}]$ is a torsion class in $H^3(M,\mathbb{Z})$. Let $\mathcal{E}$ be a horizontally $\mathcal{L}$-twisted bundle on M, cf. Section 2.6. We fix a descent datum $(U_\alpha, \mathcal{L}_{\alpha\beta}, \mu_{\alpha\beta\gamma}, \mathcal{E}_\alpha, \varphi_{\alpha\beta})$ for $\mathcal{E}$. For any $(\alpha,\beta)\in\Lambda^2$ with $U_{\alpha\beta}\neq\emptyset$ we have an isomorphism of filtered algebras, respecting the $C^\infty(U_{\alpha\beta})$-module structure:

$$\phi_{\alpha\beta}\colon \Psi(\pi^{-1}U_{\alpha\beta}|U_{\alpha\beta};\mathcal{E}_\beta) \to \Psi(\pi^{-1}U_{\alpha\beta}|U_{\alpha\beta};\mathcal{E}_\alpha), \tag{3.1}$$

It is defined as the composition

$$\Psi(\pi^{-1}U_{\alpha\beta}|U_{\alpha\beta};\mathcal{E}_\beta) \overset{\psi_{\alpha\beta}}{\to} \Psi(\pi^{-1}U_{\alpha\beta}|U_{\alpha\beta};\mathcal{E}_\beta\otimes\pi^*\mathcal{L}_{\alpha\beta}) \overset{\varphi_{\alpha\beta}}{\to} \Psi(\pi^{-1}U_{\alpha\beta}|U_{\alpha\beta};\mathcal{E}_\alpha)$$

where $\psi_{\alpha\beta} = \chi^{-1}_{\mathcal{L}_{\alpha\beta}} = \chi_{\mathcal{L}_{\beta\alpha}}$.

Recall (cf. [4]) that for every $\alpha\in\Lambda$ we have an infinite dimensional bundle $\pi_*\mathcal{E}_\alpha$ on U_α defined by $\Gamma(V,\pi_*\mathcal{E}_\alpha) = \Gamma(\pi^{-1}V,\mathcal{E}_\alpha)$, $V\subset U_\alpha$. Over $U_{\alpha\beta}$ we have isomorphisms $\pi_*\varphi_{\alpha\beta}\colon \pi_*\mathcal{E}_\alpha\otimes\mathcal{L}_{\alpha\beta}\to\pi_*\mathcal{E}_\beta$ defined by

$$\pi_*\varphi_{\alpha\beta}(\xi\otimes l) = \varphi(\xi\otimes\pi^*(l)).$$

Here $\xi\in\Gamma(U_{\alpha\beta},\pi_*\mathcal{E}_\alpha) = \Gamma(\pi^{-1}U_{\alpha\beta},\mathcal{E}_\alpha)$, $l\in\Gamma(U_{\alpha\beta},\mathcal{L}_{\alpha\beta})$.

Note that the isomorphisms $\pi_*\varphi_{\alpha\beta}\colon \pi_*\mathcal{E}_\alpha\otimes\mathcal{L}_{\alpha\beta}\to\pi_*\mathcal{E}_\beta$ induce the isomorphisms

$$(\pi_*\varphi_{\alpha\beta})^*\colon \operatorname{End}(\pi_*\mathcal{E}_\beta)\to\operatorname{End}(\pi_*\mathcal{E}_\alpha\otimes\mathcal{L}_{\alpha\beta})\cong\operatorname{End}(\pi_*\mathcal{E}_\alpha)$$

over the $U_{\alpha\beta}$. The restriction of this isomorphism to $\Psi(\pi^{-1}U_{\alpha\beta}|U_{\alpha\beta},\mathcal{E}_\beta)\subset\operatorname{End}(\pi_*\mathcal{E}_\beta)$ coincides with the isomorphism

$$\phi_{\alpha\beta}\colon \Psi(\pi^{-1}U_{\alpha\beta}|U_{\alpha\beta},\mathcal{E}_\beta)\to\Psi(\pi^{-1}U_{\alpha\beta}|U_{\alpha\beta},\mathcal{E}_\alpha).$$

Since the isomorphisms $(\pi_*\varphi_{\alpha\beta})^*$ satisfy the natural cocycle identity we have the following:

Lemma 3.2 *The isomorphisms $\phi_{\alpha\beta}$ satisfy*

$$\phi_{\alpha\beta} \circ \phi_{\beta\gamma} = \phi_{\alpha\gamma}$$

whenever $U_{\alpha\beta\gamma} \neq \emptyset$.

Recall that the isomorphisms $\varphi_{\alpha\beta}$ induce the natural isomorphisms $\varphi^*_{\alpha\beta}\colon \mathrm{End}(\mathcal{E}_\beta) \to \mathrm{End}(\mathcal{E}_\alpha)$. Then

$$\sigma_m \circ \phi_{\alpha\beta} = p^* \left(\varphi^*_{\alpha\beta}\right) \circ \sigma_m \tag{3.2}$$

Definition 3.3 A fiberwise pseudodifferential operator P of order $\leq m$ with coefficients in the horizontally $\mathcal{L}$-twisted vector bundle $\mathcal{E}$ is a collection $\{P_\alpha\}_{\alpha\in\Lambda}$, $P_\alpha \in \Psi^m(\pi^{-1}U_\alpha|U_\alpha;\mathcal{E}_\alpha)$ such that

$$P_\alpha = \phi_{\alpha\beta}(P_\beta).$$

where $\phi_{\alpha\beta}$ is defined in Equation 3.1. The space of fiberwise pseudodifferential operators of order $\leq m$, with coefficients in the $\pi^*\mathcal{L}$-twisted vector bundle $\mathcal{E}$, is denoted by $\Psi^m_{\mathcal{L}}(M|B;\mathcal{E})$.

Note that Equation (3.2) implies that if $P = \{P_\alpha\} \in \Psi^m_{\mathcal{L}}(M|B;\mathcal{E})$ then the collection $\sigma_m(P_\alpha)$ defines a section of the (untwisted) bundle $p^*\,\mathrm{End}(\mathcal{E})$. We will call this section the principal symbol of $P = \{P_\alpha\}$

Remark 3.4 We define in the same way the space $\Psi^m_{\mathcal{L}}(M|B;\mathcal{E},\mathcal{E}')$ of fiberwise pseudodifferential operators of order $\leq m$, from the horizontally $\mathcal{L}$-twisted vector bundle $\mathcal{E}$ to the horizontally $\mathcal{L}$-twisted vector bundle $\mathcal{E}'$. In particular $\Psi^m_{\mathcal{L}}(M|B;\mathcal{E},\mathcal{E}) = \Psi^m_{\mathcal{L}}(M|B;\mathcal{E})$. We also have a principal symbol map $\sigma_m\colon \Psi^m_{\mathcal{L}}(M|B;\mathcal{E},\mathcal{E}) \to p^*\,\mathrm{Hom}(\mathcal{E},\mathcal{E}')$.

We set

$$\Psi_{\mathcal{L}}(M|B;\mathcal{E}) := \bigcup_{m\in\mathbb{Z}} \Psi^m_{\mathcal{L}}(M|B;\mathcal{E}) \text{ and } \Psi^{-\infty}_{\mathcal{L}}(M|B;\mathcal{E}) := \bigcap_{m\in\mathbb{Z}} \Psi^m_{\mathcal{L}}(M|B;\mathcal{E}).$$

Introduce now a composition in $\Psi_{\mathcal{L}}(M|B;\mathcal{E})$ by

$$\{P_\alpha\} \circ \{Q_\alpha\} = \{P_\alpha Q_\alpha\}$$

Since, $\phi_{\alpha\beta}$ are algebra isomorphisms the right hand side of this equality defines an element in $\Psi_{\mathcal{L}}(M|B;\mathcal{E})$.

Proposition 3.5 *The composition of operators is $C^\infty(B)$-linear and endows $\Psi_{\mathcal{L}}(M|B;\mathcal{E})$ with the structure of associative algebra; $\Psi^{-\infty}_{\mathcal{L}}(M|B;\mathcal{E})$ is an ideal in $\Psi_{\mathcal{L}}(M|B;\mathcal{E})$.*

We can now define the algebra of forms on B with values in $\Psi_{\mathcal{L}}(M|B;\mathcal{E})$ by

$$\Omega^*\left(B, \Psi_{\mathcal{L}}(M|B;\mathcal{E})\right) = \Omega^*(B) \otimes_{C^\infty(B)} \Psi_{\mathcal{L}}(M|B;\mathcal{E}).$$

Recall that for every α and $V \subset U_\alpha$ we have a fiberwise trace $\mathrm{Tr}_\alpha\colon \Psi^{-\infty}(\pi^{-1}V|V;\mathcal{E}_\alpha) \to C^\infty(V)$. It is easy to see that for $P \in \Psi^{-\infty}(\pi^{-1}U_{\alpha\beta}|U_{\alpha\beta};\mathcal{E}_\beta)$

$$\mathrm{Tr}_\alpha\, \phi_{\alpha\beta}(P) = \mathrm{Tr}_\beta(P).$$

We therefore obtain a well-defined map $\mathrm{Tr}\colon \Psi^{-\infty}_{\mathcal{L}}(M|B,\mathcal{E}) \to C^\infty(B)$ by setting $\mathrm{Tr}\{P_\alpha\}|_{U_\alpha} = \mathrm{Tr}_\alpha(P_\alpha)$. This trace is a $C^\infty(B)$-module map satisfying $\mathrm{Tr}[A,B] = 0$. It extends naturally to define a map $\mathrm{Tr}\colon \Omega^*\left(B, \Psi^{-\infty}_{\mathcal{L}}(M|B;\mathcal{E})\right) \to \Omega^*(B)$.

If the bundle $\mathcal{E}$ is $\mathbb{Z}_2$ graded we have a similarly defined supertrace

$$\mathrm{Str}\colon \Omega^*\left(B, \Psi^{-\infty}_{\mathcal{L}}(M|B;\mathcal{E})\right) \to \Omega^*(B)\,.$$

Note that the definition of the $\Psi_{\mathcal{L}}(M|B;\mathcal{E})$ depends on the descent datum for $\mathcal{E}$. It is straightforward however to see that an isomorphism of descent data defines canonically an isomorphism of the corresponding bundles of algebras.

3.3 Analytic index.

Definition 3.6 Let $m \geq 0$ be fixed. A fiberwise pseudodifferential operator $P \in \Psi^m_{\mathcal{L}}(M|B;\mathcal{E},\mathcal{E}')$ is fiberwise elliptic if the principal symbol $\sigma_m(P) \in \Gamma(p^* \operatorname{Hom}(\mathcal{E},\mathcal{E}'))$is an isomorphism.

We say that $Q \in \Psi^{-m}_{\mathcal{L}}(M|B;\mathcal{E}',\mathcal{E})$ is a parametrix of $P \in \Psi^m_{\mathcal{L}}(M|B;\mathcal{E},\mathcal{E}')$ if $PQ - 1 \in \Psi^{-\infty}_{\mathcal{L}}(M|B;\mathcal{E}',\mathcal{E}')$ and $QP - 1 \in \Psi^{-\infty}_{\mathcal{L}}(M|B;\mathcal{E},\mathcal{E})$.

Lemma 3.7 *Every elliptic $P \in \Psi^m_{\mathcal{L}}(M|B;\mathcal{E},\mathcal{E}')$ has a parametrix.*

Proof Construct first a parametrix R_α for P_α. Let ρ_α be a partition of unity subordinate to $\{U_\alpha\}$. Then define $R'_\alpha = \sum_\beta \rho_\beta \phi_{\alpha\beta}(R_\beta)$. Then each R'_α is a parametrix for P_α and $\phi_{\alpha\beta}(R'_\beta) = R'_\alpha$. □

Let $D \in \Psi^m_{\mathcal{L}}(M|B;\mathcal{E},\mathcal{E}')$ be elliptic. Let $F \in \Psi^0_{\mathcal{L}}(M|B;\mathcal{E},\mathcal{E}')$ be such that $\sigma_0(F)|_{S^*(M|B)} = \sigma_m(D)|_{S^*(M|B)}$. Choose a parametrix R for F. Let $U_D \in \Psi^0_{\mathcal{L}}(M|B;\mathcal{E}\oplus\mathcal{E}')$ be an invertible operator such that

$$U_D - \begin{bmatrix} 0 & -R \\ F & 0 \end{bmatrix} \in \Psi^{-\infty}_{\mathcal{L}}(M|B;\mathcal{E}\oplus\mathcal{E}').$$

An explicit construction of an example of such an operator is as follows. Let $S_0 = 1 - RF$, $S_1 = 1 - FR$. Then set $U_D = \begin{bmatrix} S_0 & -(1+S_0)R \\ F & S_1 \end{bmatrix}$. With such a choice the inverse is given by an explicit formula $U_D^{-1} = \begin{bmatrix} S_0 & (1+S_0)R \\ -F & S_1 \end{bmatrix}$.

Definition 3.8 The index of D is the K-theory class of the algebra $\Psi^{-\infty}_{\mathcal{L}}(M|B;\mathcal{E}\oplus\mathcal{E}')$ defined by

$$\begin{aligned} &\operatorname{ind}(D) = [P_D - Q] \in K_0(\Psi^0_{\mathcal{L}}(M|B,\mathcal{E}\oplus\mathcal{E}'), \\ &\Psi^{-\infty}_{\mathcal{L}}(M|B,\mathcal{E}\oplus\mathcal{E}')) \cong K_0(\Psi^{-\infty}_{\mathcal{L}}(M|B,\mathcal{E}\oplus\mathcal{E}')), \end{aligned}$$

where P_D and Q are the idempotents given by $P_D = U_D \begin{bmatrix} 1 & 0 \\ 0 & 0 \end{bmatrix} U_D^{-1}$ and $Q = \begin{bmatrix} 0 & 0 \\ 0 & 1 \end{bmatrix}$.

We leave to the reader the standard K-theoretic proof that the index is well defined and is stable under homotopies of D in the class of elliptic operators in $\Psi^m_{\mathcal{L}}(M|B;\mathcal{E},\mathcal{E}')$. Assume that the horizontally $\mathcal{L}$-twisted bundles $\mathcal{E}$ and $\mathcal{E}'$ are Hermitian (and in particular $\mathcal{L}$ is unitary) and that the fibers of π are equipped with smoothly varying volume forms. In this situation for a projective family D we can define a formally adjoint projective family D^* by forming the formal adjoints for each family D_α.

Lemma 3.9 *Then, identifying $K_0(\Psi^{-\infty}_{\mathcal{L}}(M|B,\mathcal{E}\oplus\mathcal{E}'))$ with*

$$K_0(\Psi^{-\infty}_{\mathcal{L}}(M|B,\mathcal{E}'\oplus\mathcal{E}))$$

we have

$$\operatorname{ind} D^* = -\operatorname{ind} D.$$

Proof We have

$$\operatorname{ind} D^* = \left[U_{D^*} \begin{bmatrix} 0 & 0 \\ 0 & 1 \end{bmatrix} U_{D^*}^{-1} - \begin{bmatrix} 1 & 0 \\ 0 & 0 \end{bmatrix} \right] \in K_0(\Psi_{\mathcal{L}}^{-\infty}(M|B, \mathcal{E} \oplus \mathcal{E}')).$$

By deforming D we may assume that $\sigma_m(D)|_{S^*(M|B)}$ is an isometry. In this case we may choose $U_{D^*} = U_D^{-1}$, and the statement follows. □

3.4 Chern character of the index. We continue with the notations of the previous section. We assume that we are given a horizontally $\mathcal{L}$-twisted bundle $\mathcal{E}$ with a connection represented by a descent datum $(U_\alpha, \mathcal{L}_{\alpha\beta}, \mu_{\alpha\beta\gamma}, \mathcal{E}_\alpha, \varphi_{\alpha\beta})$ with connection $(\nabla_\alpha, \nabla_{\alpha\beta}, \omega_\alpha)$; see Section 2.6. Following Mathai and Stevenson [30], we describe in this section a morphism of complexes $CC_\bullet^-(\Psi_{\mathcal{L}}^{-\infty}(M|B;\mathcal{E})) \to (\Omega^*(M)[u], d_\Omega)_\bullet$.

Recall the bundles $\pi_*\mathcal{E}_\alpha$ and isomorphisms $\pi_*\varphi_{\alpha\beta}$ defined in Section 3.2. It is easy to see that $(U_\alpha, \mathcal{L}_{\alpha\beta}, \mu_{\alpha\beta\gamma}, \pi_*\mathcal{E}_\alpha, \pi_*\varphi_{\alpha\beta})$ is a descent datum for an infinite dimensional twisted bundle. We now proceed to define a connection on this descent datum.

Choose a horizontal distribution, i.e., a subbundle $\mathcal{H} \subset TM$ such that $TM = \mathcal{H} \oplus T(M|B)$. This choice together with connections $\nabla_\alpha^{\mathcal{E}}$ defines for each α a connection $\nabla_\alpha^{\mathcal{H}}$ as follows:

$$(\nabla_\alpha^{\mathcal{H}})_X \xi = (\nabla_\alpha^{\mathcal{E}})_{X^{\mathcal{H}}} \xi$$

where $X^{\mathcal{H}}$ is the horizontal lift of $X \in \Gamma(B, TB)$.

Lemma 3.10 $(\pi_*\varphi_{\alpha\beta})^* \nabla_\beta^{\mathcal{H}} = \nabla_\alpha^{\mathcal{H}} \otimes \mathrm{id} + \mathrm{id} \otimes \nabla_{\alpha\beta}$.

The curvature of the connection $\nabla_\alpha^{\mathcal{H}}$ is a 2-form $\theta_\alpha^{\mathcal{H}}$ on U_α with values in fiberwise differential operators given by

$$\theta_\alpha^{\mathcal{H}}(X,Y) = \theta_\alpha^{\mathcal{E}}(X^{\mathcal{H}}, Y^{\mathcal{H}}) + (\nabla_\alpha^{\mathcal{E}})_{T(X,Y)}.$$

where

$$T^{\mathcal{H}}(X,Y) = [X^{\mathcal{H}}, Y^{\mathcal{H}}] - [X,Y]^{\mathcal{H}}, \ X, Y \in \Gamma(B, TB). \tag{3.3}$$

Each $\nabla_\alpha^{\mathcal{H}}$ defines a filtration-preserving derivation $\partial_\alpha^{\mathcal{H}}$ of the algebra of fiberwise pseudodifferential operators

$$\partial_\alpha^{\mathcal{H}} \colon \Psi(\pi^{-1}U_\alpha|U_\alpha, \mathcal{E}_\alpha) \to \Omega^1(U_\alpha, \Psi(\pi^{-1}U_\alpha|U_\alpha, \mathcal{E}_\alpha)) \text{ defined by } \partial_\alpha^{\mathcal{H}}(D) = [\nabla_\alpha^{\mathcal{H}}, D].$$

If $D \in \Psi(\pi^{-1}U_{\alpha\beta}|U_{\alpha\beta}, \mathcal{E}_\beta)$ then the result of Lemma 3.10 implies that

$$\partial_\alpha^{\mathcal{H}}(\phi_{\alpha\beta}(D)) = \phi_{\alpha\beta}(\partial_\beta^{\mathcal{H}}(D)).$$

Therefore, if $\{D_\alpha\}$, $D_\alpha \in \Psi_{\mathcal{L}}(\pi^{-1}U_\alpha|U_\alpha, \mathcal{E}_\alpha)$, defines an element in $\Psi_{\mathcal{L}}^m(M|B, \mathcal{E})$ then $\{\partial_\alpha^{\mathcal{H}}(D_\alpha)\} \in \Omega^1(B, \Psi_{\mathcal{L}}^m(M|B, \mathcal{E}))$. We therefore obtain a derivation

$$\partial^{\mathcal{H}} \colon \Psi_{\mathcal{L}}(M|B, \mathcal{E}) \to \Omega^1(B, \Psi_{\mathcal{L}}(M|B, \mathcal{E})),$$

which extends to a derivation of the algebra $\Omega^*(B, \Psi_{\mathcal{L}}(M|B, \mathcal{E}))$.

Lemma 3.11 *There exists* $\theta^{\mathcal{H}} \in \Omega^2(B, \Psi_{\mathcal{L}}^1(M|B, \mathcal{E}))$ *such that*

$$\theta^{\mathcal{H}}|_{U_\alpha} = \theta_\alpha^{\mathcal{H}} + \pi^*\omega_\alpha$$

Proof By Lemma 3.10, $\phi_{\alpha\beta}^* \theta_\beta^{\mathcal{H}} = \theta_\alpha^{\mathcal{H}} + \pi^*(\omega_\alpha - \omega_\beta)$, and the statement follows as in Proposition 2.12. □

We have

$$(\partial^{\mathcal{H}})^2(D) = [\theta^{\mathcal{H}}, D] \text{ and } \partial^{\mathcal{H}}(\theta^{\mathcal{H}}) = 2\pi i(\pi^* \Omega).$$

where Ω is the 3-curvature form of the connection on $\mathcal{L}$.

Following Mathai and Stevenson [30] one can construct the morphism of complexes

$$\Phi_{\nabla^{\mathcal{H}}} \colon CC^-_\bullet(\Psi^{-\infty}_{\mathcal{L}}(M|B; \mathcal{E} \oplus \mathcal{E}')) \to (\Omega^*(B)[u], d_\Omega)_\bullet$$

as follows. (Note that in the nontwisted case a similar morphism was constructed in [21, 33, 34].) Denote by

$$\Delta^k := \{(t_0, \dots, t_k) \in \mathbb{R}^k \mid 0 \le t_i, \sum_{i=0}^{k} t_i = 1\}.$$

the standard k-simplex.

Define the maps $\Phi^k_{\nabla^{\mathcal{H}}} \colon C_k\left(\Psi_{\mathcal{L}}{}^{-\infty}(M|B; \mathcal{E} \oplus \mathcal{E}')\right) \to \Omega^*(M)[u]$ by

$$\Phi^k_{\nabla^{\mathcal{H}}}(A_0, \dots, A_k) := \int_{\Delta^k} \operatorname{Tr}\left(A_0 e^{-ut_0 \frac{\theta^{\mathcal{H}}}{2\pi i}} \partial^{\mathcal{H}}(A_1) e^{-ut_1 \frac{\theta^{\mathcal{H}}}{2\pi i}} \cdots e^{-ut_{k-1} \frac{\theta^{\mathcal{H}}}{2\pi i}} \partial^{\mathcal{H}}(A_k) e^{-ut_k \frac{\theta^{\mathcal{H}}}{2\pi i}}\right) dt_1 \cdots dt_k,$$

for $A_0 \otimes \cdots \otimes A_k \in C_k\left(\Psi^{-\infty}_{\mathcal{L}}(M|B; \mathcal{E} \oplus \mathcal{E}')\right)$. Then let $\Phi_{\nabla^{\mathcal{H}}} = \sum_{k=0}^{\infty} \Phi^k_{\nabla^{\mathcal{H}}}$.

Theorem 3.12 [30] *The map*

$$\Phi_{\nabla^{\mathcal{H}}} \colon (CC^-_\bullet(\Psi^{-\infty}_{\mathcal{L}}(M|B; \mathcal{E})), b + uB) \to (\Omega^*(B)[u], d_\Omega)_\bullet$$

is a morphism of complexes.

This morphism depends on the choice of horizontal distribution $\mathcal{H}$. However the results of [30] show that a different choice of $\mathcal{H}$ leads to a chain homotopic morphism.

Assume now that $D \in \Psi_{\mathcal{L}}(M|B; \mathcal{E}, \mathcal{E}')$ is an twisted elliptic family. In Definition 3.8 we defined $\operatorname{ind}(D) = [P_D - Q] \in K_0(\Psi^{-\infty}_{\mathcal{L}}(M|B; \mathcal{E}, \mathcal{E}'))$. Here the idempotents P_D and Q belong to the algebra $\overline{\Psi^{-\infty}_{\mathcal{L}}(M|B; \mathcal{E} \oplus \mathcal{E}')}$ – the unitalization of $\Psi^{-\infty}_{\mathcal{L}}(M|B; \mathcal{E} \oplus \mathcal{E}')$ with the multiplier $Q \in \Psi^0_{\mathcal{L}}(M|B; \mathcal{E} \oplus \mathcal{E}')$ adjoined. It follows that if we directly apply the formula (2.1) to $[P_D - Q]$ we obtain a 0-chain $\operatorname{Ch}[P_D - Q]$ in the relative cyclic complex

$$CC^-_\bullet\left(\overline{\Psi^{-\infty}_{\mathcal{L}}(M|B; \mathcal{E} \oplus \mathcal{E}')}, \Psi^{-\infty}_{\mathcal{L}}(M|B; \mathcal{E} \oplus \mathcal{E}')\right).$$

We have the natural morphism

$$\iota \colon CC^-_\bullet\left(\Psi^{-\infty}_{\mathcal{L}}(M|B; \mathcal{E} \oplus \mathcal{E}')\right) \to CC^-_\bullet\left(\overline{\Psi^{-\infty}_{\mathcal{L}}(M|B; \mathcal{E} \oplus \mathcal{E}')}, \Psi^{-\infty}_{\mathcal{L}}(M|B; \mathcal{E} \oplus \mathcal{E}')\right),$$

and the following equality of homology classes: $\iota(\operatorname{ch}(\operatorname{ind} D)) = [\operatorname{Ch}[P_D - Q]]$. It is straightforward to see that the map $\Phi_{\nabla^{\mathcal{H}}}$ extends to a morphism

$$\Phi_{\nabla^{\mathcal{H}}} \colon CC^-_\bullet\left(\overline{\Psi^{-\infty}_{\mathcal{L}}(M|B; \mathcal{E} \oplus \mathcal{E}')}, \Psi^{-\infty}_{\mathcal{L}}(M|B; \mathcal{E} \oplus \mathcal{E}')\right) \to (\Omega^*(B)[u], d_\Omega)_\bullet$$

defined by the same formula, and that we have an equality $\Phi_{\nabla^{\mathcal{H}}} \circ \iota = \Phi_{\nabla^{\mathcal{H}}}$. It follows that

$$\Phi_{\nabla^{\mathcal{H}}}(\operatorname{ch}(\operatorname{ind} D)) = [\Phi_{\nabla^{\mathcal{H}}}(\operatorname{Ch}[P_D - Q])].$$

4 Dirac operators and superconnections

4.1 Dirac operators. The goal of this section is to give a superconnection proof of the family index theorem for projective families of Dirac operators. We assume that the fibers of the smooth fibration $\pi : M \to B$ are even dimensional compact Riemannian manifolds.

We begin with the definition of a horizontally twisted Clifford module. Denote by $C(M|B)$ the Clifford algebra of the fiberwise cotangent bundle $T^*(M|B) = T^*M/(\ker \pi_*)^\perp$. Let $\mathcal{L}$ be a unitary gerbe on B.

Definition 4.1
- A twisted Clifford module is a horizontally $\mathcal{L}$-twisted Hermitian $\mathbb{Z}_2$-graded vector bundle $\mathcal{E} = \mathcal{E}^+ \oplus \mathcal{E}^-$ on M together with the homomorphism
$c\colon C(M|B) \to \mathrm{End}(\mathcal{E})$ of bundles of unital $\mathbb{Z}_2$-graded $*$-algebras.
- A Clifford connection $\nabla^{\mathcal{E}}$ on $\mathcal{E}$ is a Hermitian connection such that $\nabla^{\mathcal{E}}(c(a)) = c(\nabla^{M|B}(a))$.

Clifford connections on Clifford modules always exist; the proof is analogous to the proof of existence of twisted connections in Lemma 2.11.

Choose a horizontal distribution, i.e., a subbundle $\mathcal{H} \subset TM$ such that $TM = \mathcal{H} \oplus T(M|B)$. This choice together with the Riemannian metric on the fibers of π allows one to define a connection $\nabla^{M|B}$ on the fiberwise tangent bundle $T(M|B)$, see [4] Section 10.1. We denote by $R^{M|B}$ the curvature of this connection.

Set $\mathrm{End}_{C(M|B)}(\mathcal{E}) = \{A \in \mathrm{End}(\mathcal{E}) \mid [A, c(a)] = 0 \text{ for every } a \in C(M|B)\}$. Let $\Gamma \in C(M|B)$ be the chirality operator defined locally by $\Gamma = i^{k/2} e^1 \cdots e^k$ where $k = \dim M - \dim B$ and $e^1, \dots, e^k$ is the local orthonormal basis of $T^*(M|B)$. Define then the relative supertrace

$$\mathrm{str}_{\mathcal{E}/\mathcal{S}}\colon \mathrm{End}_{C(M|B)}(\mathcal{E}) \to C^\infty(M) \text{ by } \mathrm{str}_{\mathcal{E}/\mathcal{S}}(A) = 2^{-k/2}\, \mathrm{str}\, c(\Gamma) A.$$

We fix from now on a Clifford connection $\nabla^{\mathcal{E}}$ on $\mathcal{E}$ and a descent datum $(U_\alpha, \mathcal{L}_{\alpha\beta}, \mu_{\alpha\beta\gamma}, \mathcal{E}_\alpha, \varphi_{\alpha\beta})$ for the horizontally $\mathcal{L}$-twisted Clifford module $\mathcal{E}$. The connection $\nabla^{\mathcal{E}}$ defines a connection $(\nabla^{\mathcal{E}}_\alpha, \nabla_{\alpha\beta}, \omega_\alpha)$ (defined up to equivalence) on this descent datum. Each $\mathcal{E}_\alpha$ is then a Clifford module on the fibration $\pi^{-1}U_\alpha \to U_\alpha$, and each connection $\nabla^{\mathcal{E}}_\alpha$ is a Clifford connection.

Recall (see Proposition 2.12) that one defines $\theta^{\mathcal{E}} \in \Omega^2(M, \mathrm{End}(\mathcal{E}))$ by setting $\theta^{\mathcal{E}}|_{\pi^{-1}U_\alpha} = \theta^{\mathcal{E}}_\alpha + \pi^*\omega_\alpha$. Denote by $c(R^{M|B})$ the action of the 2-form with values in the Clifford algebra obtained from $R^{M|B}$ via the Lie algebra isomorphism $\mathfrak{so}(T(M|B)) \to C^2(M|B)$. Here $C^2(M|B) \subset C(M|B)$ is the subspace consisting of elements $\sum u_i v_i$, $u_i, v_i \in T^*(M|B)$ with $\sum \langle u_i, v_i \rangle = 0$. Define $\theta^{\mathcal{E}/\mathcal{S}} = \theta^{\mathcal{E}} - c(R^{M|B})$.

The argument in [4], Proposition 3.43, shows that $\theta^{\mathcal{E}/\mathcal{S}} \in \Omega^2(M, \mathrm{End}_{C(M|B)}(\mathcal{E}))$. Over an open set $\pi^{-1}U_\alpha$ we have $\theta^{\mathcal{E}/\mathcal{S}}|_{U_\alpha} = \theta^{\mathcal{E}/\mathcal{S}}_\alpha + \pi^*\omega_\alpha$ where $\theta^{\mathcal{E}/\mathcal{S}}_\alpha \in \mathrm{End}_{C(M|B)}(\mathcal{E}_\alpha)$ is defined via the equality $\theta^{\mathcal{E}}_\alpha = \theta^{\mathcal{E}/\mathcal{S}}_\alpha + c(R^{M|B})$. We can then define a differential form $\mathrm{Ch}_{\mathcal{L}}(\mathcal{E}/\mathcal{S})$ by

$$\mathrm{Ch}_{\mathcal{L}}(\mathcal{E}/\mathcal{S}) = \mathrm{str}_{\mathcal{E}/\mathcal{S}}\, e^{-\frac{u\theta^{\mathcal{E}/\mathcal{S}}}{2\pi i}} \in \Omega^*(M)[u]. \tag{4.1}$$

The proof of the following result is standard and analogous to the proof of Proposition 2.13.

Lemma 4.2 *We have $d_{\pi^*\Omega}\,\mathrm{Ch}_{\mathcal{L}}(\mathcal{E}/\mathcal{S}) = 0$ and the corresponding class in $H^*_{\pi^*\Omega}(M)$ is independent of the choice of Clifford connection $\nabla^{\mathcal{E}}$.*

We also introduce the fiberwise $\widehat{A}$-genus by $\widehat{A}(TM|B) = \widehat{A}\left(\frac{u}{2\pi i}R^{M|B}\right) \in \Omega^*(M)[u]$, where $\widehat{A}(x)$ is the power series defined by

$$\widehat{A}(x) = \det{}^{1/2}\left(\frac{x/2}{\sinh x/2}\right).$$

Using the above data, we can define on each fibration $\pi^{-1}U_\alpha \to U_\alpha$ a family of Dirac operators D_α acting on the sections of the bundle $\mathcal{E}_\alpha$. Locally $D_\alpha = \sum_i c(e^i)\left(\nabla^{\mathcal{E}}_\alpha\right)_{e_i}$, where $\{e_i\}$, $\{e^i\}$ are dual bases of $T(M|B)$ and $T^*(M|B)$ respectively. We leave it to the reader to check the following.

Lemma 4.3 *The collection $D = \{D_\alpha\}$ defines an element in $\Psi^1_{\mathcal{L}}(M|B;\mathcal{E})$.*

With respect to the decomposition $\mathcal{E} = \mathcal{E}^+ \oplus \mathcal{E}^-$ we have the decomposition $\Psi_{\mathcal{L}}(M|B;\mathcal{E}) = \Psi_{\mathcal{L}}(M|B;\mathcal{E}^+,\mathcal{E}^+) \oplus \Psi_{\mathcal{L}}(M|B;\mathcal{E}^+,\mathcal{E}^-) \oplus \Psi_{\mathcal{L}}(M|B;\mathcal{E}^-,\mathcal{E}^+) \oplus \Psi_{\mathcal{L}}(M|B;\mathcal{E}^-,\mathcal{E}^-)$. The Dirac operator then decomposes as $D = D^+ \oplus D^-$, where $D^+ \in \Psi^1_{\mathcal{L}}(M|B;\mathcal{E}^+,\mathcal{E}^-)$, $D^- \in \Psi^1_{\mathcal{L}}(M|B;\mathcal{E}^-,\mathcal{E}^+)$

Classical arguments show that D^+ is fiberwise elliptic and hence the analytic index $\mathrm{ind}(D^+)$ of D^+ is well defined in $K(\Psi^{-\infty}_{\mathcal{L}}(M|B;\mathcal{E}))$.

We then have the following result

Theorem 4.4 *The following formula holds in $H^\bullet_{\mathcal{L}}(B)$:*

$$\Phi_{\nabla^{\mathcal{H}}}(\mathrm{ch}(\mathrm{ind}\, D^+))] = u^{-\frac{\dim M - \dim B}{2}}\left[\int_{M/B}\widehat{A}\left(\frac{u}{2\pi i}R^{M|B}\right)\wedge \mathrm{Ch}_{\mathcal{L}}(\mathcal{E}/\mathcal{S})\right].$$

This index theorem is established in [28, 29] for general projective families however with the more restrictive conditions on the class $[\mathcal{L}]$ of the gerbe $\mathcal{L}$. The superconnection proof of this result occupies the rest of the paper.

4.2 Superconnections and index. We continue in the notations of the previous section.

A twisted superconnection $\mathbb{A}$ on the descent datum $(U_\alpha, \mathcal{L}_{\alpha\beta}, \mu_{\alpha\beta\gamma}, \mathcal{E}_\alpha, \varphi_{\alpha\beta})$ is a collection $(\mathbb{A}_\alpha)_{\alpha\in\Lambda}$ of superconnections on the vector bundles $\pi_*\mathcal{E}_\alpha$ over the open sets U_α such that when $U_{\alpha\beta} \neq \emptyset$,

$$(\pi_*\varphi_{\alpha\beta})^*\mathbb{A}_\beta = \mathbb{A}_\alpha \otimes \mathrm{id} + \mathrm{id}\otimes\nabla_{\alpha\beta}. \tag{4.2}$$

We say that $\mathbb{A}$ is a Bismut superconnection if each $\mathbb{A}_\alpha$ is. Specifically we have

$$\mathbb{A}_\alpha = D_\alpha + \nabla^{\mathcal{H}}_\alpha - \frac{1}{4}c(T^{\mathcal{H}})$$

where $T^{\mathcal{H}}$ is defined in (3.3).

The fact that $\nabla^{\mathcal{H}}$ is a connection on the descent datum, and $D \in \Psi^1(M|B,\mathcal{E})$, $c(T^{\mathcal{H}}) \in \Omega^2(B,\Psi^0(M|B,\mathcal{E}))$ implies that the conditions (4.2) are satisfied. From now on $\mathbb{A}$ will denote the Bismut superconnection.

We will also consider the rescaled Bismut superconnection $\mathbb{A}_s$ defined by

$$(\mathbb{A}_s)_\alpha = \mathbb{A}_{\alpha,s} := s^{1/2}D_\alpha + \nabla^{\mathcal{H}}_\alpha - \frac{1}{4}s^{-1/2}c(T^{\mathcal{H}})$$

where s is either a positive number or a multiple of the formal variable u. Denote by $\theta^{\mathbb{A}_s}_\alpha$ the curvature of the rescaled superconnection $(\mathbb{A}_s)^2_\alpha$. In particular we have forms $u\theta^{\mathbb{A}_{\alpha,u^{-1}}} \in \Omega^*(U_\alpha, \Psi(\pi^{-1}U_\alpha|U_\alpha,\mathcal{E}_\alpha))[u^{1/2}]$.

Proposition 4.5 *There exists a form* $u\theta^{\mathbb{A}_{u^{-1}}} \in \Omega^*(B, \Psi(M|B;\mathcal{E}))[u^{1/2}]$ *such that*

$$u\theta^{\mathbb{A}_{u^{-1}}}\Big|_{U_\alpha} = u(\theta^{\mathbb{A}_{\alpha,u^{-1}}} + \pi^*\omega_\alpha).$$

Proof Recall that the curvature of $\nabla_{\alpha\beta}$ is equal to $\omega_\alpha - \omega_\beta$. Therefore from Equation (4.2) we obtain

$$\phi_{\alpha\beta}\left(u\theta^{\mathbb{A}_{\beta,u^{-1}}}\right) = u(\theta^{\mathbb{A}_{\alpha,u^{-1}}} + \pi^*\omega_\alpha - \pi^*\omega_\beta).$$

The statement of the Proposition follows. □

Notice that $u\theta^{\mathbb{A}_{u^{-1}}} = D^2+$ forms of degree > 0. Therefore we can define $e^{-\frac{u\theta^{\mathbb{A}}{u^{-1}}}{2\pi i}} \in \Omega^*(B, \Psi^{-\infty}(M|B;\mathcal{E}))[u^{1/2}]$ by the usual Duhamel's formula.

Note that parity considerations as in the finite-dimensional case show that the coefficients for the nonintegral powers of u are odd with respect to the grading and hence have vanishing supertrace.

Given a superconnection $\mathbb{A} = (\mathbb{A}_\alpha)_{\alpha\in\Lambda}$ on the $\mathbb{Z}_2$-graded horizontally $\mathcal{L}$-twisted Clifford module $\mathcal{E}$, by the infinite-dimensional version of Proposition 2.15 the differential form

$$\mathrm{Str}(e^{-\frac{u\theta^{\mathbb{A}}{u^{-1}}}{2\pi i}}) \in \Omega^*(B)[u],$$

is closed with respect to the twisted de Rham differential d_Ω. The proof is identical to the proof of the first part of Proposition 2.13. The proof of the following Theorem is adapted from [22] and it uses ideas from [23]. The use of cyclic theory is inspired by [35].

Theorem 4.6 *The following equality holds in* $H^\bullet_{\mathcal{L}}(B)$*:*

$$\Phi_{\nabla^{\mathcal{H}}}(\mathrm{ch}(\mathrm{ind}\, D^+)) = \left[\mathrm{Str}(e^{-\frac{u\theta^{\mathbb{A}}{u^{-1}}}{2\pi i}})\right].$$

Proof Consider the twisted bundle $\widetilde{\mathcal{E}} = \mathcal{E} \oplus \mathcal{E}$ with the grading given by $\Gamma = \begin{bmatrix}\gamma & 0\\ 0 & -\gamma\end{bmatrix}$. The algebra $\Psi_{\mathcal{L}}(M|B;\widetilde{\mathcal{E}})$ of operators on $\widetilde{\mathcal{E}}$ is naturally $\mathbb{Z}_2$-graded. When discussing cyclic complexes of this algebra and its subalgebras we always consider it as a $\mathbb{Z}_2$-graded algebra. For an operator or (super)connection K on $\mathcal{E}$ set $\widetilde{K} = K \oplus K$; so, for example, $\widetilde{\mathbb{A}} = \mathbb{A} \oplus \mathbb{A}$, $\widetilde{\nabla}^{\mathcal{H}} = \nabla^{\mathcal{H}} \oplus \nabla^{\mathcal{H}}$, etc.

Let $F = D(1+D^2)^{-1/2}$. Then it is immediate that $F \in \Psi^0_{\mathcal{L}}(M|B;\mathcal{E})$ is odd with respect to the grading Γ and fiberwise elliptic. Construct the invertible operator $U_D \in \Psi^{-\infty}_{\mathcal{L}}(M|B;\widetilde{\mathcal{E}})$ by the same formula as before. Namely, choose a parametrix R for F. Let $S_0 = 1 - RF$, $S_1 = 1 - FR$. Then set $U_D = \begin{bmatrix}S_0 & -(1+S_0)R\\ F & S_1\end{bmatrix}$. With such a choice the inverse is given by an explicit formula $U_D^{-1} = \begin{bmatrix}S_0 & (1+S_0)R\\ -F & S_1\end{bmatrix}$. Set $P_D = U_D^{-1}\begin{bmatrix}1_{\mathcal{E}} & 0\\ 0 & 0\end{bmatrix} U_D$.

The choices in the constructions can be made so that we have $(P_D)^{\pm} = P_{D^{\pm}}$; see Definition 3.8.

Define the map $\Phi_{\widetilde{\nabla}^{\mathcal{H}}}\colon CC^{-}_{\bullet}\left(\Psi^{-\infty}_{\mathcal{L}}(M|B;\widetilde{\mathcal{E}})\right) \to (\Omega^*(M)[u], d_\Omega)$ by

$$\begin{aligned}&\Phi^k_{\widetilde{\nabla}^{\mathcal{H}}}(A_0,\ldots,A_k)\\ &:= \int_{\Delta^k} \operatorname{Str}\left(A_0 e^{-ut_0\frac{\widetilde{\theta}^{\mathcal{H}}}{2\pi i}}\widetilde{\partial}^{\mathcal{H}}(A_1)e^{-ut_1\frac{\widetilde{\theta}^{\mathcal{H}}}{2\pi i}}\cdots e^{-ut_{k-1}\frac{\widetilde{\theta}^{\mathcal{H}}}{2\pi i}}\widetilde{\partial}^{\mathcal{H}}(A_k)e^{-ut_k\frac{\widetilde{\theta}^{\mathcal{H}}}{2\pi i}}\right)dt_1\cdots dt_k\end{aligned}$$

for $A_0\otimes\cdots\otimes A_k \in C_k\left(\Psi^{-\infty}_{\mathcal{L}}(M|B;\widetilde{\mathcal{E}})\right)$ and set again $\Phi_{\widetilde{\nabla}^{\mathcal{H}}} = \sum_{k=0}^{\infty}\Phi^k_{\widetilde{\nabla}^{\mathcal{H}}}$. Notice that

$$\begin{aligned}&\left[\Phi_{\widetilde{\nabla}^{\mathcal{H}}}\left(\operatorname{Ch}\left(P_D - \begin{bmatrix}0 & 0\\ 0 & 1_{\mathcal{E}}\end{bmatrix}\right)\right)\right] = \\ &\qquad \Phi_{\nabla^{\mathcal{H}}}\left(\operatorname{ch}\left(\operatorname{ind}D^+\right)\right) - \Phi_{\nabla^{\mathcal{H}}}\left(\operatorname{ch}\left(\operatorname{ind}D^-\right)\right) = 2\Phi_{\nabla^{\mathcal{H}}}\left(\operatorname{ch}\left(\operatorname{ind}D^+\right)\right).\end{aligned} \tag{4.3}$$

Replacing in the formulas above $\widetilde{\theta}^{\mathcal{H}}$ by $\widetilde{\theta}^{\mathbb{A}_{u^{-1}}}$ and $\widetilde{\partial}^{\mathcal{H}}$ by $[\widetilde{\mathbb{A}}_{u^{-1}},\cdot]$ we obtain the definition of the morphism $\Phi_{\widetilde{\mathbb{A}}}\colon CC^{entire}_{\bullet}\left(\Psi^{-\infty}_{\mathcal{L}}(M|B;\widetilde{\mathcal{E}})\right) \to (\Omega^*(M)[u], d_\Omega)_{\bullet}$.

Lemma 4.7 *The morphisms*

$$\Phi_{\widetilde{\mathbb{A}}},\ \Phi_{\widetilde{\nabla}^{\mathcal{H}}}\colon CC^{entire}_{\bullet}\left(\Psi^{-\infty}_{\mathcal{L}}(M|B;\widetilde{\mathcal{E}})\right) \to (\Omega^*(B)[u], d_\Omega)_{\bullet}$$

are chain homotopic.

Proof This follows from the explicit formula for the chain homotopy; see Proposition 5.6 of [30]. This formula can be described as follows. Let $\mathbb{A}_{u^{-1}}(s) = s\mathbb{A}_{u^{-1}} + (1-s)\nabla^{\mathcal{H}}$. Then we can write

$$\begin{aligned}\mathbb{A}_{u^{-1}}(s) &= su^{-1/2}D + \nabla^{\mathcal{H}} - \frac{su^{1/2}}{4}c(T^{\mathcal{H}}),\\ \theta^{\mathbb{A}_{u^{-1}}(s)} &= \theta^{\mathcal{H}} + s[\nabla^{\mathcal{H}}, u^{-1/2}D - \frac{u^{1/2}}{4}c(T^{\mathcal{H}})] + s^2(u^{-1/2}D - \frac{u^{1/2}}{4}c(T^{\mathcal{H}}))^2\end{aligned}$$

and

$$\frac{d}{ds}\mathbb{A}_{u^{-1}}(s) = u^{-1/2}D - \frac{u^{1/2}}{4}c(T^{\mathcal{H}}).$$

Define $H_k\colon C_k(\Psi^{-\infty}_{\mathcal{L}}(M|B,\mathcal{E}\oplus\mathcal{E})) \to \Omega^*(B)[u]$ by

$$\begin{aligned}&H_k(A_0,\ldots,A_k)\\ &= \int_0^1 ds\left(\sum_{m=0}^{k}(-1)^m\int_{\Delta^{k+1}}dt_1\cdots dt_{k+1}\operatorname{Str}\left(A_0 e^{-ut_0\frac{\widetilde{\theta}^{\mathbb{A}_{u^{-1}}(s)}}{2\pi i}}\right.\right.\\ &\qquad [\widetilde{\mathbb{A}}_{u^{-1}}(s),A_1]e^{-ut_1\frac{\widetilde{\theta}^{\mathbb{A}_{u^{-1}}(s)}}{2\pi i}}\cdots[\widetilde{\mathbb{A}}_{u^{-1}}(s),A_m]e^{-ut_m\frac{\widetilde{\theta}^{\mathbb{A}_{u^{-1}}(s)}}{2\pi i}}\,\widetilde{\frac{d}{ds}\mathbb{A}_{u^{-1}}}\\ &\qquad \left.\left. e^{-ut_{m+1}\frac{\widetilde{\theta}^{\mathbb{A}_{u^{-1}}(s)}}{2\pi i}}\cdots e^{-ut_{k-1}\frac{\widetilde{\theta}^{\mathbb{A}_{u^{-1}}(s)}}{2\pi i}}[\widetilde{\mathbb{A}}_{u^{-1}}(s),A_k]e^{-ut_{k+1}\frac{\widetilde{\theta}^{\mathbb{A}_{u^{-1}}(s)}}{2\pi i}}\right)\right)\end{aligned} \tag{4.4}$$

and set $H = \sum H_k\colon CC^{entire}_{\bullet}\left(\Psi^{-\infty}_{\mathcal{L}}(M|B;\widetilde{\mathcal{E}})\right) \to \Omega^*(B)[u]$. Then

$$\Phi_{\widetilde{\mathbb{A}}} - \Phi_{\nabla^{\mathcal{H}}} = d_\Omega\circ H + H\circ(b+uB).$$

□

Let $\mathcal{F}$ be the algebra defined by

$$\mathcal{F} = \{F \in \Psi^0_{\mathcal{L}}(M|B;\widetilde{\mathcal{E}})^{even} \mid [D,F] \in \Psi^0_{\mathcal{L}}(M|B;\mathcal{E})\}.$$

A simple modification of the argument in [20] as done in [2], cf. also [22], shows that $\Phi_{\widetilde{\mathbb{A}}}$ extends to a morphism, also denoted $\Phi_{\widetilde{\mathbb{A}}}\colon CC^{entire}_{\bullet}(\mathcal{F}) \to (\Omega^*(M)[u], d_{\Omega})$ defined by the same formula. Note that $U_D \in \mathcal{F}$, $P_D \in \mathcal{F}$ and $\begin{bmatrix}1_{\mathcal{E}} & 0\\ 0 & 0\end{bmatrix} \in \mathcal{F}$.

Recall now that inner automorphisms act by the identity on the entire cyclic homology, see, e.g., [27] 4.1.3 or [26]. It follows that the chains $\mathrm{Ch}(P_D)$ and $\mathrm{Ch}\left(\begin{bmatrix}1_{\mathcal{E}} & 0\\ 0 & 0\end{bmatrix}\right)$ are homologous in $CC^{entire}_{\bullet}(\mathcal{F})$.

We therefore obtain

$$\begin{aligned}
\Phi\left(\mathrm{ch}\left(\mathrm{ind}\, D^+\right)\right) &= \frac{1}{2}\left[\Phi_{\widetilde{\nabla}^{\mathcal{H}}}\left(\mathrm{Ch}\left(P_D - \begin{bmatrix}0 & 0\\ 0 & 1_{\mathcal{E}}\end{bmatrix}\right)\right)\right]\\
&= \frac{1}{2}\left[\Phi_{\widetilde{\mathbb{A}}}\left(\mathrm{Ch}\left(P_D - \begin{bmatrix}0 & 0\\ 0 & 1_{\mathcal{E}}\end{bmatrix}\right)\right)\right]\\
&= \frac{1}{2}\left[\Phi_{\widetilde{\mathbb{A}}}\left(\mathrm{Ch}\left(P_D\right)\right) - \Phi_{\widetilde{\mathbb{A}}}\left(\mathrm{Ch}\left(\begin{bmatrix}0 & 0\\ 0 & 1_{\mathcal{E}}\end{bmatrix}\right)\right)\right]\\
&= \frac{1}{2}\left[\Phi_{\widetilde{\mathbb{A}}}\left(\mathrm{Ch}\left(\begin{bmatrix}1_{\mathcal{E}} & 0\\ 0 & 0\end{bmatrix}\right)\right)\right] - \frac{1}{2}\left[\Phi_{\widetilde{\mathbb{A}}}\left(\mathrm{Ch}\left(\begin{bmatrix}0 & 0\\ 0 & 1_{\mathcal{E}}\end{bmatrix}\right)\right)\right]\\
&= \left[\mathrm{Str}(e^{-\frac{u\theta^{\mathbb{A}_{u^{-1}}}}{2\pi i}})\right].
\end{aligned}$$

□

4.3 The local index theorem.

Theorem 4.8 *Let D be a projective family of Dirac operators on a horizontally $\mathcal{L}$-twisted Clifford module $\mathcal{E}$. We have the following equality of classes in $H^{\bullet}_{\mathcal{L}}(B)$:*

$$\left[\mathrm{Str}(e^{-\frac{u}{2\pi i}\theta^{\mathbb{A}_{u^{-1}}}})\right] = \left[u^{-\frac{k}{2}}\int_{M|B}\widehat{A}\left(\frac{u}{2\pi i}R^{M|B}\right)\mathrm{Ch}_{\mathcal{L}}(\mathcal{E}/\mathcal{S})\right].$$

Proof Over each open set U_α we have $\theta^{\mathbb{A}_{u^{-1}}} = \mathbb{A}^2_{\alpha,u^{-1}} + \pi^*\omega_\alpha$. Since $\mathbb{A}^2_{\alpha,u^{-1}}$ and $\pi^*\omega_\alpha$ commute, we have

$$\mathrm{Str}\, e^{-\frac{u}{2\pi i}\theta^{\mathbb{A}_{u^{-1}}}} = e^{-\frac{u}{2\pi i}\pi^*\omega_\alpha}\,\mathrm{Str}\, e^{-\frac{u}{2\pi i}\mathbb{A}^2_{\alpha,u^{-1}}}.$$

According to Bismut's local index theorem for families [5],

$$\lim_{t\to 0}\mathrm{Str}\, e^{-\frac{1}{2\pi i}\mathbb{A}^2_{\alpha,t}} = \int_{\pi^{-1}U_\alpha|U_\alpha}\widehat{A}\left(\frac{1}{2\pi i}R^{M|B}\right)\mathrm{str}_{\mathcal{E}/\mathcal{S}}\, e^{-\frac{1}{2\pi i}\theta^{\mathcal{E}/\mathcal{S}}_{\alpha}}.$$

Moreover, by the result of Bismut and Freed [6]

$$\mathrm{Str}\, e^{-\frac{1}{2\pi i}\mathbb{A}^2_{\alpha}} - \int_{\pi^{-1}U_\alpha|U_\alpha}\widehat{A}\left(\frac{1}{2\pi i}R^{M|B}\right)\mathrm{str}_{\mathcal{E}/\mathcal{S}}\, e^{-\frac{1}{2\pi i}\theta^{\mathcal{E}/\mathcal{S}}_{\alpha}} = d\int_0^1 \xi_\alpha(t)dt \tag{4.5}$$

where

$$\xi_\alpha(t) = -\frac{1}{2\pi i}\,\mathrm{Str}\,\frac{d\mathbb{A}_{\alpha,t}}{dt}e^{-\frac{1}{2\pi i}\mathbb{A}^2_{\alpha,t}} \tag{4.6}$$

is integrable at 0.

Let $s > 0$ and let δ_s^B be the operator on $\Omega^*(B)$ which multiplies the forms of degree k by $s^{k/2}$. Then $\delta_s^B \circ (\mathbb{A}_{\alpha,t}) \circ \delta_{s^{-1}}^B = s^{1/2}\mathbb{A}_{\alpha,t/s}$. We therefore obtain

$$\begin{aligned}\lim_{t\to 0} \operatorname{Str} e^{-\frac{s}{2\pi i}\mathbb{A}^2_{\alpha,t/s}} &= \delta_s^B \left(\lim_{t\to 0} \operatorname{Str} e^{-\frac{1}{2\pi i}\mathbb{A}^2_{\alpha,t}}\right)\\ &= \delta_s^B \int_{\pi^{-1}U_\alpha|U_\alpha} \widehat{A}\left(\frac{1}{2\pi i}R^{M|B}\right) \operatorname{str}_{\mathcal{E}/\mathcal{S}} e^{-\frac{1}{2\pi i}\theta_\alpha^{\mathcal{E}/\mathcal{S}}}\\ &= s^{-\frac{k}{2}} \int_{\pi^{-1}U_\alpha|U_\alpha} \widehat{A}\left(\frac{s}{2\pi i}R^{M|B}\right) \operatorname{str}_{\mathcal{E}/\mathcal{S}} e^{-\frac{s}{2\pi i}\theta_\alpha^{\mathcal{E}/\mathcal{S}}}\end{aligned}$$

where, as before, $k = \dim M - \dim B$. Since both sides are polynomials in s we deduce that

$$\lim_{t\to 0} \operatorname{Str} e^{-\frac{u}{2\pi i}\mathbb{A}^2_{\alpha,u^{-1}t}} = u^{-\frac{k}{2}} \int_{\pi^{-1}U_\alpha|U_\alpha} \widehat{A}\left(\frac{u}{2\pi i}R^{M|B}\right) \operatorname{str}_{\mathcal{E}/\mathcal{S}} e^{-\frac{u}{2\pi i}\theta_\alpha^{\mathcal{E}/\mathcal{S}}}$$

Multiplying both sides by $e^{-\frac{u}{2\pi i}\pi^*\omega_\alpha}$ we obtain

$$\lim_{t\to 0} \operatorname{Str}\exp\left(-\frac{u}{2\pi i}\theta^{\mathbb{A}_{u^{-1}t}}\right)\Big|_{U_\alpha} = u^{-\frac{k}{2}} \int_{\pi^{-1}U_\alpha|U_\alpha} \widehat{A}\left(\frac{u}{2\pi i}R^{M|B}\right) \operatorname{Ch}_{\mathcal{L}}(\mathcal{E}/\mathcal{S})$$

and therefore

$$\lim_{t\to 0} \operatorname{Str}\exp\left(-\frac{u}{2\pi i}\theta^{\mathbb{A}_{u^{-1}t}}\right) = u^{-\frac{k}{2}} \int_{M|B} \widehat{A}\left(\frac{u}{2\pi i}R^{M|B}\right) \operatorname{Ch}_{\mathcal{L}}(\mathcal{E}/\mathcal{S}).$$

Moreover by (4.5), we have

$$\begin{aligned}\operatorname{Str} e^{-\frac{u}{2\pi i}\mathbb{A}^2_{\alpha,u^{-1}}} - u^{-\frac{k}{2}} \int_{\pi^{-1}U_\alpha|U_\alpha} \widehat{A}\left(\frac{u}{2\pi i}R^{M|B}\right) \operatorname{str}_{\mathcal{E}/\mathcal{S}} e^{-\frac{u}{2\pi i}\theta_\alpha^{\mathcal{E}/\mathcal{S}}}&\\ = ud\int_0^1 u^{-\frac{1}{2}}\delta_u^B \xi_\alpha(t)dt.&\end{aligned} \tag{4.7}$$

Here the right hand side is defined as follows. Write $\xi_\alpha^{[l]} \in \Omega^l(B)$ for the component of degree l of ξ_α from Equation (4.6). Then we have

$$u^{-\frac{1}{2}}\delta_u^B \xi_\alpha(t) = \sum u^{\frac{l-1}{2}}\xi_\alpha^{[l]}(t) = \frac{1}{2\pi i}\operatorname{Str}\frac{d\mathbb{A}_{\alpha,u^{-1}t}}{dt}\exp\left(-\frac{u}{2\pi i}\mathbb{A}^2_{\alpha,u^{-1}t}\right) \in \Omega^*(B)[u].$$

Multiplying the identity (4.7) by $e^{-\frac{u}{2\pi i}\pi^*\omega_\alpha}$ and using the equality

$$e^{-\frac{u}{2\pi i}\pi^*\omega_\alpha}ud(\cdot) = (ud + u^2\Omega)(e^{-\frac{u}{2\pi i}\pi^*\omega_\alpha}\cdot)$$

we obtain

$$\operatorname{Str}\exp\left(-\frac{u}{2\pi i}\theta^{\mathbb{A}_{u^{-1}}}\right)\Big|_{U_\alpha} - u^{-\frac{k}{2}} \int_{\pi^{-1}U_\alpha|U_\alpha} \widehat{A}\left(\frac{u}{2\pi i}R^{M|B}\right) \operatorname{Ch}_{\mathcal{L}}(\mathcal{E}/\mathcal{S}) = (ud+u^2\Omega)\Xi_\alpha$$

where $\Xi_\alpha = \int_0^1 \frac{1}{2\pi i}\operatorname{Str}\frac{d\mathbb{A}_{\alpha,u^{-1}t}}{dt}\exp\left(-\frac{u}{2\pi i}\,\theta^{\mathbb{A}_{u^{-1}t}}\big|_{U_\alpha}\right)dt.$

From the equations (4.2) it follows that over $U_{\alpha\beta}$ we have

$$\frac{d\mathbb{A}_{\alpha,u^{-1}t}}{dt} = \phi_{\alpha\beta}\left(\frac{d\mathbb{A}_{\beta,u^{-1}t}}{dt}\right).$$

Hence there exists a form $\dot{\mathbb{A}}_t \in u^{-1/2}\Omega^*(B;\Psi_{\mathcal{L}}(M|B,\mathcal{E}))[u]$ such that $\dot{\mathbb{A}}_t\Big|_{U_\alpha} = \frac{d\mathbb{A}_{\alpha,u^{-1}t}}{dt}$. Therefore setting $\Xi = \int_0^1 \frac{1}{2\pi i}\operatorname{Str}\dot{\mathbb{A}}_t\exp\left(-\frac{u}{2\pi i}\theta^{\mathbb{A}_{u^{-1}t}}\right)dt \in \Omega^*(B)[u]$ we can write

$$\operatorname{Str}\exp\left(-\frac{u}{2\pi i}\theta^{\mathbb{A}_{u^{-1}}}\right) - u^{-\frac{k}{2}}\int_{M|B}\widehat{A}\left(\frac{u}{2\pi i}R^{M|B}\right)\operatorname{Ch}_{\mathcal{L}}(\mathcal{E}/\mathcal{S}) = d_{\Omega}\Xi$$

and the statement of the Theorem follows. □

Combining the results of Theorems 4.6 and 4.8 we obtain the main theorem of this paper:

Theorem 1.1 *Let D be a projective family of Dirac operators on a horizontally $\mathcal{L}$-twisted Clifford module $\mathcal{E}$ on a fibration $\pi : M \to B$. Then the following equality holds in $H^{\bullet}_{\mathcal{L}}(B)$:*

$$\left[\Phi_{\nabla^{\mathcal{H}}}(\operatorname{ch}(\operatorname{ind} D^{+}))\right] = \left[u^{-\frac{k}{2}}\int_{M|B}\widehat{A}\left(\frac{u}{2\pi i}R^{M|B}\right)\operatorname{Ch}_{\mathcal{L}}(\mathcal{E}/\mathcal{S})\right],$$

where $k = \dim M - \dim B$ is the dimension of the fibers.

References

[1] Atiyah, M. F. and Singer, I. M. *The index of elliptic operators. IV.* Ann. of Math. (2) 93 1971 119–138.

[2] Benameur, M.-T. and Carey A. *On the analyticity of the bivariant JLO cocycle.* Electron. Res. Announc. Math. Sci. 16 (2009), 37–43.

[3] Benameur, M.-T. and Heitsch, J. L. *Index theory and non-commutative geometry. II. Dirac operators and index bundles.* J. K-Theory 1 (2008), no. 2, 305–356.

[4] Berline, N., Getzler, E. and Vergne, M. *Heat kernels and Dirac operators.* Corrected reprint of the 1992 original. Grundlehren Text Editions. Springer-Verlag, Berlin, 2004.

[5] Bismut, J.-M., *The Atiyah-Singer index theorem for families of Dirac operators: two heat equation proofs.* Invent. Math. 83 (1985), no. 1, 91–151.

[6] Bismut, J.-M. and Freed, D. S., *The analysis of elliptic families. II. Dirac operators, eta invariants, and the holonomy theorem.* Comm. Math. Phys. 107 (1986), no. 1, 103–163.

[7] Bott, R. and Tu, L. W. *Differential forms in algebraic topology.* Graduate Texts in Mathematics, 82. Springer-Verlag, New York-Berlin, 1982.

[8] Breen, L. and Messing, W. *Differential Geometry of Gerbes.* Advances in Math. 198, 732846 (2005).

[9] Bressler, P., Gorokhovsky, A., Nest, R. and Tsygan B., *Deformations of gerbes on smooth manifolds.* K-theory and noncommutative geometry, 349–392, EMS Ser. Congr. Rep., European Mathematical Society, Zürich 2008.

[10] Bressler, P., Gorokhovsky, A., Nest, R. and Tsygan B., *Deformation quantization of gerbes.* Adv. Math. 214 (2007), no. 1, 230–266.

[11] Brylinski, J.-L. *Loop spaces, characteristic classes and geometric quantization.* Modern Birkhuser Classics. Birkhuser Boston, Inc., Boston, MA, 2008.

[12] Carey A. and Wang B.-L., *Riemann-Roch and index formulae in twisted K-theory.* To appear in Proceedings of NSF/CBMS Conference on "Topology, C*-algebras, and String Duality".

[13] Carey, A., Mickelsson J. and Wang B.-L., *Differential twisted K-theory and its applications,* J. Geom. Phys. 59 (2009), no. 5, 632–653.

[14] Carrillo Rouse P. and Wang B.-L., *Twisted longitudinal index theorem for foliations and wrong way functoriality* , arXiv:1005.3842.

[15] Connes, A. *Entire cyclic cohomology of Banach algebras and characters of θ-summable Fredholm modules.* K-Theory 1 (1988), no. 6, 519–548.

[16] Connes, A. *Noncommutative differential geometry.* Inst. Hautes Études Sci. Publ. Math. No. 62 (1985), 257–360.

[17] Connes, A. *Noncommutative geometry.* Academic Press, Inc., San Diego, CA, 1994.

[18] Dixmier, J. and Douady, A., *Champs continus d'espaces hilbertiens et de C^*-algèbres.* Bull. Soc. Math. France 91 1963 227–284.

[19] Donovan, P. and Karoubi, M., *Graded Brauer groups and K-theory with local coefficients.* Inst. Hautes Études Sci. Publ. Math. No. 38 (1970), 5–25

[20] Getzler, E. and Szenes, A. *On the Chern character of a theta-summable Fredholm module.* J. Funct. Anal. 84 (1989), no. 2, 343–357.

[21] Gorokhovsky, A. *Characters of cycles, equivariant characteristic classes and Fredholm modules.* Comm. Math. Phys. 208 (1999), no. 1, 1–23.

[22] Gorokhovsky, A. *Bivariant Chern character and longitudinal index.* J. of Funct. Analysis 237 (2006), 105-134.

[23] Gorokhovsky, A. and Lott, J. *Local index theory over ètale groupoids.* J. Reine Angew. Math. 560 (2003), 151–198.

[24] Hitchin, N. J. *The moduli space of complex Lagrangian submanifolds.* Surveys in differential geometry, 327–345, Surv. Differ. Geom., VII, Int. Press, Somerville, MA, 2000.

[25] Karoubi, M. *Twisted K-theory—old and new.*, K-theory and noncommutative geometry, 117–149, EMS Ser. Congr. Rep., European Mathematical Society, Zürich 2008.

[26] Khalkhali, M. On the entire cyclic cohomology of Banach algebras. Comm. Algebra 22 (1994), no. 14, 5861–5874.

[27] Loday, J.-L., *Cyclic homology.* Grundlehren der Mathematischen Wissenschaften, 301. Springer-Verlag, Berlin, 1992.

[28] Mathai, V.; Melrose, R. B. and Singer, I M. *The index of projective families of elliptic operators.* Geom. Topol. 9 (2005), 341–373.

[29] Mathai, V.; Melrose, R. B. and Singer, I M. *The index of projective families of elliptic operators: the decomposable case.* Asterisque, 327 (2009) 251-292

[30] Mathai, V. and Stevenson, D. *On a generalized Connes-Hochschild-Kostant-Rosenberg theorem.* Adv. Math. 200 (2006), no. 2, 303–335.

[31] Meyer, R., *Local and analytic cyclic homology.* EMS Tracts in Mathematics, 3. European Mathematical Society (EMS), Zurich, 2007.

[32] Murray, M. K. *Bundle gerbes.* J. London Math. Soc. (2) 54 (1996), no. 2, 403–416.

[33] Nest R. and Tsygan B. *Algebraic index theorem for families.* Adv. Math. 113 (1995), no. 2, 151–205.

[34] Nest R. and Tsygan B. *On the cohomology ring of an algebra.* Advances in geometry,337–370, Progr. Math., 172, Birkhuser Boston, Boston, MA, 1999.

[35] Nistor, V. *Super-connections and non-commutative geometry.* Cyclic cohomology and non-commutative geometry (Waterloo, ON, 1995), 115–136, Fields Inst. Commun., 17, Amer. Math. Soc., Providence, RI, 1997.

Fields Institute Communications
Volume **61**, 2011

Type III KMS States on a Class of C^*-Algebras Containing O_n and $\mathcal{Q}_{\mathbb{N}}$ and Their Modular Index

Alan L. Carey
Mathematical Sciences Institute
Australian National University
Canberra, ACT, Australia
Alan.Carey@anu.edu.au

John Phillips
Department of Mathematics and Statistics
University of Victoria
Victoria, BC Canada V8W 3R4
johnphil@uvic.ca

Ian F. Putnam
Department of Mathematics and Statistics
University of Victoria
Victoria, BC Canada V8W 3R4
putnam@math.uvic.ca

Adam Rennie
Mathematical Sciences Institute
Australian National University
Canberra, ACT, Australia
Adam.Rennie@anu.edu.au

We construct a family of purely infinite, simple, separable, nuclear C^*-algebras, $\mathcal{Q}^\lambda$, for $\lambda \in (0,1)$. These algebras are also in the class $\mathfrak{N}_{nuc}$ and therefore by results of E. Kirchberg and N. C. Phillips they are classified by their K-groups. There is an action of the circle $\mathbb{T}$ with a unique KMS state ψ on each $\mathcal{Q}^\lambda$. For $\lambda = 1/n$, $\mathcal{Q}^{1/n} \cong O_n$, with its usual $\mathbb{T}$ action and KMS state. For $\lambda = p/q$, rational in lowest terms, $\mathcal{Q}^\lambda \cong O_n$ $(n = q - p + 1)$ with UHF fixed point algebra of type $(pq)^\infty$. For any $n > 0$, $\mathcal{Q}^\lambda \cong O_n$ for infinitely many λ with distinct KMS states and UHF fixed-point algebras. However, none of the $\mathcal{Q}^\lambda$ is isomorphic to O_∞. For λ irrational the fixed point algebras are *not* AF and the $\mathcal{Q}^\lambda$ are usually *not* Cuntz algebras. For λ transcendental, $K_1(\mathcal{Q}^\lambda) \cong K_0(\mathcal{Q}^\lambda) \cong \mathbb{Z}^\infty$, so that $\mathcal{Q}^\lambda$ is Cuntz' $\mathcal{Q}_{\mathbb{N}}$, [Cu1]. If λ is algebraic (and not rational), then $K_1(\mathcal{Q}^\lambda)$ and $K_0(\mathcal{Q}^\lambda)$ have the same finite rank: $K_1 \otimes \mathbb{Q} \cong K_0 \otimes \mathbb{Q} \cong \mathbb{Q}^k$. If λ and λ^{-1} are both algebraic integers, the *only* O_n which appear are those for which $n \equiv 3 (\mathrm{mod}\, 4)$. For each λ, the representation of $\mathcal{Q}^\lambda$ defined by the KMS state ψ generates a type III_λ factor. These algebras fit into

2010 *Mathematics Subject Classification.* Primary 46L80; Secondary 19K56, 58J30, 58J42.

the framework of the modular index theory/twisted cyclic theory of [CPR2, CRT] and [CNNR].

These new examples were motivated by the 'modular index theory' of [CPR2, CNNR]. We were aiming to find examples of algebras that were not Cuntz-Krieger algebras (or the CAR algebra) and were not previously known in order to explore the possibilities opened by [CNNR]. These algebras, denoted by $\mathcal{Q}^\lambda$ for $0 < \lambda < 1$, are constructed as "corner algebras" of certain crossed product C^*-algebras.

This survey is an expanded version of a talk given at the Fields Institute in 2008: complete explanations and proofs appear elsewhere [CPPR]. The second named author would like to thank Masoud Khalkhali for organizing this conference on Noncommutative Geometry and inviting him to speak. It was a great conference.

1 The construction of the $\mathcal{Q}^\lambda$

Definition 1.1 For $0 < \lambda < 1$, let Γ_λ be the countable additive abelian subgroup of $\mathbb{R}$ defined by

$$\Gamma_\lambda = \left\{ \sum_{k=-N}^{k=N} n_k \lambda^k \;\middle|\; N \geq 0 \text{ and } n_k \in \mathbb{Z} \right\}.$$

Loosely speaking, Γ_λ consists of Laurent polynomials in λ and λ^{-1} with integer coefficients. It is not only a dense subgroup of $\mathbb{R}$, but is clearly a unital **subring** of $\mathbb{R}$. To get some idea of the structure of these groups, we prove:

Proposition 1.2 *Let* $0 < \lambda < 1$.
(1) *If* $\lambda = p/q$ *where* $0 < p < q$ *are integers in lowest terms, then* $\Gamma_\lambda = \mathbb{Z}[1/n]$, *where* $n = pq$.
(2) *If* λ *and* λ^{-1} *are both algebraic integers, then* $\Gamma_\lambda = \mathbb{Z} + \mathbb{Z}\lambda + \cdots + \mathbb{Z}\lambda^{d-1}$ *is an internal direct sum, where* $d \geq 2$ *is the degree of the minimal (monic) polynomial in* $\mathbb{Z}[x]$ *satisfied by* λ.
(3) *If* λ *is transcendental then,* $\Gamma_\lambda = \bigoplus_{k\in\mathbb{Z}} \mathbb{Z}\lambda^k$ *is an internal direct sum.*
(4) *If* $\lambda = 1/\sqrt{n}$ *with* $n \geq 2$ *a square-free positive integer, then* $\Gamma_\lambda = \mathbb{Z}[1/n] + \mathbb{Z}[1/n] \cdot \sqrt{n}$ *is an internal direct sum.*
(5) *In general, if* λ *is algebraic with minimal polynomial,* $n\lambda^d + \cdots + m = 0$ *over* $\mathbb{Z}$, *then*

$$\mathbb{Z} \oplus \mathbb{Z}\lambda \oplus \cdots \oplus \mathbb{Z}\lambda^{d-1} \subseteq \Gamma_\lambda \subseteq \mathbb{Z}[\frac{1}{mn}] \oplus \mathbb{Z}[\frac{1}{mn}]\lambda \oplus \cdots \oplus \mathbb{Z}[\frac{1}{mn}]\lambda^{d-1}.$$

Hence, $\operatorname{rank}(\Gamma_\lambda) := \dim_{\mathbb{Q}}(\Gamma_\lambda \otimes_{\mathbb{Z}} \mathbb{Q}) = d$.

Definition 1.3 Now let $C_0^\lambda(\mathbb{R})$ be the separable, commutative (AF) C^*-subalgebra of $\mathcal{L}^\infty(\mathbb{R})$ generated by the countable family of projections $\mathcal{X}_{[a,b)}$ where $a, b \in \Gamma_\lambda$. That is,

$$C_0^\lambda(\mathbb{R}) = \text{closure}\left(\left\{ \sum_{k=1}^{n} c_k \mathcal{X}_{[a_k,b_k)} \;\middle|\; c_k \in \mathbb{C};\;\; a_k, b_k \in \Gamma_\lambda \right\} \right).$$

We note that $C_0(\mathbb{R}) \subset C_0^\lambda(\mathbb{R})$, and that $C_0^\lambda(\mathbb{R})$ consists of exactly those $\mathbb{C}$-valued functions on $\mathbb{R}$ which vanish at ∞, are right-continuous at each point of Γ_λ, have finite left-hand limits at each point of Γ_λ, and are continuous at each point not in Γ_λ.

Definition 1.4 Let $G_\lambda \supset G_\lambda^0$ be the following countable discrete groups of matrices:

$$G_\lambda = \left\{ \begin{pmatrix} \lambda^n & a \\ 0 & 1 \end{pmatrix} \;\middle|\; a \in \Gamma_\lambda,\ n \in \mathbb{Z} \right\} \supset G_\lambda^0 = \left\{ \begin{pmatrix} 1 & a \\ 0 & 1 \end{pmatrix} \;\middle|\; a \in \Gamma_\lambda \right\}.$$

Of course, G_λ^0 is isomorphic to the additive group Γ_λ, and G_λ is the semidirect product of $\mathbb{Z}$ acting on $G_\lambda^0 \cong \Gamma_\lambda$. We let G_λ act on $\mathbb{R}$ as an "$ax+b$" group, noting that the action leaves Γ_λ invariant. That is,

$$\text{for } t \in \mathbb{R} \text{ and } g = \begin{pmatrix} \lambda^n & a \\ 0 & 1 \end{pmatrix} \in G_\lambda \text{ define } g \cdot t := \lambda^n t + a.$$

We use this action on $\mathbb{R}$ to define the transpose action α of G_λ on $C_0^\lambda(\mathbb{R})$:

$$\alpha_g(f)(t) = f(g^{-1}t) \text{ for } f \in C_0^\lambda(\mathbb{R}) \text{ and } t \in \mathbb{R}.$$

Definition 1.5 We define the separable C^*-algebras $A^\lambda \supset A_0^\lambda$ as the crossed products

$$A^\lambda := G_\lambda \rtimes_\alpha C_0^\lambda(\mathbb{R}) = \mathbb{Z} \rtimes (G_\lambda^0 \rtimes_\alpha C_0^\lambda(\mathbb{R})) \supset A_0^\lambda := G_\lambda^0 \rtimes_\alpha C_0^\lambda(\mathbb{R}).$$

Proposition 1.6 *The algebras A^λ, and A_0^λ are simple, separable, nuclear and stable C^*-algebras in the bootstrap class $\mathfrak{N}_{nuc}$. Moreover, A^λ is purely infinite.*

In order to obtain unital C^*-algebras we cut down these algebras by the projection, e corresponding to $\mathcal{X}_{[0,1)} \in C_0^\lambda(\mathbb{R}) \subset A_0^\lambda \subset A^\lambda$. Explicitly, we define:

Definition 1.7

$$\mathcal{Q}^\lambda := eA^\lambda e \qquad F^\lambda := eA_0^\lambda e.$$

We will also have occasion to use the dense $*$-subalgebras $\mathcal{Q}_c^\lambda$ and F_c^λ consisting of finitely supported functions $x : G^\lambda \to C_0^\lambda(\mathbb{R})$ (respectively, $x : G_0^\lambda \to C_0^\lambda(\mathbb{R})$) where the values $x(g)$ are either 0 or finite linear combinations of characteristic functions $\mathcal{X}_{[a,b)}$ with $a, b \in \Gamma_\lambda$.

Corollary 1.8 *The algebras $\mathcal{Q}^\lambda$, and F^λ are simple, separable, unital, nuclear C^*-algebras in the bootstrap class $\mathfrak{N}_{nuc}$. Moreover, $\mathcal{Q}^\lambda$ is purely infinite. In fact, the orthogonal family of projections $e_n = \mathcal{X}_{[n,n+1)} \in C_0^\lambda(\mathbb{R})$ for $n \in \mathbb{Z}$ are mutually equivalent by partial isometries in A_0^λ and the finite sums $E_N := \sum_{n=-N}^{N-1} e_n = \mathcal{X}_{[-N,N)}$ form an approximate identity for A^λ (and for A_0^λ) so that*

$$A^\lambda \cong \mathcal{Q}^\lambda \otimes \mathcal{K}(l^2(\mathbb{Z})) \text{ and } A_0^\lambda \cong F^\lambda \otimes \mathcal{K}(l^2(\mathbb{Z})).$$

2 K-theory of the algebras $C^*(\Gamma_\lambda)$, $\mathcal{Q}^\lambda$, and F^λ

We first compute the K-theory of the algebras $C^*(\Gamma_\lambda) \cong C(\hat{\Gamma}_\lambda)$ using Proposition 1.2.

Proposition 2.1 *Let $0 < \lambda < 1$.*

(1) *If $\lambda = p/q$ is rational in lowest terms so that $\Gamma_\lambda = \mathbb{Z}[1/n]$, where $n = pq$, then*

$$K_0(C(\hat{\Gamma}_\lambda)) = \mathbb{Z}[1_{\hat{\Gamma}_\lambda}] \text{ and } K_1(C(\hat{\Gamma}_\lambda)) = \mathbb{Z}[1/n].$$

(2) *If λ and λ^{-1} are both algebraic integers, so that $\Gamma_\lambda = \mathbb{Z} + \mathbb{Z}\lambda + \cdots + \mathbb{Z}\lambda^{d-1}$ is an internal direct sum as above, then*

$$K_0(C(\hat{\Gamma}_\lambda)) = \bigwedge^{even}(\Gamma_\lambda) = \bigoplus_{k=0,k\ even}^{d} \bigwedge^k(\Gamma_\lambda)$$

and

$$K_1(C(\hat{\Gamma}_\lambda)) = \bigwedge^{odd}(\Gamma_\lambda) = \bigoplus_{k=1, k\ odd}^{d} \bigwedge^{k}(\Gamma_\lambda).$$

(3) *If λ is transcendental then,*

$$K_0(C(\hat{\Gamma}_\lambda)) = \bigwedge^{even}(\Gamma_\lambda) = \bigoplus_{k=0, k\ even}^{\infty} \bigwedge^{k}(\Gamma_\lambda)$$

and

$$K_1(C(\hat{\Gamma}_\lambda)) = \bigwedge^{odd}(\Gamma_\lambda) = \bigoplus_{k=1, k\ odd}^{\infty} \bigwedge^{k}(\Gamma_\lambda).$$

(4) *If $\lambda = 1/\sqrt{n}$ with $n \geq 2$ a square-free positive integer, then*

$$K_0(C(\hat{\Gamma}_\lambda)) \cong \mathbb{Z} \oplus \mathbb{Z}[1/n] \quad \text{and} \quad K_1(C(\hat{\Gamma}_\lambda)) \cong \mathbb{Z}[1/n] \oplus \mathbb{Z}[1/n]$$

(5) *In general, if λ is algebraic with $n\lambda^d + \cdots + m = 0$ over $\mathbb{Z}$ then the composition of the inclusions*

$$\mathbb{Z} \oplus \mathbb{Z}\lambda \oplus \cdots \oplus \mathbb{Z}\lambda^{d-1} \subseteq \Gamma_\lambda \subseteq \mathbb{Z}[\frac{1}{mn}] \oplus \mathbb{Z}[\frac{1}{mn}]\lambda \oplus \cdots \oplus \mathbb{Z}[\frac{1}{mn}]\lambda^{d-1}$$

induces an inclusion on K-Theory, so that both of the following maps are one-to-one

$$\bigwedge^{even}(\mathbb{Z}^d) \cong K_0(C^*(\mathbb{Z} \oplus \cdots \mathbb{Z}\lambda^{d-1})) \hookrightarrow K_0(C(\hat{\Gamma}_\lambda))$$

and

$$\bigwedge^{odd}(\mathbb{Z}^d) \cong K_1(C^*(\mathbb{Z} \oplus \cdots \mathbb{Z}\lambda^{d-1})) \hookrightarrow K_1(C(\hat{\Gamma}_\lambda)).$$

Item (1) is proved directly, while (2) and (3) follow from the known K-Theory of (commutative) tori. Items (4) and (5) are deduced from results of C. Schochet [Sc] on the K-Theory of tensor products.

Proposition 2.2 *If $\lambda = p/q$ is rational in lowest terms, we show directly that F^λ is a UHF algebra of type n^∞ where $n = pq$, and so:*

$$K_0(F^\lambda) \cong \mathbb{Z}[1/n] \text{ and } K_1(F^\lambda) = \{0\}.$$

Since $A^\lambda = \mathbb{Z} \rtimes A_0^\lambda$ and $A_0^\lambda \cong F^\lambda \otimes \mathcal{K}$ we use the Pimsner-Voiculescu exact sequence to compute:

$$K_1(\mathcal{Q}^\lambda) = \{0\}, \text{ and } K_0(\mathcal{Q}^\lambda) \cong \mathbb{Z}[1/(pq)]/(1-\lambda)\mathbb{Z}[1/(pq)] \cong \mathbb{Z}_{(q-p)}.$$

Remarks Since the identity in $\mathcal{Q}^\lambda$ is a generator for $\mathbb{Z}_{(q-p)}$ results of Kirchberg-Phillips imply that $\mathcal{Q}^\lambda \cong O_m$ where $m = (q-p)+1$. If $\lambda = 1/n$ we can explicitly write down the n generators for $\mathcal{Q}^{1/n} \cong O_n$, and $F^{1/n}$ is (isomorphic to) the usual UHF subalgebra of O_n. However, for general rational numbers $\lambda = p/q$ the situation remains a little mysterious.

For λ **irrational** we again use the Pimsner-Voiculescu exact sequence to reduce to the K-Theory of A_0^λ. However, in this case these algebras are never AF, and we must use the theory of groupoid C^*-algebras. In particular we must use Ian Putnam's exact sequence for the K-Theory of certain principal groupoid C^*-algebras.

First we establish some notation. Let $\mathbb{R}_\lambda$ be the locally compact Gelfand spectrum of the commutative C^*-algebra $C_0^\lambda(\mathbb{R})$ so that $C_0^\lambda(\mathbb{R}) \cong C_0(\mathbb{R}_\lambda)$. Now, $C_0(\mathbb{R}) \subset C_0(\mathbb{R}_\lambda)$ and the induced proper surjection: $\mathbb{R}_\lambda \to \mathbb{R}$ has the property that points in $\mathbb{R}$ that are not in Γ_λ have a single pre-image while points $\gamma \in \Gamma_\lambda$ have exactly two pre-images in $\mathbb{R}_\lambda$, denoted by γ^- and γ^+. To simplify notation, we let $\Gamma = \Gamma_\lambda \cong G^0_\lambda$. Thus, $\Gamma \subset \mathbb{R}$ is a countable dense subgroup of $\mathbb{R}$ which acts on $\mathbb{R}$ by translations. Before looking at the crossed product of Γ acting on $C_0^\lambda(\mathbb{R}) = C_0(\mathbb{R}_\lambda)$ (which gives us A_0^λ) we first consider the crossed product of Γ acting on $C_0(\mathbb{R})$. Since Γ acts on $\mathbb{R}$ by translation we can Fourier transform to get an isomorphism:

$$\Gamma \rtimes C_0(\mathbb{R}) \cong \hat{\mathbb{R}} \rtimes C(\hat{\Gamma}).$$

Then, by Connes' Thom isomorphism we get for $i = 0, 1$:

$$K_i(\Gamma \rtimes C_0(\mathbb{R})) \cong K_i(\hat{\mathbb{R}} \rtimes C(\hat{\Gamma})) \cong K_{i+1}(C(\hat{\Gamma})).$$

We are able to identify the image of the generator of $K_1(C_0(\mathbb{R})) \cong \mathbb{Z}$ with the class of the identity function [1] in $K_0(C(\hat{\Gamma}))$.

In the notation of [Put2] we define the transformation groupoids:

$$G := \mathbb{R}_\lambda \rtimes \Gamma, \quad G' := \mathbb{R} \rtimes \Gamma, \quad \text{and} \quad H := \Gamma \rtimes \Gamma.$$

Then, $A_0^\lambda = C_r^*(G)$ is the reduced C^*-algebra of G; $\Gamma \rtimes C_0(\mathbb{R}) = C_r^*(G')$ is the reduced C^*-algebra of G'; and $\mathcal{K}(l^2(\Gamma)) = C_r^*(H)$ is the reduced C^*-algebra of H. Thus, there are two disjoint embeddings of Γ in $\mathbb{R}_\lambda$:

$$i_0, i_1 : \Gamma \to \mathbb{R}_\lambda \ : \quad i_0(\gamma) = \gamma^-, \quad i_1(\gamma) = \gamma^+.$$

Now in order to mesh with the notation of [Put2], we let $Y := \Gamma$ with the equivalence relation, "="; $X := \mathbb{R}_\lambda$, with the equivalence relation $(i_0(\gamma) \sim i_1(\gamma))$; and quotient $\pi : X \to X' := \mathbb{R}$ where $X' = X/(i_0(\gamma) \sim i_1(\gamma)) = \mathbb{R}$; while the "factor groupoid" of $G = \mathbb{R}_\lambda \times \Gamma = X \times \Gamma$ is $G' := \mathbb{R} \times \Gamma = X' \times \Gamma$.

In a natural way we represent each of these three C^*-algebras (faithfully) on $\mathcal{H} := l^2(\Gamma^+) \oplus l^2(\Gamma^-)$ where $\Gamma^\pm = \{\gamma^\pm \mid \gamma \in \Gamma\}$. Then from the six-term exact sequence of [Put2] we get

$$\begin{array}{ccccc}
K_1(C_r^*(H)) & \longrightarrow & K_0(C_r^*(G')) & \longrightarrow & K_0(C_r^*(G)) \\
\uparrow & & & & \downarrow \\
K_1(C_r^*(G)) & \longleftarrow & K_1(C_r^*(G')) & \longleftarrow & K_0(C_r^*(H))
\end{array}$$

In our set-up this becomes

$$\begin{array}{ccccc}
\{0\} & \longrightarrow & K_0(\Gamma \rtimes C_0(\mathbb{R})) & \longrightarrow & K_0(\Gamma \rtimes C_0(\mathbb{R}_\lambda)) \\
\uparrow & & & & \downarrow \\
K_1(\Gamma \rtimes C_0(\mathbb{R}_\lambda)) & \longleftarrow & K_1(\Gamma \rtimes C_0(\mathbb{R})) & \longleftarrow & \mathbb{Z}
\end{array}$$

Which by Connes' Thom isomorphism becomes

$$\begin{array}{ccccc}
\{0\} & \longrightarrow & K_1(C(\hat{\Gamma})) & \longrightarrow & K_0(A_0^\lambda) \\
\uparrow & & & & \downarrow \\
K_1(A_0^\lambda) & \longleftarrow & K_0(C(\hat{\Gamma})) & \longleftarrow & \mathbb{Z}
\end{array}$$

With a little more effort we obtain the following result on the K-theory of A_0^λ and hence of F^λ.

Proposition 2.3 *For $0 < \lambda < 1$,*

$$K_0(F^\lambda) = K_0(A_0^\lambda) \cong K_1(C(\hat{\Gamma}_\lambda)) \quad \text{and} \quad K_1(F^\lambda) = K_1(A_0^\lambda) \cong K_0(C(\hat{\Gamma}_\lambda))/[1]\mathbb{Z}.$$

In particular, if λ is irrational, then $K_1(A_0^\lambda) \neq \{0\}$ so that A_0^λ is not an AF-algebra.

Remarks The last statement of the Proposition follows from items (3) and (5) of Proposition 2.1 since $K_0(C(\hat{\Gamma}_\lambda))$ is not singly generated when λ is irrational.

Now, for λ **irrational** we can use the Pimsner-Voiculescu exact sequence to calculate the K-theory of $\mathcal{Q}^\lambda$ in cases where we can effectively compute the action of λ on the K-theory of A_0^λ. For example:

If λ is transcendental, then $K_0(\mathcal{Q}^\lambda) \cong \mathbb{Z}^\infty \cong K_1(\mathcal{Q}^\lambda)$. Then by classification theory $\mathcal{Q}^\lambda$ is isomorphic to Cuntz's algebra, $\mathcal{Q}_{\mathbb{N}}$.

If $\lambda = 1/\sqrt{n}$ where n is square-free, then $K_0(\mathcal{Q}^\lambda) \cong \mathbb{Z}/(n-1)\mathbb{Z} \cong K_1(\mathcal{Q}^\lambda)$. In particular, if $n = 2$ then $\mathcal{Q}^{1/\sqrt{2}} \cong O_2$, but if $n \geq 3$ then $\mathcal{Q}^{1/\sqrt{n}}$ is not even a Cuntz-Krieger algebra.

If λ and λ^{-1} are both algebraic with minimal polynomial of degree d, then by Proposition 2.1 $K_0(F^\lambda) \cong \mathbb{Z}^{2^{d-1}}$ and $K_1(F^\lambda) \cong \mathbb{Z}^{(2^{d-1}-1)}$ with natural bases, so that one can write down explicit integer matrices for the actions of λ. By row and column reducing these matrices to obtain the Smith Normal Form one can exactly compute the K-theory of $\mathcal{Q}^\lambda$ in this case.

For example, if $\lambda^2 + a\lambda - 1 = 0$ with $a > 0$ so that $\lambda = (1/2)(\sqrt{a^2+4} - a)$ then $K_0(\mathcal{Q}^\lambda) = \mathbb{Z}/a\mathbb{Z}$ and $K_1(\mathcal{Q}^\lambda) = \mathbb{Z}/2\mathbb{Z}$. None of these algebras are Cuntz-Krieger algebras. In particular, if $a = 1$ then λ is the inverse of the golden mean and $K_0(\mathcal{Q}^\lambda) = \{0\}$ and $K_1(\mathcal{Q}^\lambda) = \mathbb{Z}/2\mathbb{Z}$.

As another example, suppose $\lambda^2 + a\lambda + 1 = 0$ where $a \leq -3$. Then, $K_0(\mathcal{Q}^\lambda) = \mathbb{Z} \oplus (\mathbb{Z}/(a+2)\mathbb{Z})$ and $K_1(\mathcal{Q}^\lambda) = \mathbb{Z}$. For these algebras, $\mathcal{Q}^\lambda$ has the correct K-theory to be a Cuntz-Krieger algebra, and is therefore stably isomorphic to one. When $a = -3$ so that $\lambda = (1/2)(3 - \sqrt{5}))$ we have $K_0(\mathcal{Q}^\lambda) = \mathbb{Z} = K_1(\mathcal{Q}^\lambda)$.

As a final specific example, suppose $\lambda^3 + m\lambda^2 + (m-1)\lambda + 1 = 0$ with $m \geq 0$. Then $K_0(\mathcal{Q}^\lambda) = \mathbb{Z}/(4m+2)\mathbb{Z}$ and $K_1(\mathcal{Q}^\lambda) = \{0\}$. Then, $\mathcal{Q}^\lambda \cong O_{(4m+3)}$. This example is not completely random, as the next result shows.

Proposition 2.4 *If λ and λ^{-1} are both algebraic integers and $\mathcal{Q}^\lambda$ is stably isomorphic to a Cuntz algebra O_n then the minimal polynomial of λ has odd degree and constant term $+1$; $n \equiv 3 (mod 4)$; and all such Cuntz algebras appear this way.*

3 The action of $\mathbb{T}$ on $\mathcal{Q}^\lambda$ and its unique KMS state

Now $C_0^\lambda(\mathbb{R})$ has a faithful semifinite (norm) lower semicontinuous trace given by integration on positive elements. This trace is clearly invariant under $G_0^\lambda \cong \Gamma_\lambda$, which acts by translations. Therefore $A_0^\lambda = G_0^\lambda \rtimes_\alpha C_0^\lambda(\mathbb{R})$ also has a faithful semifinite (norm) lower semicontinuous trace obtained by composing the trace on $C_0^\lambda(\mathbb{R})$ with the conditional expectation: $A_0^\lambda \to C_0^\lambda(\mathbb{R})$. Finally we obtain a faithful semifinite (norm) lower semicontinuous weight on $\mathcal{Q}^\lambda$ by writing $A^\lambda = \mathbb{Z} \rtimes A_0^\lambda$ and composing the trace on A_0^λ with the conditional expectation $A^\lambda \to A_0^\lambda$. Now the projection $e \in A_0^\lambda$ obtained from the element $\mathcal{X}_{[0,1]} \in C_0^\lambda(\mathbb{R})$ clearly has trace $= 1$ and therefore has weight $= 1$ as an element of A^λ. Therefore the weight on A^λ restricted to $\mathcal{Q}^\lambda = eA^\lambda e$ defines a state ψ on $\mathcal{Q}^\lambda$.

Since $A^\lambda = \mathbb{Z} \rtimes A_0^\lambda$ there is a natural dual action of $\mathbb{T}$ on A^λ and the fixed point algebra under this action of $\mathbb{T}$ is A_0^λ. Since $e \in A_0^\lambda$, it is fixed by the action of

$\mathbb{T}$ and therefore the action restricts to an action of $\mathbb{T}$ on $\mathcal{Q}^\lambda = eA^\lambda e$. We call this the **gauge action** of $\mathbb{T}$ on $\mathcal{Q}^\lambda$. The fixed-point subalgebra of $\mathcal{Q}^\lambda$ under this action is exactly $F^\lambda = eA_0^\lambda e$, and ψ restricts to a (faithful, finite) trace τ on F^λ. We have the following theorem.

Theorem 3.1 *The state ψ on $\mathcal{Q}^\lambda$ is invariant under the gauge action of $\mathbb{T}$. Moreover, considered as an action of $\mathbb{R}$, the state ψ is a* KMS_β *state for $\beta = \log(\lambda^{-1})$. Furthermore, ψ is the unique* KMS *state for this action regardless of β. The GNS representation $\pi : \mathcal{Q}^\lambda \to \mathcal{B}(\mathcal{H}_\psi)$ afforded by ψ generates a type III_λ factor in $\mathcal{B}(\mathcal{H}_\psi)$.*

Remarks It is not difficult to show that the conditional expectation, $A^\lambda \to A_0^\lambda$ restricts to a conditional expectation $\Phi : \mathcal{Q}^\lambda \to F^\lambda$ and that the KMS state $\psi = \tau \circ \Phi$. We let $\mathcal{D}$ denote the generator of the one-parameter unitary group which implements the action of $\mathbb{T}$ on $\mathcal{H}_\psi$ and let Δ denote the modular operator on $\mathcal{H}_\psi$ for the state ψ. We have the following result.

Proposition 3.2 *With the left action of $\mathcal{Q}^\lambda$ on $\mathcal{H}_\psi$, the action of $\mathbb{T}$ is unitarily implemented on $\mathcal{H}_\psi$ and the generator $\mathcal{D}$ of this unitary group is related to the modular operator Δ for ψ by:*

$$\Delta = \lambda^{\mathcal{D}} \quad \text{or} \quad e^{it\mathcal{D}} = \Delta^{it/\log(\lambda)}.$$

4 The modular spectral triple for $\mathcal{Q}^\lambda$

Definition 4.1 To simplify notation let $\mathcal{Q} = \mathcal{Q}^\lambda$ and $F = F^\lambda$. Then $\mathcal{Q}$ becomes a right pre-Hilbert F-module with F-valued inner product $(a|b) := \Phi(a^*b)$. We let X denote the Hilbert F-module obtained by completing $\mathcal{Q}$ in the module norm $||x||_X = ||\Phi(x^*x)||^{1/2}$.

Now the left action of $\mathcal{Q}$ on itself extends uniquely to an action of $\mathcal{Q}$ as adjointable operators on X. Moreover the action of $\mathbb{T}$ on $\mathcal{Q}$ extends uniquely to a representation of $\mathbb{T}$ as unitaries on X. If we denote this representation by $z \mapsto u_z$, then for each $k \in \mathbb{Z}$ we define the k-th spectral subspace of this representation to be

$$X_k := \{x \in X \mid u_z(x) = z^k x \ \text{ for all } \ z \in \mathbb{T}\}.$$

The projection operator Φ_k from X onto X_k exists, is adjointable and is defined by

$$\Phi_k(x) = \frac{1}{2\pi} \int_{\mathbb{T}} z^{-k} u_z(x) d\theta, \quad z = e^{i\theta}, \quad x \in X.$$

For $x \in X$ we use the notation $x_k = \Phi_k(x)$ and then define $X_{\mathcal{D}} \subset X$ to be the linear subspace

$$X_{\mathcal{D}} = \left\{ x = \sum_{k \in \mathbb{Z}} x_k \in X \mid \ \| \sum_{k \in \mathbb{Z}} k^2 (x_k | x_k) \ \| < \infty \right\}.$$

For $x \in X_{\mathcal{D}}$ we define $\mathcal{D}(x) = \sum_{k \in \mathbb{Z}} k x_k$.

Proposition 4.2 *The operator $\mathcal{D} : X_{\mathcal{D}} \to X$ is a self-adjoint, regular operator on X with discrete spectrum $= \mathbb{Z}$ and spectral projections Φ_k for each $k \in \mathbb{Z}$. Moreover, the finite sums $\sum_{|k| \le N} \Phi_k$ converge strictly to the identity operator on X.*

Now, if $a, b \in \mathcal{Q}$ then the "rank one" operator $\theta_{a,b} : \mathcal{Q} \to \mathcal{Q}$ defined by $\theta_{a,b}(c) = a(b|c) = a\Phi(b^*c)$ is in fact a bounded linear operator in the Hilbert space norm on $\mathcal{H}_\psi$, with bound at most $\|a\|\,\|b\|$. This is a straightforward calculation using the inequalities $S^*T^*TS \leq \|T\|^2 S^*S$, and $\Phi(x^*)\Phi(x) \leq \Phi(x^*x)$, since Φ is 2-positive. So we regard these operators as operators in $\mathcal{B}(\mathcal{H}_\psi)$.

Definition 4.3 We define $\mathcal{N}$ to be the von Neumann algebra on $\mathcal{H}_\psi$ generated by all the rank one operators $\theta_{a,b}$ with a, b in $\mathcal{Q}$.

Proposition 4.4 *There is a faithful, normal, semifinite trace $\tilde{\tau}$ on $\mathcal{N}$ such that all rank one operators are in the ideal of definition of $\tilde{\tau}$ and*

$$\tilde{\tau}(\theta_{a,b}) = \tau \circ \Phi(b^*a) = \psi(b^*a), \quad a, b \in \mathcal{Q}.$$

Moreover, $\pi(\mathcal{Q}) \subset \mathcal{N}$, $\mathcal{D}$ is affiliated to $\mathcal{N}$; and $\tilde{\tau}(\Phi_k) = \lambda^{-k}$.

Since $\tilde{\tau}(\Phi_k) = \lambda^{-k}$ we see that $\mathcal{D}$ does not satisfy any finite summability criterion. We use $\Delta = \lambda^{\mathcal{D}}$ to define a new weight on $\mathcal{N}^+$. For $T \in \mathcal{N}^+$ let $\tau_\Delta(T) = sup_N \tilde{\tau}(\Delta_N T)$ where $\Delta_N T = \Delta(\sum_{|k|\leq N} \Phi_k)$.

Proposition 4.5 *τ_Δ is a faithful, normal, semifinite weight on $\mathcal{N}^+$.*

We now give another way to define τ_Δ which is not only conceptually useful but also makes a number of important properties straightforward to verify.

Notation. Let $\mathcal{M}$ be the relative commutant in $\mathcal{N}$ of the operator Δ. Equivalently, $\mathcal{M}$ is the relative commutant of the set of spectral projections $\{\Phi_k | k \in \mathbb{Z}\}$ of $\mathcal{D}$. Clearly, $\mathcal{M} = \sum_{k\in\mathbb{Z}} \Phi_k \mathcal{N} \Phi_k$.

Definition 4.6 As $\tilde{\tau}$ restricted to each $\Phi_k \mathcal{N} \Phi_k$ is a faithful finite trace with $\tilde{\tau}(\Phi_k) = \lambda^{-k}$, we define $\widehat{\tau}_k$ on $\Phi_k \mathcal{N} \Phi_k$ to be λ^k times the restriction of $\tilde{\tau}$. Then, $\widehat{\tau} := \sum_k \widehat{\tau}_k$ on $\mathcal{M} = \sum_{k\in\mathbb{Z}} \Phi_k \mathcal{N} \Phi_k$ is a faithful normal semifinite trace $\widehat{\tau}$ with $\widehat{\tau}(\Phi_k) = 1$ for all k.

We now use $\widehat{\tau}$ to give an alternative expression for τ_Δ :

Lemma 4.7 *An element $m \in \mathcal{N}$ is in $\mathcal{M}$ if and only if it is in the fixed point algebra of the action, $\sigma_t^{\tau_\Delta}$ on $\mathcal{N}$ defined for $T \in \mathcal{N}$ by $\sigma_t^{\tau_\Delta}(T) = \Delta^{it} T \Delta^{-it}$. Both $\pi(F)$ and the projections Φ_k belong to $\mathcal{M}$. The map $\Psi : \mathcal{N} \to \mathcal{M}$ defined by $\Psi(T) = \sum_k \Phi_k T \Phi_k$ is a conditional expectation onto $\mathcal{M}$ and $\tau_\Delta(T) = \widehat{\tau}(\Psi(T))$ for all $T \in \mathcal{N}^+$. That is, $\tau_\Delta = \widehat{\tau} \circ \Psi$ so that $\widehat{\tau}(T) = \tau_\Delta(T)$ for all $T \in \mathcal{M}^+$. Finally, if one of $A, B \in \mathcal{M}$ is $\widehat{\tau}$-trace-class and $T \in \mathcal{N}$ then $\tau_\Delta(ATB) = \tau_\Delta(A\Psi(T)B) = \widehat{\tau}(A\Psi(T)B)$.*

We now have the key lemma, but we omit the proof:

Lemma 4.8 *Suppose g is a function on $\mathbb{R}$ such that $g(\mathcal{D})$ is τ_Δ trace-class in $\mathcal{M}$; then for all $f \in F$ we have*

$$\tau_\Delta(\pi(f)g(\mathcal{D})) = \tau_\Delta(g(D))\tau(f) = \tau(f)\sum_{k\in\mathbb{Z}} g(k).$$

Proposition 4.9 *(i) We have $(1+\mathcal{D}^2)^{-1/2} \in \mathcal{L}^{(1,\infty)}(\mathcal{M}, \tau_\Delta)$ in the sense that $\tau_\Delta((1+\mathcal{D}^2)^{-s/2}) < \infty$ for all $s > 1$. Moreover, for all $f \in F^\lambda$*

$$\lim_{s\to 1^+} (s-1)\tau_\Delta(\pi(f)(1+\mathcal{D}^2)^{-s/2}) = 2\tau(f)$$

so that $\pi(f)(1+\mathcal{D}^2)^{-1/2}$ is a measurable operator in the sense of [C].

(ii) For $\pi(a) \in \pi(\mathcal{Q}^\lambda) \subset \mathcal{N}$ the following (ordinary) limit exists and

$$\widehat{\tau}_\omega(\pi(a)) = \frac{1}{2} \lim_{s \to 1^+} (s-1)\tau_\Delta(\pi(a)(1+\mathcal{D}^2)^{-s/2}) = \tau \circ \Phi(a) = \psi(a),$$

the original KMS *state $\psi = \tau \circ \Phi$ on $\mathcal{Q}^\lambda$.*

Proof (i) This proof is identical to [CPR2, Proposition 3.12].
(ii) This proof is the same as [CPR2, Proposition 3.14] with $\mathcal{Q}^\lambda, F^\lambda$ replacing O_n, F. □

Definition 4.10 Let $\mathcal{A} = \mathcal{Q}_c^\lambda$ and let γ be the gauge represntation of $\mathbb{T}$ as automorphism of $\mathcal{Q}^\lambda$. The triple $(\mathcal{A}, \mathcal{H}, \mathcal{D})$ along with γ, ψ, $\mathcal{N}$, τ_Δ satisfying properties (0) to (3) below is called the **modular spectral triple** of the dynamical system $(\mathcal{Q}^\lambda, \gamma, \psi)$
0) The $*$-subalgebra $\mathcal{A}$ of $\mathcal{Q}^\lambda$ is faithfully represented in $\mathcal{N}$ acting on the Hilbert space $\mathcal{H}_\psi$,
1) there is a faithful normal semifinite weight τ_Δ on $\mathcal{N}$ such that the modular automorphism group of τ_Δ is an inner automorphism group $\sigma_t = \mathrm{Ad}(\Delta^{it})$ (for $t \in \mathbb{C}$) of (the Tomita algebra of) $\mathcal{N}$ with $\sigma_i|_\mathcal{A} = \sigma$ in the sense that $\sigma_i(\pi(a)) = \pi(\sigma(a))$, where σ is the automorphism $\sigma(a) = \Delta^{-1}(a)$ on $\mathcal{A}$,
2) τ_Δ restricts to a faithful semifinite trace $\widehat{\tau}$ on $\mathcal{M} = \mathcal{N}^\sigma$, with a faithful normal projection $\Psi : \mathcal{N} \to \mathcal{M}$ satisfying $\tau_\Delta = \widehat{\tau} \circ \Psi$ on $\mathcal{N}$,
3) with $\mathcal{D}$ the generator of the one-parameter group implementing the gauge action of $\mathbb{T}$ on $\mathcal{H}$ we have: $[\mathcal{D}, \pi(a)]$ extends to a bounded operator (in $\mathcal{N}$) for all $a \in \mathcal{A}$ and for λ in the resolvent set of $\mathcal{D}$, $(\lambda - \mathcal{D})^{-1} \in \mathcal{K}(\mathcal{M}, \tau_\Delta)$, where $\mathcal{K}(\mathcal{M}, \tau_\Delta)$ is the ideal of compact operators in $\mathcal{M}$ relative to τ_Δ. In particular, $\mathcal{D}$ is affiliated to $\mathcal{M}$.

For matrix algebras $\mathcal{A} = \mathcal{Q}_c^\lambda \otimes M_k$ over $\mathcal{Q}_c^\lambda$, $(\mathcal{Q}_c^\lambda \otimes M_k, \mathcal{H} \otimes M_k, \mathcal{D} \otimes \mathrm{Id}_k)$ is also a modular spectral triple in the obvious fashion.

4.1 Modular K_1. We now make appropriate modifications to [CPR2, Section 4]) using [CNNR] introducing elements of these modular spectral triples $(\mathcal{A}, \mathcal{H}, \mathcal{D})$ (where $\mathcal{A}$ is a matrix algebra of analytic elements over $\mathcal{Q}^\lambda$) that will have a well defined pairing with our Dixmier functional $\widehat{\tau}_\omega$. Let $A = \mathcal{Q}^\lambda$. Following [HR] we say that a unitary (invertible, projection,...) in the $n \times n$ matrices over $\mathcal{Q}^\lambda$ for some n is a unitary (invertible, projection,...) over $\mathcal{Q}^\lambda$. We will also use σ_t for the inflated automorphism $\sigma_t \otimes \mathrm{Id}_n$ of $\mathcal{A}$.

Definition 4.11 Let v be a partial isometry in the $*$-algebra $\mathcal{A}$. We say that v satisfies the **modular condition** with respect to σ if the operators $v\sigma_t(v^*)$ and $v^*\sigma_t(v)$ are in the fixed point algebra $F \subset \mathcal{A}$ for all $t \in \mathbb{R}$. Of course, any partial isometry in F is a modular partial isometry.

Lemma 4.12 [CPR2, Lemma 4.8] *Let $v \in \mathcal{A}$ be a modular partial isometry. Then we have*

$$u_v = \begin{pmatrix} 1 - v^*v & v^* \\ v & 1 - vv^* \end{pmatrix}$$

is a modular unitary over $\mathcal{A}$. Moreover there is a modular homotopy $u_v \sim u_{v^*}$.

Note that in [CPR2] we used a different approach which is implied by the one given here. In [CPR2] we defined modular unitaries in terms of the regular automorphism

$$\pi(\sigma(a)) = \pi(\Delta^{-1}(a)) = \Delta^{-1}\pi(a)\Delta = \sigma_i(\pi(a)).$$

That is, we said that a unitary in $\mathcal{A}$ is modular if $u\sigma(u^*)$ and $u^*\sigma(u)$ are in the fixed point algebra. Define the modular K_1 group as follows.

Definition 4.13 Let $K_1(A,\sigma)$ be the abelian group with one generator $[v]$ for each partial isometry v over A satisfying the modular condition and with the following relations:

1) $[v]=0$ if v is over F,
2) $[v]+[w]=[v\oplus w]$,
3) if v_t, $t\in[0,1]$, is a continuous path of modular partial isometries in some matrix algebra over A then $[v_0]=[v_1]$.

One could use modular unitaries as in [CPR2] in place of these modular partial isometries.

The following can now be seen to hold.

Lemma 4.14 (Compare [CPR2, Lemma 4.9]) *Let $(\mathcal{A},\mathcal{H},\mathcal{D})$ be our modular spectral triple relative to $(\mathcal{N},\tau_\Delta)$ and set $F=\mathcal{A}^\sigma$ and $\sigma:\mathcal{A}\to\mathcal{A}$. Let $L^\infty(\Delta)=L^\infty(\mathcal{D})$ be the von Neumann algebra generated by the spectral projections of Δ then $L^\infty(\Delta)\subset\mathcal{Z}(\mathcal{M})$. Let $v\in\mathcal{A}$ be a partial isometry with vv^*, $v^*v\in F$. Then $\pi(v)Q\pi(v^*)\in\mathcal{M}$ and $\pi(v^*)Q\pi(v)\in\mathcal{M}$ for all spectral projections Q of $\mathcal{D}$, if and only if v is modular. That is, $\pi(v)\Delta\pi(v^*)$ and $\pi(v^*)\Delta\pi(v)$ (or $\pi(v)\mathcal{D}\pi(v^*)$ and $\pi(v^*)\mathcal{D}\pi(v)$) are both affiliated to $\mathcal{M}$ if and only if v is modular.*

Thus we see that modular partial isometries conjugate Δ to an operator affiliated to $\mathcal{M}$, and so $v\Delta v^*$ commutes with Δ (and $v\mathcal{D}v^*$ commutes with $\mathcal{D}$).

We will next show that there is an analytic pairing between (part of) modular K_1 and modular spectral triples. To do this, we are going to use the analytic formulae for spectral flow in [CP2].

4.2 A local index formula for the algebras $\mathcal{Q}^\lambda$. Using the fact that we have full spectral subspaces we know from [CNNR] that there is a formula for spectral flow which is analogous to the local index formula in noncommutative geometry. We remind the reader that $\tau_\Delta=\hat\tau\circ\Psi$ where $\Psi:\mathcal{N}\to\mathcal{M}$ is the canonical expectation, so that τ_Δ **restricted to** $\mathcal{M}$ is $\hat\tau$.

Theorem 4.15 *(Compare [CPR2, Theorem 5.5]) Let $(\mathcal{A},\mathcal{H},\mathcal{D})$ be the $(1,\infty)$-summable, modular spectral triple for the algebra $\mathcal{Q}^\lambda$ we have constructed previously. Then for any modular partial isometry v and for any Dixmier trace $\hat\tau_{\tilde\omega}$ associated to $\hat\tau$, we have spectral flow as an actual limit*

$$\begin{aligned} sf_{\hat\tau}(vv^*\mathcal{D},v\mathcal{D}v^*) &= \frac{1}{2}\lim_{s\to1^+}(s-1)\hat\tau(v[\mathcal{D},v^*](1+\mathcal{D}^2)^{-s/2}) \\ &= \frac{1}{2}\hat\tau_{\tilde\omega}(v[\mathcal{D},v^*](1+\mathcal{D}^2)^{-1/2})=\tau\circ\Phi(v[\mathcal{D},v^*]). \end{aligned}$$

The functional on $\mathcal{A}\otimes\mathcal{A}$ defined by

$$a_0\otimes a_1\mapsto\frac{1}{2}\lim_{s\to1^+}(s-1)\tau_\Delta(a_0[\mathcal{D},a_1](1+\mathcal{D}^2)^{-s/2})$$

is a σ-twisted (b,B)-cocycle.

Remark Spectral flow in this setting is independent of the path joining the endpoints of unbounded self-adjoint operators affiliated to $\mathcal{M}$, however it is not obvious that this is enough to show that it is constant on homotopy classes of modular unitaries. This latter fact is true but the proof is lengthy, so we refer to [CNNR].

Theorem 4.16 *We let* $(\mathcal{Q}_c^\lambda \otimes M_2, \mathcal{H}\otimes\mathbb{C}^2, \mathcal{D}\otimes 1_2)$ *be the modular spectral triple of* $(\mathcal{Q}_c^\lambda \otimes M_2)$.
(1) *For any* $k, j \in \mathbb{Z}$ *we can define a modular unitary* u *of the form in Lemma 4.12 so that the spectral flow is positive, being given by*

$$sf_{\tau_\Delta}(\mathcal{D}, u\mathcal{D}u^*) \;=\; (k-j)(\lambda^j - \lambda^k) \in \mathbb{Z}[\lambda] \subset \Gamma_\lambda.$$

(2) *If* λ^{-k} *and* λ^{-j}; *are not integers we let* $m_k = [\lambda^{-k}]$ *and* $m_j = [\lambda^{-j}]$ *we can define modular unitaries* u *of the form in Lemma 4.12 where the spectral flow is given by*

$$sf_{\tau_\Delta}(\mathcal{D}, u\mathcal{D}u^*) \;=\; (k-j)[\lambda^j(\lambda^{-k} - m_k) - \lambda^k(\lambda^{-j} - m_j)] \in \Gamma_\lambda.$$

Remarks The actual form of the modular unitaries in the previous theorem is a little complicated so we omit it.

Remarks The observation of [CPR2] that the twisted residue cocycle formula for spectral flow is calculating Araki's relative entropy of two KMS states [Ar] also applies to the examples in this subsection.

Acknowledgements We would like to thank Nigel Higson, Ryszard Nest, Sergey Neshveyev, Marcelo Laca, Iain Raeburn and Peter Dukes for advice and comments. The first and fourth named authors were supported by the Australian Research Council. The second and third named authors acknowledge the support of NSERC (Canada).

References

[Ar] H. Araki, *Relative entropy of states of von Neumann algebras*, Publ. RIMS, Kyoto Univ., **11** (1976) 809–833 and *Relative entropy for states of von Neumann algebras II*, Publ. RIMS, Kyoto Univ., **13** (1977) 173–192.

[APS3] M.F. Atiyah, V.K. Patodi, I.M. Singer, *Spectral asymmetry and Riemannian geometry. III*, Math. Proc. Camb. Phil. Soc. **79** (1976) 71–99.

[CNNR] A.L. Carey, R. Nest, S. Neshveyev, A. Rennie, *Twisted cyclic theory, equivariant KK-theory and KMS states*, J. Reine Angew. Math., **650** (2011), 161–191.

[CP1] A. L. Carey, J. Phillips, *Unbounded Fredholm modules and spectral flow*, Canadian J. Math., **50** (4) (1998) 673–718.

[CP2] A. L. Carey, J. Phillips, *Spectral flow in θ-summable Fredholm modules, eta invariants and the JLO cocycle*, K-Theory, **31** (2004) 135–194.

[CPPR] A. L. Carey, J. Phillips, I. F. Putnam, A. Rennie, *Families of type III KMS states on a class of C^*-algebras containing $\mathcal{O}_n$ and $\mathcal{Q}_{\mathbf{N}}$*, J. Functional Analysis, **260** (2011) 1637–1681.

[CPR1] A.L. Carey, J. Phillips, A. Rennie, *A noncommutative Atiyah-Patodi-Singer index theorem in KK-theory*, J. Reine Angew. Math., **643** (2010) 59–109.

[CPR2] A.L. Carey, J. Phillips, A. Rennie, *Twisted cyclic theory and an index theory for the gauge invariant* KMS *state on Cuntz algebras*, K-Theory, **6** (2010) 339–380.

[CPS2] A.L. Carey, J. Phillips, F. Sukochev, *Spectral flow and Dixmier traces*, Adv. Math, **173** (2003) 68–113.

[CPRS2] A.L. Carey, J. Phillips, A. Rennie, F. Sukochev, *The local index formula in semifinite von Neumann algebras I: Spectral Flow*, Adv. in Math. **202** (2006) 451–516.

[CRSS] A. L. Carey, A. Rennie, A. Sedaev, F. Sukochev, *Dixmier traces and asymptotics of zeta functions*, J. Functional Analysis **249** (2007) 253–283.

[CRT] A.L. Carey, A. Rennie, K. Tong, *Spectral flow invariants and twisted cyclic theory from the Haar state on $SU_q(2)$*, J. Geom. Phys. **59** (2009) 1431–1452.

[C0] A. Connes, *Une classification des facteurs de type III*, Annales Scientifiques de le'Ecole Norm. Sup., 4em serie t. 6 (1973) 18–252.

[C] A. Connes, *Noncommutative geometry*, Academic Press, 1994.

[Cu] J. Cuntz, *Simple C^*-algebras generated by isometries*, Commun. Math. Phys, **57** (1977) 173–189.

[Cu1] J. Cuntz, *C^*-algebras associated with the $ax + b$-semigroup over $\mathbb{N}$*, in *K-Theory and Noncommutative Geometry*, Eds, G. Cortinas, J. Cuntz, M Karoubi, R. Nest, C.A. Weibel, EMS Series of Congress Reports, Volume 2.

[D] J. Dixmier, *von Neumann algebras*, North-Holland, 1981.

[E] G. Elliott, *Some simple C^*-algebras constructed as crossed products with discrete outer automorphism groups*, Publ. RIMS Kyoto Univ., **16** (1980) 299-311.

[HR] N. Higson, J. Roe, *Analytic K-homology*, Oxford University Press, 2000.

[KNR] J. Kaad, R. Nest, A. Rennie, *KK-theory and spectral flow in von Neumann algebras*, arXive:math.OA/0701326.

[L] E. C. Lance, *Hilbert C^*-modules*, Cambridge University Press, Cambridge, 1995.

[PR] D. Pask, A. Rennie, *The noncommutative geometry of graph C^*-algebras I: The index theorem*, Journal of Functional Analysis, **233** (2006) 92–134.

[PhR] J. Phillips, I. Raeburn, *Semigroups of isometries, Toeplitz algebras and twisted crossed products*, J. Int. Equat. and Op. Th., **17** (1993) 579-602.

[Ped] G. K. Pedersen, *C^*-algebras and their automorphism groups*, London Math. Soc. monographs **14**, Academic Press, London 1979.

[Put1] I. Putnam, *An excision theorem for the K-theory of C^*-algebras*, J. Operator Theory, **38** (1997) 151–171.

[Put2] I. Putnam, *On the K-theory of C^*-algebras of principal groupoids*, Rocky Mountain J. Math., **28** no. 4 (1998) 1483–1518.

[RS] M. Rørdam and E. Størmer, *Classification of nuclear C^*-algebras. Entropy in operator algebras*, Encyclopedia of Mathematical Sciences, **126** (2002), Springer, Berlin.

[Sc] C. Schochet, *Topological methods for C^*-algebras II: geometric resolutions and the Künneth formula*, Pac. J. Math., **98** No. 2 (1982) 443–458.

[Ta] M. Takesaki, *Tomita's theory of modular Hilbert algebras and its applications*, Lecture Notes in Mathematics, **128** (1970), Springer, Berlin.

Fields Institute Communications
Volume **61**, 2011

Duality, Correspondences and the Lefschetz Map in Equivariant KK-Theory: A Survey

Heath Emerson
Department of Mathematics and Statistics
University of Victoria
Victoria, BC Canada V8W 3R4
hemerson@math.uvic.ca

Abstract. We survey work by the author and Ralf Meyer on equivariant KK-theory. Duality plays a key role in our results. We have organized this survey around the objective of computing a certain homotopy invariant of a space equipped with a (generally proper) action of a groupoid. This invariant is called the Lefschetz map. The Lefschetz map associates an equivariant K-homology class to an equivariant Kasparov self-morphism of a space X. We will compute the Lefschetz map explicitly for a bundle of smooth manifolds over the base space of a proper groupoid, in which groupoid elements act by diffeomorphisms between fibres. We will use the topological model of equivariant KK-theory – using correspondences – originating in work of Paul Baum, Alain Connes and Georges Skandalis in the 1980's, and worked out fully in several of our recent articles, to do this. We include a discussion of this 'topological KK-theory' and the issue of when it agrees with the analytic theory. We also describe several other computations of the Lefschetz map: in the situation of a simplicial action of a group on a simplicial complex we compute the Lefschetz invariant of an equivariant cellular self-map, and in the situation of a smooth Riemannian manifold with isometric action of a group we compute the Lefschetz invariant of a smooth equivariant self-map satisfying a weak transversality condition.

The papers [12], [16], [11], [15] present a study of the equivariant Kasparov groups $\mathrm{KK}^{\mathcal{G}}\big(\mathcal{C}_0(X), \mathcal{C}_0(Y)\big)$ where $\mathcal{G}$ is a locally compact Hausdorff groupoid with Haar system and X and Y are $\mathcal{G}$-spaces, usually with X a proper $\mathcal{G}$-space. This program builds on work of Kasparov, Connes and Skandalis done mainly in the 1980's. At that point, the main interest was the index theorem of Atiyah and Singer and its generalisations, and later, the Dirac dual-Dirac method and the Novikov conjecture. In the work we are surveying, the goal is to develop Euler characteristics

---

2010 *Mathematics Subject Classification.* Primary 19K35; Secondary 46L80, 19K56.

Heath Emerson was supported by a National Science and Engineering Council of Canada (NSERC) Discovery grant.

and Lefschetz formulas in equivariant KK-theory. These are *topological* features of the subject. Via the Baum-Connes isomorphism – when it applies – this contributes to noncommutative topology and geometry. Our program started in [12] where we found the *Lefschetz map* in connection with a C*-algebra K-theory problem. The Lefschetz map is a certain map, defined for a suitable class of $\mathcal{G}$-spaces X, having the form

$$\mathrm{Lef}\colon \mathrm{KK}_*^{\mathcal{G}\ltimes X}\big(\mathcal{C}_0(X\times_Z X),\mathcal{C}_0(X)\big)\to \mathrm{KK}_*^{\mathcal{G}}\big(\mathcal{C}_0(X),\mathcal{C}_0(Z)\big), \tag{0.1}$$

where we always denote by Z the base space of the groupoid. This map is defined under certain somewhat technical circumstances, but, again, these normally involve proper $\mathcal{G}$-spaces X. The domain of the Lefschetz map is very closely related to the simpler-looking group $\mathrm{KK}_*^{\mathcal{G}}\big(\mathcal{C}_0(X),\mathcal{C}_0(X)\big)$: the latter group maps in a natural way to the domain in (0.1) and this map is an isomorphism when the anchor map $X\to Z$ is a proper map. This means that the Lefschetz map can be used to assign an invariant, which is an equivariant K-homology class, to an equivariant Kasparov self-morphism of X. We call this class the *Lefschetz invariant* of the map. The Lefschetz invariant of the identity map is called the *Euler characteristic* of X. Lefschetz invariants and Euler characteristics are interesting even when $\mathcal{G}$ is the trivial groupoid, and the reader can do worse than to consider this case to begin with, although by doing so one misses the applications to noncommutative topology.

The definition of (0.1) uses the notion of an *abstract dual* for X. Abstract duals for a given $\mathcal{G}$-space X are not unique but the Lefschetz map does not depend on the choice of a dual, only on the existence of one. Abstract duals do not always exist either: a Cantor set X doesn't have an abstract dual even if $\mathcal{G}$ is trivial. But if X is a smooth $\mathcal{G}$-manifold, with $\mathcal{G}$ acting smoothly and isometrically on X, then X has an abstract dual, and if $\mathcal{G}$ is a group, then any $\mathcal{G}$-simplicial complex has an abstract dual due to [12]. (Note that in both examples, the action need not be proper, but at least *factors* through a proper action.)

The Lefschetz map is functorial for $\mathcal{G}$-maps $X\to X'$ in a way made explicit in §1.9 (see Theorem 1.2). In brief, it is a *homotopy invariant* of the $\mathcal{G}$-space X. In particular, since Lef doesn't depend on the dual used to to compute it, one can try to compute the Lefschetz invariant of a given morphism using two different duals and thereby get an identity in equivariant K-homology (see §2). Such examples (worked out in [12] and [13]), seemed to us interesting enough to support making a systematic study of the Lefschetz map. However, to get started on this question one obviously has to first describe the morphisms f themselves in some kind of satisfactory way. To this end, we have extended the theory of correspondences, initiated by Baum, Connes, Skandalis and others, to the equivariant situation, in the paper [15]. This extension presents some new features, and we will devote a part of this survey to explaining them. Of course the theory of correspondences is useful and important in its own right. But it is *designed* for intersection theory because of the way correspondences can be composed using coincidence spaces and transversality.

The theory of correspondences, which is purely topological, requires the groupoid $\mathcal{G}$ to be *proper*. If $\mathcal{G}$ is a group, this means that $\mathcal{G}$ is compact. The Baum-Connes apparatus allows us to reduce the general case to this one in certain circumstances, e.g., that of a proper action of a non-proper groupoid (like a proper action of an infinite, discrete group.) There is a geometrically defined equivalence relation on correspondences that makes the set of equivalence classes of $\mathcal{G}$-equivariant correspondences from X to Y the morphism set $\widehat{KK}^*_{\mathcal{G}}(X,Y)$ in a $\mathbb{Z}/2$-graded category $\widehat{KK}^{\mathcal{G}}$ which maps naturally to $\mathrm{KK}^{\mathcal{G}}$ and even isomorphically in many circumstances.

The fact that correspondences can be composed in an essentially topological fashion together with the definition of the Lefschetz map implies that for smooth $\mathcal{G}$-manifolds X, (0.1) can also be described in purely topological or geometric terms. Roughly speaking, a self-correspondence of X is a pair of maps $W \xleftarrow{b} W \xrightarrow{f} X$, for some smooth manifold W, and under a transversality assumption, the coincidence space $F := \{w \in W \mid b(w) = f(w)\}$ is a smooth manifold of dimension $\dim(W) - \dim(X)$, mapping naturally to X. The orientation on the correspondence induces one on F, and the whole construction works equivariantly, subject to a small point. The 'general position' caveat is non-trivial: in the equivariant setting, it may not be possible to perturb a pair of equivariant maps to make them transverse. The framework of equivariant correspondences allows us to treat this difficulty using Bott Periodicity, but we do not discuss this much here; another useful improvement uses a so-called *excess intersection* formula due to Baum and Block [4], but we do not discuss it either. The main point is that, say, for compact groups acting smoothly on compact spaces, the Lefschetz map is a purely *geometric* or topological invariant of the space.

The simplest example is where $\mathcal{G}$ is a compact group, X a smooth and compact manifold with a smooth action of $\mathcal{G}$, and we are given a smooth map $b\colon X \to X$ in general position in the usual sense. This yields an equivariant self-correspondence $X \xleftarrow{b} X \xrightarrow{\mathrm{id}} X$, and the coincidence space F defined above is the fixed-point set of the map, a finite set of points permuted by $\mathcal{G}$, oriented by an equivariant line bundle over this finite set. This data constitutes a (finite) K-oriented $\mathcal{G}$-manifold, mapping canonically to X via the inclusion, and hence a (zero-dimensional) Baum-Douglas cycle for X. If $\mathcal{G}$ is the trivial group, the index of this cycle is the algebraic fixed-point set of the map, in the usual sense. One might thus reasonably expect that Lefschetz invariants fit into a generalized Lefschetz fixed-point theorem, applying in the equivariant situation, and applying to correspondences instead of just maps. We allude to this idea again briefly below.

The notion of *duality* in Kasparov theory is central to this whole framework, and accordingly we begin the article with a discussion of it. It is well-known from the work of Kasparov and Connes and Skandalis (see [21] and [8] respectively) that if X is a smooth manifold, then there is a canonical isomorphism

$$\mathrm{KK}(\mathcal{C}_0(TX), \mathbb{C}) \cong \mathrm{RK}^*(X) := \mathrm{KK}^X(\mathcal{C}_0(X), \mathcal{C}_0(X))$$

where the groupoid equivariant KK group on the right is equivariant *representable* K-theory, or K-theory with *locally finite support*, denoted $\mathrm{RK}^*(X)$ by Kasparov. There is a generalisation of this duality to the equivariant situation if $\mathcal{G}$ is a groupoid acting smoothly and properly on a *bundle* $X \to Z$ of smooth manifolds over the base Z of $\mathcal{G}$, and furthermore, the roles of X and TX can be in a sense reversed, so that one can establish a pair of natural (in a technical sense) isomorphisms

$$\mathrm{KK}^{\mathcal{G}}_*(\mathcal{C}_0(TX) \otimes A, B) \cong \mathrm{KK}^{\mathcal{G} \ltimes X}_*(\mathcal{C}_0(X) \otimes A, \mathcal{C}_0(X) \otimes B) \tag{0.2}$$

and

$$\mathrm{KK}^{\mathcal{G}}_*(\mathcal{C}_0(X) \otimes A, B) \cong \mathrm{KK}^{\mathcal{G} \ltimes X}_*(\mathcal{C}_0(X), \mathcal{C}_0(TX) \otimes B) \tag{0.3}$$

for all $\mathcal{G}$-C*-algebras A and B. (The tensor products are all in the category of $\mathcal{G}$-C*-algebras.) These results are proved in [16], along with variants of them for manifolds with boundary. (See also [32]). It is the first kind of duality (0.2) which is relevant for the Lefschetz map, and the second (0.3) that is used to prove that the map $\widehat{KK}^{\mathcal{G}} \to \mathrm{KK}^{\mathcal{G}}$ is an isomorphism in certain cases.

In §1.6 these results are placed in a more abstract context using the notion of an *abstract dual* for a $\mathcal{G}$-space X, which is a $\mathcal{G}$-C*-algebra P together with a given, natural family of isomorphisms

$$\mathrm{KK}^{\mathcal{G}}_*(P \otimes A, B) \xrightarrow{\cong} \mathrm{KK}^{\mathcal{G} \ltimes X}(\mathcal{C}_0(X) \otimes A, \mathcal{C}_0(X) \otimes B), \tag{0.4}$$

one for each A, B (as in (0.2) above, with $P = \mathcal{C}_0(TX)$). Given the existence of an abstract dual for a given $\mathcal{G}$-space X, the Lefschetz map is then defined for X, as in (0.1), by manipulating certain ingredients of the dual. This definition is rather abstract and is not useful for computation, but it does have a certain essentially obvious functoriality (for example the first thing one wants to check is that the definition is independent of the dual chosen.)

To compute the Lefschetz map in specific cases one generally wants a dual with an additional property: a *Kasparov* dual is an abstract dual in which P is assumed to carry the structure of a $\mathcal{G} \ltimes X$-C*-algebra, rather than just a $\mathcal{G}$-C*-algebra. This is the case in the classical example $P = \mathcal{C}_0(TX)$. But there are other examples: $\mathcal{G}$-simplicial complexes also have rather natural Kasparov duals. Kasparov duals have a local, geometric aspect not present in abstract duals, which facilitates computation, especially of the Lefschetz map. Kasparov duals are defined in §1.6. The Lefschetz map is defined in §1.9.

In §2 we describe some examples of duals and computations of the Lefschetz map for equivariant *self-maps* (special kinds of cycles for $\mathrm{KK}^{\mathcal{G}}_*(\mathcal{C}_0(X), \mathcal{C}_0(X))$.) We focus on smooth manifolds with proper isometric actions of groups, and on simplicial complexes, with simplicial actions, also of groups. We are not restricted to proper groupoids in this discussion, because the computations we survey are done in analytic Kasparov theory. In §3 we begin to discuss Lefschetz invariants of equivariant correspondences; topological equivariant KK-theory requires a proper groupoid, so we assume this. We assume that X is a smooth bundle of manifolds over the base of $\mathcal{G}$ and compute the Lefschetz invariants of *arbitrary* equivariant self-correspondences of X, along the lines described above; since, under some mild hypotheses, every equivariant Kasparov morphism $X \to X$ is represented by a unique equivariant self-correspondence of X, up to geometric equivalence, it can be said that we have provided a complete computation of Lef in this situation.

1 Abstract duality and the Lefschetz map

Throughout this paper, *groupoid* shall mean locally compact Hausdorff groupoid with Haar system. All topological spaces will be assumed paracompact, locally compact and Hausdorff. For the material in this section, see [16]. For source material on equivariant KK-theory for groupoids, see [23]. One seems to be forced to consider groupoids, as opposed to groups, in equivariant Kasparov theory, even if one is ultimately only interested in groups. This will be explained later.

Therefore we will work more or less uniformly with groupoids when discussing general theory. When we discuss topological equivariant Kasparov theory, we will further assume that all groupoids are *proper*. This restriction is needed for various geometric constructions. The additional assumption of properness involves no serious loss of generality for our purposes because the Baum-Connes isomorphism, when it applies, gives a method of replacing non-proper groupoids by proper ones.

1.1 Equivariant Kasparov theory for groupoids. Let $\mathcal{G}$ be a groupoid. We let Z denote the base space. A $\mathcal{G}$-C*-algebra is in particular a C*-algebra over Z. This means that there is given a non-degenerate equivariant *-homomorphism from $\mathcal{C}_0(Z)$ to central multipliers of A. This identifies A with the section algebra of a continuous bundle of C*-algebras over Z. For a groupoid action we require in addition an isomorphism $r^*(A) \to s^*(A)$ which is compatible with the structure of $r^*(A)$ and $s^*(A)$ as C*-algebras over $\mathcal{G}$. Here $r\colon \mathcal{G} \to Z$ and $s\colon \mathcal{G} \to Z$ are the range and source map of the groupoid, and r^* (and similarly s^*) denotes the usual pullback operation of bundles. From the bundle point of view, all of this means that groupoid elements g with $s(g) = x$ and $r(g) = y$ induce *-homomorphisms $A_x \to A_y$ between the fibres of A at x and y.

In particular, if A is commutative, then A is the C*-algebra of continuous functions on a locally compact $\mathcal{G}$-space X, equipped with a map $\varrho_X : X \to Z$ called the *anchor map* for X, and a homeomorphism

$$\mathcal{G} \times_{Z,s} X \to \mathcal{G} \times_{Z,r} X, \qquad (g,x) \mapsto (g, gx)$$

where the domain and range of this homeomorphism (by abuse of notation) are respectively

$$\mathcal{G} \times_{Z,s} X := \{(g,x) \in \mathcal{G} \times X \mid s(g) = \varrho_X(x)\},$$

and similarly for $\mathcal{G} \times_{Z,r} X$ using r instead of s.

1.2 Tensor products. The category of $\mathcal{G}$-C*-algebras has a symmetric monoidal structure given by tensor products. We describe this very briefly (see [16][Section 2] for details).

Let A and B be two $\mathcal{G}$-C*-algebras. Since they are each C*-algebras over Z, their ordinary tensor product $A \otimes B$ is a C*-algebra over $Z \times Z$. (All tensor products used in Kasparov theory are *minimal*, or 'spatial' tensor products.) We can then restrict the C*-algebra $A \otimes B$ over $Z \times Z$ to a C*-algebra over the diagonal $Z \subset Z \times Z$. The result is called the *tensor product of A and B over Z*. The tensor product of A and B over Z carries a diagonal action of $\mathcal{G}$. We leave it to the reader to check that we obtain a $\mathcal{G}$-C*-algebra in this way.

In order not to complicate notation, *we write just $A \otimes B$ for the tensor product of A and B in the category of $\mathcal{G}$-C*-algebras.* We emphasize that the tensor product is over Z; this is not the same as the tensor product in the category of C*-algebras.

For commutative C*-algebras, *i.e.* for $\mathcal{G}$-spaces, say X and Y, with anchor maps as usual denoted $\varrho_X : X \to Z$ and $\varrho_Y : Y \to Z$, the tensor product is Gelfand dual to the operation which forms from X and Y the fibre product

$$X \times_Z Y := \{(x,y) \in X \times Y \mid \varrho_X(x) = \varrho_Y(y)\}.$$

The required anchor map $\varrho_{X \times_Z Y} : X \times_Z Y \to Z$ is of course the composition of the first coordinate projection and the anchor map for X (or the analogue using Y; they are equal). Of course groupoid elements act diagonally in the obvious way. Such *coincidence spaces* as the one just described will appear again and again in the theory of correspondences.

Finally, for the record, we supply the following important definition.

Definition 1.1 Let $\mathcal{G}$ be a groupoid. A $\mathcal{G}$-space X is *proper* if the map

$$\mathcal{G} \times_Z X \to X,\ (g, x) \mapsto (gx, x)$$

is a proper map, where $\mathcal{G} \times_Z X := \{(g, x) \in \mathcal{G} \times X \mid s(g) = \varrho_X(x)\}$.

A groupoid is itself called *proper* if it acts properly on its base space Z. Explicitly, the map

$$\mathcal{G} \to X \times_Z X, \qquad g \mapsto (r(g), s(g))$$

is required to be proper.

1.3 Equivariant Kasparov theory. Le Gall has defined $\mathcal{G}$-equivariant KK-theory in [23]. We briefly sketch the definitions. Let A and B be (possibly $\mathbb{Z}/2$-graded) $\mathcal{G}$-C*-algebras.

Then a cycle for $\mathrm{KK}^{\mathcal{G}}(A, B)$ is given by a $\mathbb{Z}/2$-graded $\mathcal{G}$-equivariant Hilbert B-module $\mathcal{E}$, together with a $\mathcal{G}$-equivariant grading-preserving *-homomorphism from A to the C*-algebra of bounded, adjointable operators on $\mathcal{E}$, and an essentially $\mathcal{G}$-equivariant self-adjoint operator F on $\mathcal{E}$ which is graded odd and satisfies that $[a, F]$ and $a(F^2 - 1)$ are compact operators (essentially zero operators) for all $a \in A$.

Modulo an appropriate equivalence relation, the set of equivalence classes of cycles can be identified with the morphism set $\mathrm{KK}^{\mathcal{G}}(A, B)$ in an additive, symmetric monoidal category. Higher KK-groups are defined using Clifford algebras, and since these are 2-periodic, there are only two up to isomorphism. We denote by $\mathrm{KK}^{\mathcal{G}}_*(A, B)$ the sum of these two groups.

If A and B are $\mathcal{G}$-C*-algebras, then the group $\mathrm{RKK}^{\mathcal{G}}(X; A, B)$ is by definition the groupoid-equivariant Kasparov group $\mathrm{KK}^{\mathcal{G} \ltimes X}(\mathcal{C}_0(X) \otimes A, \mathcal{C}_0(X) \otimes B)$. The tensor products are in the category of $\mathcal{G}$-C*-algebras. This group differs from $\mathrm{KK}^{\mathcal{G}}(A, \mathcal{C}_0(X) \otimes B)$ only in the support condition on cycles. For example, if $\mathcal{G}$ is trivial and $A = B = \mathbb{C}$ then $\mathrm{KK}^{\mathcal{G}}(A, \mathcal{C}_0(X) \otimes B)$ is the ordinary (compactly supported) K-theory of X and $\mathrm{RKK}^{\mathcal{G}}(X; A, B)$ is the representable K-theory of X (a non-compactly supported, 'locally-finitely supported' theory.) We discuss these groups in more detail in the next section. Of course similar remarks hold for higher $\mathrm{RKK}^{\mathcal{G}}$-groups.

In general, cycles for $\mathrm{RKK}^{\mathcal{G}}_*(X; A, B)$ are $\mathcal{G}$-equivariant continuous families of $\mathrm{KK}_*(A, B)$-cycles, parameterized by the points of X.

In particular, there is a canonical map

$$p_X^*\colon \mathrm{KK}^{\mathcal{G}}_*(A, B) \to \mathrm{RKK}^{\mathcal{G}}_*(X; A, B) \tag{1.1}$$

which associates to a cycle the corresponding constant family of cycles over X. We often refer to this map as the 'inflation map.'

1.4 Equivariant K-theory. In this section, we present an exceedingly brief overview of equivariant K-theory, roughly sufficient for the theory of equivariant correspondences. For more details see [11].

Let X be a proper $\mathcal{G}$-space. Recall that a $\mathcal{G} \ltimes X$-space consists of a $\mathcal{G}$-space Y together with a $\mathcal{G}$-equivariant map $\varrho_Y : Y \to X$ serving as the anchor map for the $\mathcal{G} \ltimes X$-action.

Definition 1.2 Let Y be a $\mathcal{G} \ltimes X$-space. The *$\mathcal{G}$-equivariant representable* K*-theory of* Y *with* X*-compact supports* is the group

$$\mathrm{RK}^{-*}_{\mathcal{G},X}(Y) := \mathrm{KK}^{\mathcal{G} \ltimes X}_*\big(\mathcal{C}_0(X), \mathcal{C}_0(Y)\big).$$

The $\mathcal{G}$*-equivariant representable* K*-theory of* Y is

$$\mathrm{RK}^*_{\mathcal{G}}(Y) := \mathrm{RK}^*_{\mathcal{G},Y}(Y).$$

Cycles for $\mathrm{KK}^{\mathcal{G}\ltimes X}\big(\mathcal{C}_0(X),\mathcal{C}_0(Y)\big)$ consist of pairs $(\mathcal{H},F)$ where $\mathcal{E}$ is a countably generated $\mathbb{Z}/2$-graded $\mathcal{G}\ltimes X$-equivariant right Hilbert $\mathcal{C}_0(X)$-module equipped with a $\mathcal{G}\ltimes X$-equivariant non-degenerate *-homomorphism from $\mathcal{C}_0(X)$ to the C*-algebra of bounded, adjointable operators on $\mathcal{E}$, and F is a bounded, odd, self-adjoint, essentially $\mathcal{G}$-equivariant adjointable operator on $\mathcal{H}$ such that $f(F^2-1)$ is a compact operator, for all $f\in\mathcal{C}_0(X)$. The properness of $\mathcal{G}$ implies that F may be averaged to be actually $\mathcal{G}$-equivariant, so we assume this in the following.

The Hilbert $\mathcal{C}_0(Y)$-module $\mathcal{E}$ is the space of continuous sections of a continuous field of $\mathbb{Z}/2$-graded Hilbert spaces $\{\mathcal{H}_y \mid y\in Y\}$ over Y. Since F must be $\mathcal{C}_0(Y)$-linear, it consists of a continuous family $\{F_y \mid y\in Y\}$ of odd operators on these graded Hilbert spaces such that F_y^2-1 is a compact operator on $\mathcal{H}_y$ for all $y\in Y$.

By $\mathcal{G}\ltimes X$-equivariance, the representation of $\mathcal{C}_0(X)$ on $\mathcal{E}$ must factor through the *-homomorphism $\mathcal{C}_0(X)\to\mathcal{C}_0(Y)$ Gelfand dual to the anchor map $\varrho_Y\colon Y\to X$. Therefore F commutes with the action of $\mathcal{C}_0(X)$ as well; in fact the induced representation of $\mathcal{C}_0(X)$ on each Hilbert space $\mathcal{H}_y$ sends a continuous function $f\in\mathcal{C}_0(X)$ to the operator of multiplication by the complex number $f\big(\varrho_Y(y)\big)$.

In particular, the only role of the representation of $\mathcal{C}_0(X)$ is to relax the support condition on the compact operator valued function F^2-1 from requiring it to vanish at ∞ of Y to only requiring it to vanish at infinity along the fibres of $\varrho_Y: Y\to X$.

If $\varrho_Y: Y\to X$ is a *proper* map then $\mathrm{RK}^*_{\mathcal{G},X}(Y) = \mathrm{RK}^*_{\mathcal{G}}(X) := \mathrm{RK}^*_{\mathcal{G},X}(X)$; these two groups have exactly the same cycles.

Example 1.1 Any $\mathcal{G}$-equivariant complex vector bundle V on Y, or, more generally, a $\mathbb{Z}/2$-graded $\mathcal{G}$-equivariant complex vector bundle, yields a cycle for $\mathrm{RK}^0_{\mathcal{G}}(Y)$ by choosing a $\mathcal{G}$-invariant Hermitian metric on V and forming the corresponding $\mathcal{G}\ltimes Y$-equivariant $\mathbb{Z}/2$-graded right Hilbert $\mathcal{C}_0(Y)$ module of sections, where the grading is the trivial one. We set the operator equal to zero.

Example 1.2 Let X be a $\mathcal{G}$-space and let V be a $\mathcal{G}$-equivariantly K-oriented vector bundle over X of (real) dimension n. The $\mathcal{G}$-equivariant vector bundle projection $\pi_V: V\to X$ gives V the structure of a space over X, so that V becomes a $\mathcal{G}\ltimes X$-space. Then the *Thom isomorphism* provides an invertible *Thom class*

$$t_V\in \mathrm{RK}^{\dim V}_{\mathcal{G},X}(V) := \mathrm{KK}^{\mathcal{G}\ltimes X}_{-\dim V}\big(\mathcal{C}_0(X),\mathcal{C}_0(V)\big).$$

In the case $\mathcal{G} = \mathrm{Spin}^c(\mathbb{R}^n)$ and $X=\star$ and $V:=\mathbb{R}^n$ with the representation $\mathrm{Spin}^c(\mathbb{R}^n)\to\mathrm{Spin}(\mathbb{R}^n)\to\mathrm{O}(n,\mathbb{R})$, the class $t_{\mathbb{R}^n}$ is the 'Bott' class figuring in equivariant Bott Periodicity.

Certain further normalizations can be made in order to describe the groups $\mathrm{RK}^*_{\mathcal{G},X}(Y)$. A standard one is to replace the $\mathbb{Z}/2$-grading on $\mathcal{E}$ by the standard even grading, so that $\mathcal{E}$ consists of the sum of two copies of the same Hilbert module. This means that F can be taken to be of the form $\begin{pmatrix}0 & F_1^*\\ F_1 & 0\end{pmatrix}$ and the conditions involving F are replaced by ones involving F_1 and F_1^*; in what follows, we denote F_1 simply by F. Then the Fredholm conditions are that $f(FF^*-1)$ and $f(F^*F-1)$ are compact for all $f\in\mathcal{C}_0(X)$. In other words, $y\mapsto F_y$ takes essentially unitary values in $\mathbb{B}(\mathcal{H}_y)$ for all $y\in Y$ and the compact operator valued functions FF^*-1 and F^*F-1 vanish at infinity along the fibres of $\varrho\colon Y\to X$

(so in particular if $Y \to X$ is a proper map (e.g., the identity map) the support condition becomes vacuous and we just have maps to Fredholm operators.)

The equivariant stabilization theorem for Hilbert modules implies that we may take $\mathcal{H}$ to have the special form $L^2(\mathcal{G})^\infty \otimes_{\mathcal{C}_0(Z)} \mathcal{C}_0(Y)$, where $L^2(\mathcal{G})$ is the $\mathcal{G}$-equivariant right Hilbert $\mathcal{C}_0(Z)$-module defined using the Haar system of $\mathcal{G}$, and the superscript indicates the sum of countably many copies of $L^2(\mathcal{G})$. Thus, the corresponding field of Hilbert spaces has fibre $L^2(\mathcal{G}^{\varrho_Y(y)})^\infty$ at $y \in Y$.

All of this leads to a description of $\mathrm{RK}^0_{\mathcal{G}}(Y) := \mathrm{RK}^0_{\mathcal{G},Y}(Y)$ as homotopy classes of $\mathcal{G}$-equivariant maps from Y to the space of Fredholm operators on $L^2(\mathcal{G})^\infty$, but topologizing this space is somewhat delicate. Similarly, the relative groups $\mathrm{RK}^*_{\mathcal{G},X}(Y)$ are maps to Fredholm operators with compact vertical support with respect to the map $Y \to X$, where the *support* of a map to Fredholm operators is by definition the complement of the set where the map takes invertible values. See [11] for details.

Remark 1.1 If $\mathcal{G}$ acts properly and *co-compactly* on X, A is a trivial $\mathcal{G}$-C*-algebra and B is a $\mathcal{G} \ltimes X$-C*-algebra, then there is a canonical isomorphism

$$\mathrm{KK}^{\mathcal{G}\ltimes X}(\mathcal{C}_0(X) \otimes A, B) \cong \mathrm{KK}(A, \mathcal{G} \ltimes B).$$

In particular, the $\mathcal{G}$-equivariant representable K-theory of X agrees with the K-theory of the corresponding cross-product. Under this identification, classes in $\mathrm{RK}^0_{\mathcal{G}}(X)$ which are represented by equivariant vector bundles on X correspond to classes in $\mathrm{K}_0(\mathcal{G} \ltimes \mathcal{C}_0(X))$ which are represented by projections in the stabilisation of $\mathcal{G} \ltimes \mathcal{C}_0(X)$. The question of whether or not all $\mathrm{RK}^0_{\mathcal{G}}$-classes are so represented is highly dependent on the groupoid – see [11].

Thus, even if the reader is only interested in groups, or the trivial group, it is convenient to introduce groupoids to some extent in order to describe cohomology theories with different support conditions.

1.5 Tensor and forgetful functors. The following simple functor will play an important role. If P is a $\mathcal{G} \ltimes X$-algebra, we denote by T_P the map

$$\mathrm{RKK}^{\mathcal{G}}(X; A, B) := \mathrm{KK}^{\mathcal{G}\ltimes X}(\mathcal{C}_0(X) \otimes A, \mathcal{C}_0(X) \otimes B) \to \mathrm{KK}^{\mathcal{G}}(A \otimes P, B \otimes P)$$

which sends a $\mathcal{G} \ltimes X$-equivariant right Hilbert $\mathcal{C}_0(X) \otimes B$-Hilbert module $\mathcal{E}$ to $\mathcal{E} \otimes_X P$, the tensor product being *in the category of $\mathcal{G} \ltimes X$-algebras* (we accordingly use a subscript for emphasis) and sends $F \in \mathbb{B}(\mathcal{E})$ to the operator $F \otimes_X \mathrm{id}_P$. This definition makes sense since F commutes with the $\mathcal{C}_0(X)$-structure on $\mathcal{E}$.

The functor T_P is the composition of external product

$$\_ \otimes_X 1_P\colon \mathrm{KK}^{\mathcal{G}\ltimes X}(\mathcal{C}_0(X) \otimes A, \mathcal{C}_0(X) \otimes B) \to \mathrm{KK}^{\mathcal{G}\ltimes X}(A \otimes_X P, B \otimes_X P)$$

(where the X-structure on $A \otimes P$ etc. is on the P factor), and the forgetful map

$$\mathrm{KK}^{\mathcal{G}\ltimes X}(A \otimes_X P, B_X \otimes P) \to \mathrm{KK}^{\mathcal{G}}(A \otimes_X P, B \otimes_X P)$$

which maps a $\mathcal{G} \ltimes X$-algebra or Hilbert module to the underlying $\mathcal{G}$-algebra, or Hilbert module, thus forgetting the X-structure.

1.6 Kasparov duals. We begin our discussion of duality by formalizing some duality calculations of Kasparov, *cf.* [21, Theorem 4.9]. Explicit examples will be discussed later.

For convenience of notation we will often write $\mathbb{1} := \mathcal{C}_0(Z)$. This notation expresses the fact that $\mathcal{C}_0(Z)$ is the tensor unit in the tensor category of $\mathcal{G}$-C*-algebras. Similarly, if $\mathcal{G}$ acts on a space X then we sometimes denote by $\mathbb{1}_X$ the $\mathcal{G}$-C*-algebra $\mathcal{C}_0(X)$; thus $\mathbb{1}_X$ is the tensor unit in the category of $\mathcal{G} \ltimes X$ C*-algebras, X being the base of $\mathcal{G} \ltimes X$. This notation is consistent with the source of this material (see [16].)

Definition 1.3 Let $n \in \mathbb{Z}$. An *n-dimensional $\mathcal{G}$-equivariant Kasparov dual* for the $\mathcal{G}$-space X is a triple (P, D, Θ), where

- P is a (possibly $\mathbb{Z}/2$-graded) $\mathcal{G} \ltimes X$-C*-algebra,
- $D \in \mathrm{KK}^{\mathcal{G}}_{-n}(P, \mathbb{1})$, and
- $\Theta \in \mathrm{RKK}^{\mathcal{G}}_n(X; \mathbb{1}, P)$,

subject to the following conditions:

(1) $\Theta \otimes_P D = \mathrm{id}_{\mathbb{1}}$ in $\mathrm{RKK}^{\mathcal{G}}_0(X; \mathbb{1}, \mathbb{1})$;
(2) $\Theta \otimes f = \Theta \otimes_P T_P(f)$ in $\mathrm{RKK}^{\mathcal{G}}_{*+n}(X; A, B \otimes P)$ for all $\mathcal{G}$-C*-algebras A and B and all $f \in \mathrm{RKK}^{\mathcal{G}}_*(X; A, B)$;
(3) $T_P(\Theta) \otimes_{P \otimes P} \Phi_P = (-1)^n T_P(\Theta)$ in $\mathrm{KK}^{\mathcal{G}}_n(P, P \otimes P)$, where Φ_P is the flip automorphism on $P \otimes P$.

Remark 1.2 Condition (2) above is somewhat confusing (but is necessary). Both compositions $\Theta \otimes f$ and $\Theta \otimes_P T_P(f)$ are in the category $\mathrm{RKK}^{\mathcal{G}}(X)$. But they are quite different; the composition $\Theta \otimes_P T_P(f)$ involves first applying to f the functor $T_P\colon \mathrm{RKK}^{\mathcal{G}}(X; A, B) \to \mathrm{KK}^{\mathcal{G}}(A \otimes P, B \otimes P)$ – recall that T_P factors through a forgetful functor (see the discussion at the beginning of §1.5) – and then applying to the resulting class the inflation map $p_X^*\colon \mathrm{KK}^{\mathcal{G}}_*(A \otimes P, B \otimes P) \to \mathrm{RKK}^{\mathcal{G}}_*(X; A \otimes P, B \otimes P)$ to obtain a morphism in $\mathrm{RKK}^{\mathcal{G}}_*(X; A \otimes P, B \otimes P)$, which is then composed with $\Theta \in \mathrm{RKK}^{\mathcal{G}}_n(X; \mathbb{1}, P)$. The composition $\Theta \otimes f$ is simply the external product in $\mathrm{RKK}^{\mathcal{G}}(X)$ of $\Theta \in \mathrm{RKK}^{\mathcal{G}}_n(X; \mathbb{1}, P)$ and $f \in \mathrm{RKK}^{\mathcal{G}}(X; A, B)$. It turns out that (2) is automatic when X is a universal proper $\mathcal{G}$-space. In general, the idea behind proving (2) in examples involves constructing a homotopy between the restriction of the two projection maps $X \times_Z X \to X$ on the support of the class Θ.

The following theorem is proved in [16].

Theorem 1.1 *Let $n \in \mathbb{Z}$, let P be a $\mathcal{G} \ltimes X$-C*-algebra, $D \in \mathrm{KK}^{\mathcal{G}}_{-n}(P, \mathbb{1})$, and $\Theta \in \mathrm{RKK}^{\mathcal{G}}_n(X; \mathbb{1}, P)$. Define two natural transformations*

$$\mathrm{PD}\colon \mathrm{KK}^{\mathcal{G}}_{i-n}(P \otimes A, B) \to \mathrm{RKK}^{\mathcal{G}}_i(X; A, B), \qquad f \mapsto \Theta \otimes_P f,$$
$$\mathrm{PD}^*\colon \mathrm{RKK}^{\mathcal{G}}_i(X; A, B) \to \mathrm{KK}^{\mathcal{G}}_{i-n}(P \otimes A, B), \qquad g \mapsto (-1)^{in} T_P(g) \otimes_P D,$$

These two are inverse to each other if and only if (P, D, Θ) is an n-dimensional $\mathcal{G}$-equivariant Kasparov dual for X.

1.7 Abstract duals. The reader may have noticed that the only place the $\mathcal{C}_0(X)$-structure on P comes into play in the conditions listed in Definition 1.3, and in the statement of Theorem 1.1, is via the functor T_P. In particular, if one has a Kasparov dual (P, D, Θ) and if one changes the $\mathcal{C}_0(X)$-structure on P, for example by composing it with a $\mathcal{G}$-equivariant homeomorphism of X, then the map PD of Theorem 1.1 does not change; since by the theorem PD^* is its inverse map, it would not change either, strangely, since its definition uses T_P. In fact it turns out that the functor T_P can be reconstructed from PD if one knows that PD is an isomorphism. This is an important idea in connection with the Lefschetz map and suggests the following useful definition.

Definition 1.4 An n-dimensional *abstract dual* for X is a pair (P, Θ), where P is a $\mathcal{G}$-C*-algebra and $\Theta \in \mathrm{RKK}^{\mathcal{G}}_n(X; \mathbb{1}, P)$, such that the map PD defined as in Theorem 1.1 is an isomorphism for all $\mathcal{G}$-C*-algebras A and B.

This definition is shorter, does not involve any X-structure on the dual algebra P, and, as mentioned, is useful for theoretical reasons, but it seems like it should be difficult to check in practise because it is not clear what map to use as a potential inverse to PD.

In any case, it is clear from Theorem 1.1 that a pair (P, Θ) is an abstract dual if it is part of a Kasparov dual (P, D, Θ).

Proposition 1.1 *An abstract dual for a space X is unique up to a canonical $\mathrm{KK}^{\mathcal{G}}$-equivalence if it exists, and even covariantly functorial in the following sense.*

Let X and Y be two $\mathcal{G}$-spaces and let $f\colon X \to Y$ be a $\mathcal{G}$-equivariant continuous map. Let (P_X, Θ_X) and (P_Y, Θ_Y) be abstract duals for X and Y of dimensions n_X and n_Y, respectively. Then there is a unique $P_f \in \mathrm{KK}^{\mathcal{G}}_{n_Y - n_X}(P_X, P_Y)$ with $\Theta_X \otimes_{P_X} P_f = f^*(\Theta_Y)$. Given two composable maps between three spaces with duals, we have $P_{f\circ g} = P_f \circ P_g$. If $X = Y$, $f = \mathrm{id}_X$, and $(P_X, \Theta_X) = (P_Y, \Theta_Y)$, then $P_f = \mathrm{id}_{P_X}$. If only $X = Y$, $f = \mathrm{id}_X$, then P_f is a $\mathrm{KK}^{\mathcal{G}}$-equivalence between the two duals of X.

Although the map $f\colon X \to Y$ appearing in Proposition 1.1 does not have to be proper, it nonetheless yields a morphism P_f in $\mathrm{KK}^{\mathcal{G}}$.

1.8 Duality co-algebra. Let (P, Θ) be an n-dimensional abstract dual for a $\mathcal{G}$-space X. It follows from the Yoneda Lemma that another abstract dual (P', Θ') also for X and say of dimension n' is related to (P, Θ) by an invertible element

$$\psi \in \mathrm{KK}^{\mathcal{G}}_{n'-n}(P, P'), \text{ such that } \Theta \otimes_P \psi = \Theta'. \tag{1.2}$$

We repeat for emphasis that since (P, Θ) is only an *abstract* dual, we are not assuming that there is a $\mathcal{G} \ltimes X$-structure on P. However, we are going to attempt to reconstruct what we might consider to be a $\mathcal{G} \ltimes X$-structure on P *at the level of* KK-*theory.* Along the way we will keep track of how the change in dual from (P, Θ) to (P', Θ') affects our constructions.

Define $D \in \mathrm{KK}^{\mathcal{G}}_{-n}(P, \mathbb{1})$ by the requirement

$$\mathrm{PD}(D) := \Theta \otimes_P D = 1_{\mathbb{1}} \qquad \text{in } \mathrm{RKK}^{\mathcal{G}}_0(X; \mathbb{1}, \mathbb{1}). \tag{1.3}$$

as in the first condition in Definition 1.3). The class D should thus play the role of the class named D in a Kasparov dual. It is routine to check that when we change the dual, as above, D is replaced by $\psi^{-1} \otimes_P D$.

We call D the *counit of the duality* because it plays the algebraic role of a counit in the theory of adjoint functors (see [16] and also Remark 3.3 below).

Define $\nabla \in \mathrm{KK}^{\mathcal{G}}_n(P, P \otimes P)$ by the requirement that

$$\mathrm{PD}(\nabla) := \Theta \otimes_P \nabla = \Theta \otimes_X \Theta \qquad \text{in } \mathrm{RKK}^{\mathcal{G}}_{2n}(X; \mathbb{1}, P \otimes P).$$

We call ∇ the *comultiplication of the duality.* When we change the dual, ∇ is replaced by

$$(-1)^{n(n'-n)}\psi^{-1} \otimes_P \nabla \otimes_{P\otimes P} (\psi \otimes \psi) \in \mathrm{KK}^{\mathcal{G}}_{n'}(P', P' \otimes P').$$

Remark 1.3 If $n = 0$ then the object P of $\mathrm{KK}^{\mathcal{G}}$ with counit D and comultiplication ∇ is a cocommutative, counital coalgebra object in the tensor category $\mathrm{KK}^{\mathcal{G}}$:

$$\nabla \otimes_{P\otimes P} (\nabla \otimes 1_P) = \nabla \otimes_{P\otimes P} (1_P \otimes \nabla), \tag{1.4}$$

$$\nabla \otimes_{P\otimes P} \Phi_P = \nabla, \tag{1.5}$$

$$\nabla \otimes_{P\otimes P} (D \otimes 1_P) = 1_P = \nabla \otimes_{P\otimes P} (1_P \otimes D). \tag{1.6}$$

Equation (1.4) holds in $\mathrm{KK}^{\mathcal{G}}_{2n}(P, P^{\otimes 3})$, equation (1.5) holds in $\mathrm{KK}^{\mathcal{G}}_{n}(P, P \otimes P)$, and (1.6) holds in $\mathrm{KK}^{\mathcal{G}}_{0}(P, P)$.

Now, for $\mathcal{G}$-C*-algebras A and B, we define

$$T'_P\colon \mathrm{RKK}^{\mathcal{G}}_*(X; A, B) \to \mathrm{KK}^{\mathcal{G}}_*(P \otimes A, P \otimes B), \qquad f \mapsto \nabla \otimes_P \mathrm{PD}^{-1}(f),$$

where PD is the duality isomorphism, ∇ is the comultiplication of the duality, and $\otimes_P$ operates on the *second* copy of P in the target $P \otimes P$ of ∇. A computation yields that

$$\mathrm{PD}\bigl(T'_P(f)\bigr) = \Theta \otimes_X f \qquad \text{in } \mathrm{RKK}^{\mathcal{G}}_{i+n}(X; A, P \otimes B) \tag{1.7}$$

for all $f \in \mathrm{RKK}^{\mathcal{G}}_i(X; A, B)$. It follows that

$$T'_P(f) = T_P(f)$$

if (P, Θ) is part of a Kasparov dual, and thus T_P is in fact independent of the $\mathcal{G} \ltimes X$-structure on P, verifying our guess above.

When we change the dual, we replace T'_P by the map

$$\mathrm{RKK}^{\mathcal{G}}_i(X; A, B) \ni f \mapsto (-1)^{i(n-n')}\psi^{-1} \otimes_P T_P(f) \otimes_P \psi \in \mathrm{KK}^{\mathcal{G}}_i(P' \otimes A, P' \otimes B). \tag{1.8}$$

In fact, one can check that the maps T'_P above define a functor

$$T'_P\colon \mathrm{RKK}^{\mathcal{G}}(X) \to \mathrm{KK}^{\mathcal{G}}.$$

This is a $\mathrm{KK}^{\mathcal{G}}$-functor in the sense that it is compatible with the tensor products $\otimes$, and it is left adjoint to the 'inflation map' $p_X^*\colon \mathrm{KK}^{\mathcal{G}} \to \mathrm{RKK}^{\mathcal{G}}$.

It follows that we can write the inverse duality map involved in an abstract dual (P, Θ) as

$$\mathrm{PD}^{-1}(f) = (-1)^{in} T'_P(f) \otimes_P D \qquad \text{in } \mathrm{KK}^{\mathcal{G}}_{i-n}(P \otimes A, B) \tag{1.9}$$

for $f \in \mathrm{RKK}^{\mathcal{G}}_i(X; A, B)$. By the above discussion this formula agrees with the map PD^* when we have a Kasparov dual.

1.9 The Lefschetz map. The formal computations summarized in the previous section allow us to single out an interesting invariant of a $\mathcal{G}$-space X, at least under the hypothesis that X has *some* abstract dual.

For any $\mathcal{G}$-space X the diagonal embedding $X \to X \times_Z X$ is a proper map and hence induces a *-homomorphism

$$\mathbb{1}_X \otimes \mathbb{1}_X \cong \mathcal{C}_0(X \times_Z X) \to \mathcal{C}_0(X) = \mathbb{1}_X.$$

This map is $\mathcal{G} \ltimes X$-equivariant and hence yields

$$\Delta_X \in \mathrm{RKK}^{\mathcal{G}}(X; \mathbb{1}_X, \mathbb{1}) \cong \mathrm{KK}^{\mathcal{G}\ltimes X}\big(\mathcal{C}_0(X \times_Z X), \mathcal{C}_0(X)\big).$$

We call this the *diagonal restriction class.* It yields a canonical map

$$\text{\textvisiblespace} \otimes_{\mathbb{1}_X} \Delta_X \colon \mathrm{KK}^{\mathcal{G}}(\mathbb{1}_X \otimes A, \mathbb{1}_X \otimes B) \to \mathrm{RKK}^{\mathcal{G}}(X; \mathbb{1}_X \otimes A, B). \tag{1.10}$$

In particular, this contains a map $\mathrm{KK}^{\mathcal{G}}(\mathbb{1}_X, \mathbb{1}_X) \to \mathrm{RKK}^{\mathcal{G}}(X; \mathbb{1}_X, \mathbb{1})$.

Example 1.3 If $f\colon X \to X$ is a proper, continuous, $\mathcal{G}$-equivariant map, then

$$[f] \otimes_{\mathbb{1}_X} \Delta_X \in \mathrm{RKK}^{\mathcal{G}}(X; \mathbb{1}_X, \mathbb{1})$$

is the class of the *-homomorphism induced by $(\mathrm{id}_X, f)\colon X \to X \times_Z X$.

Now drop the assumption that f be proper. Then (id_X, f) is still a proper, continuous, $\mathcal{G}$-equivariant map. The class of the *-homomorphism it induces is equal to $f^*(\Delta_X)$, where we use the maps

$$f^*\colon \mathrm{RKK}^{\mathcal{G}}_*(X; A, B) \to \mathrm{RKK}^{\mathcal{G}}_*(X; A, B)$$

for $A = \mathbb{1}_X$, $B = \mathbb{1}$ induced by $f\colon X \to X$ (the functor $X \mapsto \mathrm{RKK}^{\mathcal{G}}(X; A, B)$ is functorial with respect to arbitrary $\mathcal{G}$-maps, not just proper ones.) This suggests that we can think of $\mathrm{RKK}^{\mathcal{G}}(X; \mathbb{1}_X, \mathbb{1})$ as generalized, possibly *non-proper* self-maps of X.

In fact if the anchor map $X \to Z$ is a proper map, so that X is a bundle of compact spaces over Z, then $\text{\textvisiblespace} \otimes_{\mathbb{1}_X} \Delta_X$ is an isomorphism (an easy exercise in the definitions.)

Now let T'_P be the tensor functor and Δ_X the diagonal restriction class of an abstract dual. We define the *multiplication class* of P by

$$[m] := T'_P(\Delta_X) \in \mathrm{KK}^{\mathcal{G}}_0(P \otimes \mathbb{1}_X, P). \tag{1.11}$$

A change of dual as in (1.2) replaces $[m]$ by $\psi^{-1} \otimes_P [m] \otimes_P \psi$.

Lemma 1.1 *Let (P, D, Θ) be a Kasparov dual. Then $[m]$ is the class in $\mathrm{KK}^{\mathcal{G}}$ of the multiplication homomorphism $\mathcal{C}_0(X) \otimes_Z P \to P$ that describes the X-structure on P (up to commuting the tensor factors).*

We now have enough theoretical development to define the Lefschetz map and sketch the proof of its homotopy invariance.

Let X be a $\mathcal{G}$-space and (P, Θ) an n-dimensional abstract dual for X, PD and PD^{-1} the duality isomorphisms. As before, we write $\mathbb{1} := \mathcal{C}_0(Z)$, $\mathbb{1}_X := \mathcal{C}_0(X)$ and $\Delta_X \in \mathrm{RKK}^{\mathcal{G}}(X; \mathbb{1}_X, \mathbb{1}) = \mathrm{KK}^{\mathcal{G}\ltimes X}(\mathbb{1}_X \otimes \mathbb{1}_X, \mathbb{1}_X)$ the diagonal restriction class and

$$\bar{\Theta} := \mathrm{forget}_X(\Theta) \in \mathrm{KK}^{\mathcal{G}}_n(\mathbb{1}_X, P \otimes \mathbb{1}_X).$$

Definition 1.5 The equivariant *Lefschetz map*

$$\mathrm{Lef}\colon \mathrm{RKK}^{\mathcal{G}}_*(X; \mathbb{1}_X, \mathbb{1}) \to \mathrm{KK}^{\mathcal{G}}_*(\mathbb{1}_X, \mathbb{1})$$

for a $\mathcal{G}$-space X is defined as the composite map

$$\mathrm{RKK}^{\mathcal{G}}_i(X; \mathbb{1}_X, \mathbb{1}) \xrightarrow{\mathrm{PD}^{-1}} \mathrm{KK}^{\mathcal{G}}_{i-n}(P \otimes \mathbb{1}_X, \mathbb{1}) \xrightarrow{\bar{\Theta} \otimes_{P \otimes \mathbb{1}_X} \text{\textvisiblespace}} \mathrm{KK}^{\mathcal{G}}_i(\mathbb{1}_X, \mathbb{1}).$$

The equivariant *Euler characteristic* of X is

$$\mathrm{Eul}_X := \mathrm{Lef}(\Delta_X) \in \mathrm{KK}^{\mathcal{G}}_0(\mathbb{1}_X, \mathbb{1}) = \mathrm{KK}^{\mathcal{G}}_0\big(\mathcal{C}_0(X), \mathcal{C}_0(Z)\big).$$

Let $f \in \mathrm{RKK}^{\mathcal{G}}_i(X; \mathbb{1}_X, \mathbb{1})$. Equations (1.9) and (1.11) yield

$$\mathrm{Lef}(f) = (-1)^{in}\bar{\Theta} \otimes_{P\otimes \mathbb{1}_X} T'_P(f) \otimes_P D, \tag{1.12}$$

$$\mathrm{Eul}_X = (-1)^{in}\bar{\Theta} \otimes_{P\otimes \mathbb{1}_X} [m] \otimes_P D. \tag{1.13}$$

We have already established that if (P, Θ) is part of a Kasparov dual, then $T'_P = T_P$ and $[m]$ is the KK-class of the multiplication *-homomorphism $\mathcal{C}_0(X, P) \to P$, so that (1.12) yields explicit formulas for $\mathrm{Lef}(f)$ and Eul_X. This is extremely important because otherwise it would not be possible to compute these invariants.

Let X and X' be $\mathcal{G}$-spaces, and let $f\colon X \to X'$ be a $\mathcal{G}$-homotopy equivalence. Then f induces an equivalence of categories $\mathrm{RKK}^{\mathcal{G}}(X') \cong \mathrm{RKK}^{\mathcal{G}}(X)$, that is, we get invertible maps

$$f^*\colon \mathrm{RKK}^{\mathcal{G}}_*(X'; A, B) \to \mathrm{RKK}^{\mathcal{G}}_*(X; A, B)$$

for all $\mathcal{G}$-C*-algebras A and B. Now assume, in addition, that f is proper; we do not need the inverse map or the homotopies to be proper. Then f induces a *-homomorphism $f^!\colon \mathcal{C}_0(X') \to \mathcal{C}_0(X)$, which yields $[f^!] \in \mathrm{KK}^{\mathcal{G}}\bigl(\mathcal{C}_0(X'), \mathcal{C}_0(X)\bigr)$. We write $[f^!]$ instead of $[f^*]$ to better distinguish this from the map f^* above. Unless f is a *proper* $\mathcal{G}$-homotopy equivalence, $[f^!]$ need not be invertible.

Theorem 1.2 *Let X and X' be $\mathcal{G}$-spaces with abstract duals, and let $f\colon X \to X'$ be both a proper map and a $\mathcal{G}$-homotopy equivalence. Then*

$$[f^!] \otimes_{\mathcal{C}_0(X)} \mathrm{Eul}_X = \mathrm{Eul}_{X'} \qquad \text{in } \mathrm{KK}^{\mathcal{G}}_0(\mathcal{C}_0(X'), \mathbb{1})$$

and the Lefschetz maps for X and X' are related by a commuting diagram

$$\begin{array}{ccccc}
\mathrm{RKK}^{\mathcal{G}}_*(X;\mathcal{C}_0(X),\mathbb{1}) & \xleftarrow[\cong]{f^*} & \mathrm{RKK}^{\mathcal{G}}_*(X';\mathcal{C}_0(X),\mathbb{1}) & \xrightarrow{[f^!]^*} & \mathrm{RKK}^{\mathcal{G}}_*(X';\mathcal{C}_0(X'),\mathbb{1}) \\
\downarrow{\scriptstyle \mathrm{Lef}_X} & & & & \downarrow{\scriptstyle \mathrm{Lef}_{X'}} \\
\mathrm{KK}^{\mathcal{G}}_*(\mathcal{C}_0(X),\mathbb{1}) & & \xrightarrow{[f^!]^*} & & \mathrm{KK}^{\mathcal{G}}_*(\mathcal{C}_0(X'),\mathbb{1}),
\end{array}$$

where $[f^!]^*$ denotes composition with $[f^!]$.

In particular, Eul_X *and the map* Lef_X *do not depend on the chosen dual.*

The proof relies on the discussion preceding the theorem.

Theorem 1.2 implies that the Lefschetz maps for properly $\mathcal{G}$-homotopy equivalent spaces are equivalent because then $[f^!]$ is invertible, so that all horizontal maps in the diagram in Theorem 1.2 are invertible. In this sense, the Lefschetz map and the Euler class are invariants of the proper $\mathcal{G}$-homotopy type of X.

The construction in Example 1.3 associates a class $[\Delta_f] \in \mathrm{RKK}^{\mathcal{G}}_0(X; \mathcal{C}_0(X), \mathbb{1})$ to any continuous, $\mathcal{G}$-equivariant map $f\colon X \to X$; it does not matter whether f is proper. We abbreviate

$$\mathrm{Lef}(f) := \mathrm{Lef}([\Delta_f])$$

and call this the Lefschetz invariant of f. Of course, equivariantly homotopic self-maps induce the same class in $\mathrm{RKK}^{\mathcal{G}}(X; \mathcal{C}_0(X), \mathbb{1})$ and therefore have the same Lefschetz invariant. We have $\mathrm{Lef}(\mathrm{id}_X) = \mathrm{Eul}_X$.

More generally, specializing (1.10) gives a map

$$\text{\textvisiblespace} \otimes_{\mathbb{1}_X} \Delta_X \colon \mathrm{KK}^{\mathcal{G}}_*\bigl(\mathcal{C}_0(X), \mathcal{C}_0(X)\bigr) \to \mathrm{RKK}^{\mathcal{G}}_*(X; \mathcal{C}_0(X), \mathbb{1}),$$

which we compose with the Lefschetz map; abusing notation, we still denote this composition by

$$\mathrm{Lef}\colon\ \mathrm{KK}^{\mathcal{G}}_*(\mathcal{C}_0(X), \mathcal{C}_0(X)) \to \mathrm{KK}^{\mathcal{G}}_*(\mathcal{C}_0(X), \mathbb{1})$$

Finally, we record that Lefschetz invariants for elements of $\mathrm{RKK}^{\mathcal{G}}_*(X; \mathcal{C}_0(X), \mathbb{1})$ can be arbitrarily complicated: the Lefschetz map is rather easily seen to be split surjective. The splitting is given by specializing the *inflation map*

$$p_X^*\colon\ \mathrm{KK}^{\mathcal{G}}_*(A, B) \to \mathrm{KK}^{\mathcal{G}\ltimes X}(\mathbb{1}_X \otimes A, \mathbb{1}_X \otimes B) \tag{1.14}$$

to $A := \mathbb{1}_X$ and $B := \mathbb{1}$.

The fundamental example of a Kasparov dual is provided by the vertical tangent space to a bundle of smooth manifolds over the base Z of a groupoid, in which morphisms act smoothly. We come back to this in §4.

2 Examples of computations of the Lefschetz map

In this section we will give some examples of computations of the Lefschetz map for various instances of spaces with duals and for equivariant self-morphisms coming from actual *maps*; throughout, $\mathcal{G}$ will be a *group*. The problem of computing the Lefschetz invariants of more general Kasparov self-morphisms for the next section is a central problem for us, and will be treated in §4 once we have available the theory of equivariant correspondences. In this setting, we will also be able to treat the groupoid cases.

Most of the examples of $\mathcal{G}$-spaces X we consider here are quite close to proper actions, but they do not quite have to be proper. The point is that an abstract (or Kasparov) dual for a $\mathcal{G}$-space X also yields one for X regarded as a $\mathcal{G}'$-space where $\mathcal{G}'$ is a (not necessarily closed) subgroup (or subgroupoid) of $\mathcal{G}$. Even if X is proper as a $\mathcal{G}$-space, it need not be proper as a $\mathcal{G}'$-space since $\mathcal{G}'$ may not be closed.

Known examples of duals are of two types; they are both Kasparov duals. If $\mathcal{G}$ is a locally compact group acting as simplicial automorphisms of a finite-dimensional simplicial complex then X has a Kasparov dual by [12] (based on ideas of Kasparov and Skandalis in [22]). This does not quite imply that the action of $\mathcal{G}$ is proper, since the action of $\mathcal{G}$ could be trivial. Neither does it quite imply that $\mathcal{G}$ must be discrete, though since the connected component of the identity of $\mathcal{G}$ must act trivially, the only interesting examples must involve disconnected groups. For instance, the non-discrete group $\mathrm{SL}(2, \mathbb{Q}_p)$ acts simplicially on a tree.

If X is a complete Riemannian manifold then X is a proper $\mathcal{G}$-space, where $\mathcal{G}$ is the Lie group of isometries of X, and either the Clifford algebra of X or the C*-algebra of $\mathcal{C}_0$-functions on the tangent bundle TX of X is part of a Kasparov dual for X (see, e.g., [12] or [16]), and also §4. Hence if $\mathcal{G}$ is *any* group of isometries of a Riemannian manifold X with finitely many components, then the $\mathcal{G}$-space X also has a Kasparov dual. If the tangent bundle TX admits an equivariant K-orientation, then $\mathcal{C}_0(TX)$ can be replaced by $\mathcal{C}_0(X)$. For instance, the circle with the group $\mathbb{Z}$ acting by an irrational rotation has a Kasparov dual of dimension 1, given by $(\mathcal{C}(\mathbb{T}), D, \Theta)$ where D is the class of the Dirac operator on the circle.

The Lefschetz invariants of equivariant self-maps of X will in both these situations turn out to be in some sense *zero-dimensional* in the sense that they are built out of point-evaluation classes. As we will see later, the Lefschetz invariants of more general Kasparov self-morphisms are more complicated, higher-dimensional objects.

2.1 The combinatorial Lefschetz map. Let X be a finite-dimensional simplicial complex and let $\mathcal{G}$ be a locally compact group acting smoothly and simplicially on X (that is, stabilisers of points are open). To emphasise that $\mathcal{G}$ is a group, we will denote it by G throughout.

We follow [12] and [13]. Assume that X admits a colouring (that is, X is typed) and that G preserves the colouring. This ensures that if $g \in G$ maps a simplex to itself, then it fixes that simplex pointwise.

Let SX be the set of (non-degenerate) simplices of X and let $S_dX \subseteq SX$ be the subset of d-dimensional simplices. The group G acts on the discrete set SX preserving the decomposition $SX = \bigsqcup S_dX$. Decompose SX into G-orbits. For each orbit $\dot{\sigma} \subseteq SX$, choose a representative $\sigma \in SX$ and let $\xi_\sigma \in X$ be its barycentre and $\mathrm{Stab}(\sigma) \subseteq G$ its stabiliser. Restriction to the orbit $\dot{\sigma}$ defines a G-equivariant *-homomorphism

$$\xi_{\dot{\sigma}}\colon \mathcal{C}_0(X) \to \mathcal{C}_0\bigl(G/\mathrm{Stab}(\sigma)\bigr) \to \mathbb{K}\bigl(\ell^2(G/\mathrm{Stab}(\sigma))\bigr), \tag{2.1}$$

where the second map is the representation by pointwise multiplication operators. We let $[\xi_{\dot{\sigma}}]$ be its class in $\mathrm{KK}_0^G(\mathcal{C}_0(X), \mathbb{1})$.

Let $\varphi\colon X \to X$ be a G-equivariant self-map of X. Since φ is G-equivariantly homotopic to a G-equivariant cellular map, we may assume without loss of generality that φ is itself cellular. Hence it induces a G-equivariant chain map

$$\varphi\colon C_\bullet(X) \to C_\bullet(X),$$

where $C_\bullet(X)$ is the chain complex of *oriented* simplices of X. A basis for $C_\bullet(X)$ is given by the set of (un-)oriented simplices, by arbitrarily choosing an orientation on each simplex. We may describe the chain map φ by its matrix coefficients $\varphi_{\sigma\tau} \in \mathbb{Z}$ with respect to this basis; thus the subscripts are unoriented simplices. For example, if φ maps a simplex to itself, and reverses orientation, then $\varphi_{\sigma,\sigma} = -1$. Since φ is G-equivariant, $\varphi_{g(\sigma),g(\sigma)} = \varphi_{\sigma\sigma}$. So the following makes sense.

Notation 2.1 For $\dot{\sigma} \in G\backslash S_dX$, let $\mathbf{n}(\varphi, \dot{\sigma}) := (-1)^d \varphi_{\sigma\sigma} \in \mathbb{Z}$ for any choice of representative $\sigma \in \dot{\sigma}$.

The following theorem is proved in [13] using the simplicial dual developed in [12], and inspired by ideas of Kasparov and Skandalis in [22].

Theorem 2.1 *Let X be a finite-dimensional coloured simplicial complex and let G be a locally compact group that acts smoothly and simplicially on X, preserving the colouring. Let $\varphi\colon X \to X$ be a G-equivariant self-map. Define $\mathbf{n}(\varphi, \dot{\sigma}) \in \mathbb{Z}$ and $[\xi_{\dot{\sigma}}] \in \mathrm{KK}^G(\mathcal{C}_0(X), \mathbb{1})$ for $\dot{\sigma} \in G\backslash SX$ as above. Then*

$$\mathrm{Lef}(\varphi) = \sum_{\dot{\sigma} \in G\backslash SX} \mathbf{n}(\varphi, \dot{\sigma})[\xi_{\dot{\sigma}}] \in \mathrm{KK}^G(\mathcal{C}_0(X), \mathbb{1}). \tag{2.2}$$

If $G\backslash X$ is compact, the sum in (2.2) is finite. But even in the general case it 'converges' in a suitable sense: namely that the corresponding sum of cycles converges to a cycle for the K-homology class for X. (The sum is *locally* finite, in an obvious sense.)

If G is trivial, X is connected and has only a finite number of simplices, then the formula just given reduces to the ordinary Lefschetz number of φ by a standard argument.

2.2 The smooth Lefschetz map. Let X be a smooth manifold and G a group acting by isometric diffeomorphisms of X. Then we can build a Kasparov dual for X using the Clifford algebra. The real (resp. complex) Clifford algebra of a Euclidean vector space V is generated as a real (resp. complex) unital *-algebra by an orthonormal basis $\{e_i\}$ for V with the relations that e_i is self-adjoint and $e_ie_j + e_je_i = 2\delta_{ij}$. In the complex case this produces a finite-dimensional $\mathbb{Z}/2$-graded C*-algebra. More generally, if V is a Euclidean vector bundle, this construction applies and produces a locally trivial bundle of finite-dimensional $\mathbb{Z}/2$-graded C*-algebras over X. If $V := T^*X$ for a Riemannian manifold X then the *Clifford algebra* of X is the corresponding C*-algebra of sections vanishing at infinity. It is denoted $\mathscr{C}_\tau(X)$. This C*-algebra carries a canonical action of the group of isometries of X and hence likewise for any subgroup.

We discuss the specific mechanics of the Clifford Kasparov dual to the following extent. Let d be the de Rham differential on X, acting on L^2-forms on X. This Hilbert space carries a unitary action of G and both d and its adjoint d^* are G-equivariant, and $d + d^*$ is an elliptic operator called the *Euler* (or *de Rham*) operator on X. If ω is a section of T^*X, exterior product with ω defines an operator $\lambda_\omega\colon L^2(\wedge T^*X) \to L^2(\wedge T^*X)$ and if λ_ω^* is its adjoint then the assignment $\omega \mapsto \lambda_\omega + \lambda_\omega^*$ determines a representation of $\mathscr{C}_\tau(X)$ which graded commutes modulo bounded operators with $d + d^*$. Hence we get a cycle for $\mathrm{KK}^G(\mathscr{C}_\tau(X), \mathbb{C})$. It represents the class D appearing in the Kasparov dual.

We start by recording one of the easiest computations of the Lefschetz map. The interested reader can easily prove it for himself or herself after looking briefly at the definition of the class Θ (see [12]) and reviewing the definition of the Lefschetz map. Recall that the Euler class of X is the Lefschetz invariant of the identity map of X.

Proposition 2.1 *The Euler class of X is the class of the Euler operator on X.*

Notice that the functoriality result 1.2 combines with Proposition 2.1 to imply the homotopy invariance of the class of the Euler operator:

$$f_*(\mathrm{Eul}_X) = \mathrm{Eul}_{X'}$$

for a proper $\mathcal{G}$-equivariant homotopy equivalence $f\colon X \to X'$ between smooth and proper $\mathcal{G}$-manifolds. This fact is not trivial.

We now describe the Lefschetz invariant of a more general smooth G-equivariant self-map of X. This requires a preliminary discussion.

Let Y be a locally compact space and G be a locally compact group acting continuously on Y, and let $\pi\colon E \to Y$ be a G-equivariant Euclidean $\mathbb{R}$-vector bundle over E. Let $A\colon E \to E$ be a G-equivariant vector bundle automorphism, that is, a continuous map $E \to E$ over Y that restricts to $\mathbb{R}$-vector space isomorphisms on the fibres of E. We are going to define a G-equivariant $\mathbb{Z}/2$-graded real line bundle $\mathrm{sign}(A)$ over Y. (Since we work with complex K-theory we will only use its complexification.)

If Y is a point, then G-equivariant real vector bundles over Y correspond to real orthogonal representations of G. The endomorphism A becomes in this case an invertible linear map $A\colon \mathbb{R}^n \to \mathbb{R}^n$ commuting with G. The sign is a virtual 1-dimensional representation of G and hence corresponds to a pair (χ, n), where $n \in \{0, 1\}$ is the parity (we are referring to the grading, either even or odd) of the line bundle and $\chi\colon G \to \{-1, +1\}$ is a real-valued character. The overall parity will turn out to be 0 if A preserves orientation and 1 if A reverses orientation (see Example 2.1). In this sense, our invariant will refine the orientation of A.

As above, let $\mathrm{Cliff}(E)$ be the bundle of *real* Clifford algebras associated to E. We can also define in an analogous way $\mathrm{Cliff}(E)$ if E carries an indefinite bilinear form and it is a well-known fact from algebra that if the index of the bilinear form on E is divisible by 8, then the fibres of $\mathrm{Cliff}(E)$ are isomorphic to matrix algebras. In this case, a *G-equivariant spinor bundle* for E is a $\mathbb{Z}/2$-graded real vector bundle S_E together with a grading preserving, G-equivariant *-algebra isomorphism $c\colon \mathrm{Cliff}(E) \to \mathrm{End}(S_E)$. This representation is determined uniquely by its restriction to $E \subseteq \mathrm{Cliff}(E)$, which is a G-equivariant map $c\colon E \to \mathrm{End}(S_E)$ such that $c(x)$ is odd and symmetric and satisfies $c(x)^2 = \|x\|^2$ for all $x \in E$.

The spinor bundle is unique up to tensoring with a G-equivariant real line bundle L: if $c_t\colon E \to S_t$ for $t = 1, 2$ are two G-equivariant spinor bundles for E, then we define a G-equivariant real line bundle L over Y by

$$L := \mathrm{Hom}_{\mathrm{Cliff}(E)}(S_1, S_2),$$

and the evaluation isomorphism $S_1 \otimes L \xrightarrow{\cong} S_2$ intertwines the representations c_1 and c_2 of $\mathrm{Cliff}(E)$.

Definition 2.1 Let $A\colon E \to E$ be a real G-equivariant vector bundle automorphism and let $A = T \circ (A^*A)^{1/2}$ be its polar decomposition with an orthogonal vector bundle automorphism $T\colon E \to E$.

Let F be another G-equivariant vector bundle over Y with a non-degenerate bilinear form, such that the signature of $E \oplus F$ is divisible by 8, so that $\mathrm{Cliff}(E \oplus F)$ is a bundle of matrix algebras over $\mathbb{R}$. We assume that $E \oplus F$ has a G-equivariant spinor bundle, that is, there exists a G-equivariant linear map $c\colon E \oplus F \to \mathrm{End}(S)$ that induces an isomorphism of graded *-algebras

$$\mathrm{Cliff}(E \oplus F) \cong \mathrm{End}(S).$$

Then

$$c'\colon E \oplus F \to \mathrm{End}(S), \qquad (\xi, \eta) \mapsto c(T(\xi), \eta)$$

yields another G-equivariant spinor bundle for $E \oplus F$. We let

$$\mathrm{sign}(A) := \mathrm{Hom}_{\mathrm{Cliff}(E\oplus F)}\big((S, c'), (S, c)\big).$$

This is a G-equivariant $\mathbb{Z}/2$-graded real line bundle over Y.

It is not hard to check that $\mathrm{sign}(A)$ is well-defined and a homotopy invariant. Furthermore, $\mathrm{sign}(A_1 \circ A_2) \cong \mathrm{sign}(A_1) \otimes \mathrm{sign}(A_2)$ for two equivariant automorphisms $A_1, A_2\colon E \rightrightarrows E$ of the same bundle, and $\mathrm{sign}(A_1 \oplus A_2) \cong \mathrm{sign}(A_1) \otimes \mathrm{sign}(A_2)$ for two equivariant vector bundle automorphisms $A_1\colon E_1 \to E_1$ and $A_2\colon E_2 \to E_2$.

If Y is a point and $\mathcal{G}$ is trivial, then $\mathrm{sign}(A) = \mathbb{R}$ for orientation-preserving A and $\mathrm{sign}(A) = \mathbb{R}^{\mathrm{op}}$ for orientation-reversing A, as claimed above.

Example 2.1 Consider $G = \mathbb{Z}/2$. Let $\tau\colon G \to \{1\}$ be the trivial character and let $\chi\colon G \to \{+1, -1\}$ be the non-trivial character. Let $\mathbb{R}_\chi$ denote the real representation of G on $\mathbb{R}$ with character χ. This can be considered trivially graded; let $\mathbb{R}_\chi^{\mathrm{op}}$ denote the same representation but with the opposite grading (the whole vector space is considered odd.)

Consider $A\colon \mathbb{R}_\chi \to \mathbb{R}_\chi$, $t \mapsto -t$, so A commutes with G. Then $\mathrm{sign}(A) \cong \mathbb{R}_\chi^{\mathrm{op}}$ carries a non-trivial representation.

To see this, let F be $\mathbb{R}_\chi$ with negative definite metric. Thus the Clifford algebra of $\mathbb{R}_\chi \oplus \mathbb{R}_\chi$ is $\mathrm{Cliff}_{1,1} \cong \mathbb{M}_{2\times 2}(\mathbb{R})$. Explicitly, the map

$$c(x,y) = \begin{pmatrix} 0 & x-y \\ x+y & 0 \end{pmatrix}$$

induces the isomorphism. We equip $\mathbb{R}^2$ with the representation $\tau \oplus \chi$, so that c is equivariant.

Twisting by A yields another representation

$$c'(x,y) := c(-x,y) = Sc(x,y)S^{-1} \qquad \text{with} \quad S = S^{-1} = \begin{pmatrix} 0 & 1 \\ -1 & 0 \end{pmatrix}.$$

Since S reverses the grading and exchanges the representations τ and χ, it induces an isomorphism $(\mathbb{R}_\tau \oplus \mathbb{R}_\chi^{\mathrm{op}}) \otimes \mathbb{R}_\chi^{\mathrm{op}} \xrightarrow{\cong} \mathbb{R}_\tau \oplus \mathbb{R}_\chi^{\mathrm{op}}$. Hence $\mathrm{sign}(A) = \mathbb{R}_\chi^{\mathrm{op}}$.

Now let X be a smooth Riemannian manifold and assume that G acts on X isometrically and continuously.

Let $\varphi\colon X \to X$ be a G-equivariant self-map of X. In order to write down an explicit local formula for $\mathrm{Lef}(\varphi)$, we impose the following restrictions on φ:

- φ is smooth;
- the fixed point subset $\mathrm{Fix}(\varphi)$ of φ is a submanifold of X;
- if $(p,\xi) \in TX$ is fixed by the derivative $D\varphi$, then ξ is tangent to $\mathrm{Fix}(\varphi)$.

The last two conditions are automatic if φ is isometric with respect to *some* Riemannian metric (not necessarily the given one) and this of course applies by averaging the given metric if φ has finite order.

In the simplest case, φ and id_X are transverse, that is, $\mathrm{id} - D\varphi$ is invertible at each fixed point of φ; this implies that φ has isolated fixed points.

To describe the Lefschetz invariant, we abbreviate $Y := \mathrm{Fix}(\varphi)$. This is a closed submanifold of X by assumption. Let ν be the normal bundle of Y in X. Since the tangent space of Y is left fixed by the derivative $D\varphi$, it induces a linear map $D_\nu\varphi\colon \nu \to \nu$. By assumption, the map $\mathrm{id}_\nu - D_\nu\varphi\colon \nu \to \nu$ is invertible.

Theorem 2.2 *Let X be a complete smooth Riemannian manifold, let G be a locally compact group that acts on X smoothly and by isometries, and let $\varphi\colon X \to X$ be a self-map satisfying the three conditions enumerated above. Let ν be the normal bundle of Y in X and let $D_\nu\varphi\colon \nu \to \nu$ be induced by the derivative of φ as above. Let $r_Y\colon \mathcal{C}_0(X) \to \mathcal{C}_0(Y)$ be the restriction map and let $\mathrm{Eul}_Y \in \mathrm{KK}_0^G(\mathcal{C}_0(Y), \mathbb{1})$ be the equivariant Euler characteristic of Y. Then*

$$\mathrm{Lef}(\varphi) = r_Y \otimes_{\mathcal{C}_0(Y)} \mathrm{sign}(\mathrm{id}_\nu - D_\nu\varphi) \otimes_{\mathcal{C}_0(Y)} \mathrm{Eul}_Y\,.$$

Furthermore, Eul_Y *is the equivariant* K*-homology class of the* de Rham *operator on Y.*

In brief, the Lefschetz invariant of G is the Euler characteristic of the fixed-point set, twisted by an appropriate equivariant line bundle depending on orientation data.

If φ and id_X are transverse then the fixed point subset Y is discrete. A discrete set is a manifold and its Euler characteristic – a degenerate case of Proposition 2.1 – is represented by the Kasparov cycle in which the Hilbert space is $L^2(\Lambda_{\mathbb{C}}^*(T^*Y)) := \ell^2(Y)$ equipped with the representation $\mathcal{C}_0(Y) \to \mathbb{K}(\ell^2 Y)$ by pointwise multiplication operators, and the zero operator. The group permutes the points of Y and so acts by unitaries on $\ell^2(Y)$.

The normal bundle ν to Y in X in this case is the restriction of the vector bundle TX to the subset Y. For $p \in Y$, let n_p be $+1$ if $\mathrm{id}_{T_pX} - D_p\varphi$ preserves orientation, and -1 otherwise. The graded equivariant line bundle $\mathrm{sign}(\mathrm{id}_\nu - D_\nu\varphi)$ in Theorem 2.2 is determined by pairs (n_p, χ_p) for $p \in Y$, where n_p is the parity of the representation at p and χ_p is a certain real-valued character $\chi_p\colon \mathrm{Stab}(p) \to \{-1, +1\}$ that depends on $\mathrm{id}_{T_pX} - D_p\varphi$ and the representation of the stabiliser $\mathrm{Stab}(p) \subseteq G$ on T_pX. Equivariance implies that n_p is constant along G-orbits, whereas χ_p behaves like $\chi_{g\cdot p} = \chi_p \circ \mathrm{Ad}(g^{-1})$. Let $\ell^2_\chi(Gp)$ be the representation of the crossed product $G \ltimes \mathcal{C}_0\big(G/\mathrm{Stab}(p)\big)$ obtained by inducing the representation χ_p from $\mathrm{Stab}(p)$, and let $\mathcal{C}_0(X)$ act on $\ell^2_\chi(Gp)$ by restriction to $G/\mathrm{Stab}(p) \cong Gp$. This defines a G-equivariant *-homomorphism

$$\xi_{Gp,\chi}\colon \mathcal{C}_0(X) \to \mathbb{K}(\ell^2_\chi G).$$

Theorem 2.2 asserts the following:

Corollary 2.1 *If the graph of φ is transverse to the diagonal in $X \times X$, then*

$$\mathrm{Lef}(\varphi) = \sum_{Gp \in G\backslash \mathrm{Fix}(\varphi)} n_p[\xi_{Gp,\chi}] \in \mathrm{KK}^G_0(\mathcal{C}_0(X), \mathbb{1})]$$

where $[\xi_{Gp,\chi}] \in \mathrm{KK}^G_0(\mathcal{C}_0(X), \mathbb{1})$ and the multiplicities n_p are explained above.

Furthermore, the character $\chi\colon \mathrm{Stab}_G(p) \to \{-1, +1\}$ at a fixed point p has the explicit formula

$$\chi(g) = \mathrm{sign}\det\big(\mathrm{id} - D_p\varphi\big) \cdot \mathrm{sign}\det\big(\mathrm{id} - D_p\varphi_{\mathrm{Fix}(g)}\big).$$

The sum may be infinite if $G\backslash X$ is not compact, but it still converges in a suitable sense (in the sense, for instance, of defining a K-homology class.)

If, in addition, G is trivial and X is connected, then $\xi_{Gp,\chi} = \mathrm{ev}_p$ for all $p \in Y$; moreover, all point evaluations have the same K-homology class – denote it just [ev] – because they are homotopic. Hence we get the classical Lefschetz data multiplied by the K-homology class [ev] of a point

$$\mathrm{Lef}(\varphi) = \Big(\sum_{p \in \mathrm{Fix}(\varphi)} \mathrm{sign}(\mathrm{id}_{T_pX} - D\varphi_p)\Big) \cdot [\mathrm{ev}]$$

as asserted above. This sum is finite if X is compact.

We include the following for the benefit of the reader; it enables he or she to verify our computations by direct inspection.

Lemma 2.1 *Let $H \subseteq G$ be compact and open, let $p, q \in X^H$ belong to the same path component of the fixed point subspace X^H, and let $\chi \in \mathrm{Rep}(H)$. Then*

$$\big[\xi_{Gp,\mathrm{Ind}_H^{\mathrm{Stab}(Gp)}(\chi)}\big] = \big[\xi_{Gq,\mathrm{Ind}_H^{\mathrm{Stab}(Gq)}(\chi)}\big] \qquad \textit{in } \mathrm{KK}_0(\mathcal{C}_0(X), \mathbb{1}).$$

Remark 2.1 If the identity map $\mathrm{id}\colon X \to X$ can be equivariantly perturbed to be in general position in the sense explained above, then combining Proposition 2.1 and Corollary 2.1 proves that the class of the de Rham operator in $\mathrm{KK}^G(\mathcal{C}_0(X), \mathbb{C})$ is a sum of point-evaluation classes. Lück and Rosenberg show in [25] that this can always be achieved when G is discrete and acts properly on X.

The next two explicit examples involve isolated fixed points.

Example 2.2 Let $G \cong \mathbb{Z} \ltimes \mathbb{Z}/2\mathbb{Z}$ be the infinite dihedral group, identified with the group of affine transformations of $\mathbb{R}$ generated by $u(x) = -x$ and $w(x) = x + 1$. Then G has exactly two conjugacy classes of finite subgroups, each isomorphic to $\mathbb{Z}/2$. Its action on $\mathbb{R}$ is proper, and the closed interval $[0, \frac{1}{2}]$ is a fundamental domain. There are two orbits of fixed point in $\mathbb{R}$, those of 0 and $\frac{1}{2}$, and their stabilisers represent the two conjugacy classes of finite subgroups.

Now we use some notation from Example 2.1. Each copy of $\mathbb{Z}/2$ acting on the tangent space at the fixed point acts by multiplication by -1 on tangent vectors. Therefore, the computations in Example 2.1 show that for any nonzero real number A, viewed as a linear transformation of the tangent space that commutes with $\mathbb{Z}/2$, we have

$$\operatorname{sign}(A) = \begin{cases} \mathbb{R}_\chi^{\mathrm{op}} & \text{if } A < 0, \text{ and} \\ \mathbb{R}_\tau & \text{if } A > 0. \end{cases}$$

Let φ be a small G-equivariant perturbation of the identity map $\mathbb{R} \to \mathbb{R}$ with the following properties. First, φ maps the interval $[0, \frac{1}{2}]$ to itself. Secondly, its fixed points in $[0, \frac{1}{2}]$ are the end points 0, $\frac{1}{2}$, and $\frac{1}{4}$; thirdly, its derivative is bigger than 1 at both endpoints and between 0 and 1 at $\frac{1}{4}$. Such a map clearly exists. Furthermore, it is homotopic to the identity map, so that $\mathrm{Lef}(\varphi) = \mathrm{Eul}_{\mathbb{R}}$.

By construction, there are three fixed points modulo G, namely, the orbits of 0, $\frac{1}{4}$ and $\frac{1}{2}$. The isotropy groups of the first and third orbit are non-conjugate subgroups isomorphic to $\mathbb{Z}/2$; from Example 2.1, each of them contributes $\mathbb{R}_\chi^{\mathrm{op}}$. The point $\frac{1}{4}$ contributes the trivial character of the trivial subgroup. Hence

$$\mathrm{Lef}(\varphi) = -[\xi_{\mathbb{Z},\chi}] - [\xi_{\mathbb{Z}+\frac{1}{2},\chi}] + [\xi_{\mathbb{Z}+\frac{1}{4}}].$$

On the other hand, suppose we change φ to fix the same points but to have zero derivative at 0 and $\frac{1}{2}$ and large derivative at $\frac{1}{4}$. This is obviously possible. Then we get contributions of $\mathbb{R}_\tau$ at 0 and $\frac{1}{2}$ and a contribution of $-[\xi_{\frac{1}{4}}]$ at $\frac{1}{4}$. Hence

$$\mathrm{Lef}(\varphi) = [\xi_{\mathbb{Z},\tau}] + [\xi_{\mathbb{Z}+\frac{1}{2},\tau}] - [\xi_{\mathbb{Z}+\frac{1}{4}}].$$

Combining the two formulas yields the identity

$$[\xi_{\mathbb{Z},\tau}] + [\xi_{\mathbb{Z}+\frac{1}{2},\tau}] - [\xi_{\mathbb{Z}+\frac{1}{4}}] = -[\xi_{\mathbb{Z},\chi}] - [\xi_{\mathbb{Z}+\frac{1}{2},\chi}] + [\xi_{\mathbb{Z}+\frac{1}{4}}]. \tag{2.3}$$

By the way, the left-hand side is the description of $\mathrm{Eul}_{\mathbb{R}}$ we get from the combinatorial dual with the obvious G-invariant triangulation of $\mathbb{R}$ with vertex set $\mathbb{Z} + \frac{1}{2}\mathbb{Z} \subset \mathbb{R}$.

Using Lemma 2.1 one can check (2.3) by direct computation.

3 Geometric KK-theory

In the previous chapter we explained our computation of the Lefschetz map for self-*maps* of a $\mathcal{G}$-space X in several relatively simple situations. In these cases, $\mathcal{G}$ was in each case a group (a groupoid with trivial base). Obviously, not all equivariant Kasparov self-morphisms $\mathrm{KK}_*^{\mathcal{G}}(\mathcal{C}_0(X), \mathcal{C}_0(X))$ are represented by maps. We have organized this survey around the problem of computing Lefschetz invariants of more general equivariant Kasparov self-morphisms. This requires describing the morphisms themselves in some geometric way. The theory of *correspondences* of Baum, Connes and Skandalis (see [3] [8]) would seem ideal for this purpose. Since we are working in the equivariant setting, to use it would necessitate checking that the pseudodifferential calculus which plays such a prominent role in [8] works equivariantly with respect to an action of a group (or groupoid), as well as proving the main functoriality result for K-oriented maps and the fact that every class in $\mathrm{KK}_*^{\mathcal{G}}(C_0(X), C_0(Y))$ is represented by a unique correspondence up to a suitable notion of geometric equivalence.

Although it seems plausible that such an extension could be carried out in a fairly direct fashion, in reality there is a problem with composing correspondences using transversality in the equivariant situation (we explain this below). A trick of Baum and Block (see §3.8 and [3]) is useful in this connection, but in order for it to work, some hypotheses on vector bundles are necessary. These extra hypotheses cannot be avoided.

Our approach is rather different. To address the difficulty with $\mathcal{G}$-spaces with an insufficient supply of equivariant vector bundles, we build some vector bundles into the definition of a correspondence. To address difficulties with composing correspondences in the equivariant setting, we work only with wrong-way maps with a certain built-in factorization. In the end, we obtain a bivariant theory having the same basic formal properties of $\mathrm{KK}^{\mathcal{G}}$, but which is defined in a purely topological manner, *i.e.* involves essentially no analysis at all. Taking this approach necessitates restricting attention to *proper* groupoids (like compact groups); we first show how to reduce to this case using the Baum-Connes conjecture (see, for example, [16] for a discussion of the mechanics of Baum-Connes, duality and the Dirac dual-Dirac method.)

3.1 Using Baum-Connes to reduce to proper groupoids. Let $\mathcal{G}$ be a locally compact group (or groupoid). The *classifying space $\mathcal{E}\mathcal{G}$ for proper actions of $\mathcal{G}$* is the proper $\mathcal{G}$-space with the universal property that if X is any proper $\mathcal{G}$-space, then there is a $\mathcal{G}$-equivariant classifying map $\chi\colon X \to \mathcal{E}\mathcal{G}$ which is unique up to $\mathcal{G}$-homotopy. If $\mathcal{G}$ is a proper groupoid to begin with, then $\mathcal{E}\mathcal{G} = Z$ gives a simple model for $\mathcal{E}\mathcal{G}$, for it is proper as a $\mathcal{G}$-space and has the required universal property. In particular, if $\mathcal{G}$ is a compact group, then $\mathcal{E}\mathcal{G}$ is a point.

For $\mathcal{G}$ and $\mathcal{G}$-spaces X and Y, the *inflation map* (1.1)

$$\begin{aligned} p_{\mathcal{E}\mathcal{G}}^*\colon \mathrm{KK}_*^{\mathcal{G}}(C_0(X), C_0(Y)) &\to \mathrm{RKK}_*^{\mathcal{G}}(\mathcal{E}\mathcal{G}; C_0(X), C_0(Y)) \\ &:= \mathrm{KK}_*^{\mathcal{G}\ltimes\mathcal{E}\mathcal{G}}(C_0(X\times\mathcal{E}\mathcal{G}), C_0(Y\times\mathcal{E}\mathcal{G})) \quad (3.1) \end{aligned}$$

is an isomorphism as soon as the $\mathcal{G}$ action on X is topologically amenable (by Tu's generalization [32] of the Higson-Kasparov theorem [19]) and in particular as soon as it is proper. An abstract (respectively Kasparov) dual for the $\mathcal{G}$-space X pulls back to one for $X \times \mathcal{E}\mathcal{G}$ as a $\mathcal{G} \ltimes \mathcal{E}\mathcal{G}$-space, and the diagram

$$\begin{array}{ccc} \mathrm{KK}_*^{\mathcal{G}}(C_0(X), C_0(X)) & \xrightarrow{\mathrm{Lef}} & \mathrm{KK}_*^{\mathcal{G}}(C_0(X), \mathbb{C}) \\ \cong\downarrow p_{\mathcal{E}\mathcal{G}}^* & & \cong\downarrow p_{\mathcal{E}\mathcal{G}}^* \\ \mathrm{KK}_*^{\mathcal{G}\ltimes\mathcal{E}\mathcal{G}}(C_0(X\times\mathcal{E}\mathcal{G}), C_0(X\times\mathcal{E}\mathcal{G})) & \xrightarrow{\mathrm{Lef}} & \mathrm{KK}_*^{\mathcal{G}\ltimes\mathcal{E}\mathcal{G}}(C_0(X\times\mathcal{E}\mathcal{G}), C_0(\mathcal{E}\mathcal{G})) \end{array} \quad (3.2)$$

commutes. Hence the Lefschetz map for $\mathcal{G}$ acting on X is isomorphic to the Lefschetz map of $\mathcal{G} \ltimes \mathcal{E}\mathcal{G}$ acting on $X \times \mathcal{E}\mathcal{G}$.

This replaces the non-proper groupoid $\mathcal{G}$ by the proper groupoid $\mathcal{G} \times \mathcal{E}\mathcal{G}$ at no loss of information.

In order to compose correspondences we need a sufficient supply of 'trivial' vector bundles, but for this too, the fact that we have a proper groupoid makes a big difference. If $\mathcal{G}$ is a groupoid acting on a space, then a $\mathcal{G}$-equivariant vector bundle (see §3.2) over that space is *trivial* if it is pulled back from the unit space of $\mathcal{G}$ using the anchor map for the space. In the case where $\mathcal{G}$ is a group then its base is a point, so that a trivial $\mathcal{G}$-vector bundle over X is equivalent to a finite-dimensional representation of $\mathcal{G}$; if $\mathcal{G}$ is not compact, there may be very few such representations.

3.2 Equivariant vector bundles. As per above, $\mathcal{G}$ shall be a *proper* groupoid until further notice. If $\mathcal{G}$ is a group, this means that it must be compact. We will frequently consider this case in examples.

If X is a $\mathcal{G}$-space, we remind the reader that the anchor map for the action is denoted $\varrho_X\colon X \to Z$. A *$\mathcal{G}$-equivariant vector bundle on X* is a vector bundle on X which is also a $\mathcal{G}$-space such that elements of $\mathcal{G}$ map fibres to fibres linearly. There is an obvious notion of isomorphic $\mathcal{G}$-equivariant vector bundles.

If $\mathcal{G}$ is a compact group, then a $\mathcal{G}$-equivariant vector bundle on a point is a finite-dimensional linear representation of $\mathcal{G}$.

Notation 3.1 If V is an equivariant vector bundle over X then we denote by $\pi_V\colon V \to X$ the vector bundle projection and $\zeta_V\colon X \to V$ the zero section. We frequently denote by $|V|$ the total space of V, and denote by $\mathrm{VK}_{\mathcal{G}}(X)$ the Grothendieck group of the monoid of isomorphism classes of $\mathcal{G}$-equivariant vector bundles over X.

Given that $\mathcal{G}$ is assumed proper, equivariant vector bundles behave in some ways just like ordinary vector bundles. For example, if $Y \subset X$ is a closed, $\mathcal{G}$-invariant subset of a $\mathcal{G}$-space X, and if V is a $\mathcal{G}$-equivariant vector bundle on X, then any equivariant section of V defined on Y can be extended to an equivariant section defined on an open $\mathcal{G}$-invariant neighbourhood of Y. This involves an averaging procedure (see [15]). As a consequence, if $f_i\colon X \to Y$, $i = 0, 1$ are two $\mathcal{G}$-equivariantly homotopic maps, and if V is a $\mathcal{G}$-equivariant vector bundle on Y, then $f_0^*(V)$ and $f_1^*(V)$ are $\mathcal{G}$-equivariantly isomorphic.

On the other hand, some new problems appear in connection with equivariant vector bundles. We first formalize our notion of triviality and the corresponding notion of *subtriviality* in a definition.

Definition 3.1 Let X be a $\mathcal{G}$-space. A $\mathcal{G}$-vector bundle over X is *trivial* if it is pulled back from Z under the anchor map $\varrho_X\colon X \to Z$. We denote the pull-back of a $\mathcal{G}$-vector bundle E over Z to the $\mathcal{G}$-space X by E^X. A $\mathcal{G}$-equivariant vector bundle is *subtrivial* if it is a direct summand of a trivial $\mathcal{G}$-vector bundle.

Example 3.1 If $\mathcal{G}$ is a compact group, then a trivial $\mathcal{G}$-equivariant vector bundle over X has the form $X \times \mathbb{R}^n$ where $\mathbb{R}^n$ carries a linear representation of $\mathcal{G}$, and where $\mathcal{G}$ acts on $X \times \mathbb{R}^n$ diagonally.

We will make repeated use of the following basic fact about representations of compact groups (see [27]).

Lemma 3.1 *Let $\mathcal{G}$ be a compact group and $\mathcal{G}' \subset \mathcal{G}$ be a closed subgroup. Then any finite-dimensional representation of $\mathcal{G}'$ is contained in the restriction to $\mathcal{G}'$ of a finite-dimensional representation of $\mathcal{G}$.*

Even if $\mathcal{G}$ is a compact group, due to the notion of 'trivial' vector bundle we are using, not every $\mathcal{G}$-vector bundle is locally trivial in the category of $\mathcal{G}$-vector bundles, *i.e.* locally isomorphic to a trivial $\mathcal{G}$-vector bundle. But it is not hard to prove the following and it is a good exercise for understanding equivariant vector bundles. We leave the proof to the reader, but see §3.3 for some important ingredients in the argument.

Lemma 3.2 *Let X be a smooth $\mathcal{G}$-manifold (see §3.4). Then every $\mathcal{G}$-equivariant vector bundle on X is locally subtrivial in the sense that for every $x \in X$ there is a $\mathcal{G}$-equivariant vector bundle E on Z, a $\mathcal{G}$-equivariant neighbourhood U of x, and an equivariant embedding $\varphi\colon V_{|_U} \to E^U$ of equivariant vector bundles.*

Improving this local result to a global one is not possible, however, without further assumptions.

Example 3.2 Let $X := \mathbb{Z}$ with the trivial action of the compact group $\mathcal{G} := \mathbb{T}$. Then the 1-dimensional complex vector bundle $\mathbb{Z} \times \mathbb{C}$ with the action of $z \in \mathcal{G}$ in the fibre over n by the character $z \mapsto z^n$ is not subtrivial, since it contains infinitely many distinct irreducible representations of $\mathcal{G}$.

Definition 3.2 Let X be a $\mathcal{G}$-space.

- The space X *has enough $\mathcal{G}$-equivariant vector bundles* if whenever we are given $x \in X$ and a finite-dimensional representation of the compact isotropy group $\mathcal{G}_x^x$ of x, there is a $\mathcal{G}$-equivariant vector bundle over X whose restriction to x contains the given representation of $\mathcal{G}_x^x$.
- The space X has a *full* vector bundle if there is a *fixed* vector bundle V over X such that any irreducible representation of any $\mathcal{G}_x^x$ is contained in the representation of $\mathcal{G}_x^x$ on V_x (and we say that such a vector bundle V is full.)

Trivially, a compact group acting on a point has enough vector bundles, but it does not have a full vector bundle unless it is finite, because a compact group with a finite dual has finite-dimensional $L^2(G)$ by the Peter-Weyl theorem. For the same reason, if a $\mathcal{G}$-space X has a full vector bundle then all stabilizers $\mathcal{G}_x^x$ must be finite.

It is easy to check that if $f\colon X \to Y$ is a $\mathcal{G}$-equivariant map then X has enough equivariant vector bundles if Y does, and $f^*(V)$ is a full vector bundle on X if V is a full vector bundle on Y. Both of these assertions use the basic fact Lemma 3.1.

Example 3.3 The $\mathcal{G}$-space described in Example 3.2 does not have a full vector bundle, although it obviously has enough vector bundles.

The following example is more subtle. It is due to Julianne Sauer (see [30]).

Example 3.4 Let $X = \mathbb{R}$ and K be the compact group $K := \prod_{n\in\mathbb{Z}} \mathbb{Z}/2$ acting *trivially* on $\mathbb{R}$. Since X is K-equivariantly contractible, and by the homotopy invariance of equivariant vector bundles (mentioned above at the beginning of §3.2), any K-equivariant vector bundle V on X is trivial, and hence all the representations of K on the fibres V_x are equivalent (in a continuous fashion, but we do not need this.)

Now let $\sigma\colon K \to K$ be the shift automorphism and consider the group $G := \mathbb{Z} \ltimes_\sigma K$; it acts on $\mathbb{R}$ by letting $\sigma(x) := x + 1$ and K continue to act trivially. This is a proper action. Of course as a groupoid G is not proper, but we repair this below.

Any G-vector bundle on X yields, at each point of $\mathbb{R}$, a representation π_x of K on the fibre V_x. We will show that the only representations that appear are trivial ones. This will of course show that X does not have enough G-vector bundles.

There must be a vector bundle map T_σ of V corresponding to the element $\sigma \in G$, since V is a G-vector bundle. By the covariance rule for the semi-direct product, the vector bundle maps $\sigma(k)$ and $T_\sigma k T_\sigma^{-1}$ agree. Thus, in particular,

$$\pi_{x+1}\big(\sigma(k)\big) = T_\sigma \pi_x(k) T_{\sigma^{-1}}$$

holds for all $x \in \mathbb{R}$. This asserts the equivalence of the representations π_x and $\pi_{x+1} \circ \sigma$ on V_{x+1}. As noted above, the K-equivariant contractibility of $\mathbb{R}$ already implies that π_{x+1} and π_x are equivalent. Thus we have $\pi_{x+1} \circ \sigma \sim \pi_x \sim \pi_{x+1}$ for any $x \in \mathbb{R}$. Hence the map $\hat{\sigma}\colon \mathrm{Rep}(K) \to \mathrm{Rep}(K)$ induced by σ fixes the element of this ring obtained by decomposing π_{x+1} into irreducibles.

We have $\mathrm{Rep}(K) \cong \bigoplus_n \widehat{\mathbb{Z}/2}$ with the discrete topology. The action of $\hat{\sigma}\colon \mathrm{Rep}(K) \to \mathrm{Rep}(K)$ is by the shift, whence the only $\hat{\sigma}$-fixed element is zero. This argument shows that any $\hat{\sigma}$-fixed representation of K is a multiple of the trivial representation.

Specializing to our situation, we have that all the representations π_x are trivial, thus K acts trivially on every fibre of any G-equivariant vector bundle V over $\mathbb{R}$; $\mathbb{R}$ does not have enough $\mathcal{G}$-equivariant vector bundles.

A Morita-equivalent approach is via a mapping cyclinder construction and produces a compact *groupoid* acting on a compact space without enough vector bundles. Take $[0,1] \times K$ modulo the relation $(1,k) \sim (0,\sigma(k))$. This results in a bundle of compact, totally disconnected groups over the circle which can be shown to be a locally compact groupoid with Haar system.

Let $\mathcal{G}$ be this groupoid: it is proper. Its base Z is the circle. By a holonomy argument similar to the one just given, any $\mathcal{G}$-equivariant vector bundle over Z must restrict in each fibre to a trivial representation of K. Thus, there are not enough $\mathcal{G}$-equivariant vector bundles on Z.

Example 3.5 If G is a discrete group with a G-compact model for $\mathcal{E}G$, then Lück and Oliver have shown in [24] that there is a full $\mathcal{G}$-equivariant vector bundle on Z, where $\mathcal{G} := G \ltimes \mathcal{E}G$, (so that $Z = \mathcal{E}G$.)

3.3 The topological index of Atiyah-Singer. We now indicate why the condition of having enough vector bundles, or having a full vector bundle, is important for describing analytic equivariant KK-groups topologically.

We start with the problem, famously treated by Atiyah and Singer, of describing the equivariant (analytic) index of a $\mathcal{G}$-equivariant elliptic operator in topological terms, where $\mathcal{G}$ is a compact group, keeping in mind that an equivariant elliptic operator is an important example of a cycle for equivariant KK-theory.

Let X be a smooth manifold with a smooth action of the compact group $\mathcal{G}$. The symbol of an equivariant elliptic operator on X is an equivariant K-theory class for TX. If X is not compact, it is a class with vertically compact support with respect to the bundle projection $TX \to X$ (see §1.4).

The idea of Atiyah and Singer for defining the topological index of the operator is to smoothly embed X in a finite-dimensional (linear) representation of $\mathcal{G}$ on $\mathbb{R}^n$. The derivative of this embedding gives a smooth embedding of TX in $\mathbb{R}^{2n}$, where TX has the induced action of $\mathcal{G}$. Note that this action preserves the complex structure on $\mathbb{R}^{2n}$ in which the first n coordinates are real and the last n imaginary.

Since TX is an equivariantly K-oriented manifold, as is $\mathbb{R}^{2n}$, the normal bundle ν to the embedding $TX \to \mathbb{R}^{2n}$ is a $\mathcal{G}$-equivariantly K-oriented vector bundle on TX. The tubular neighbourhood embedding identifies its total space with a $\mathcal{G}$-invariant neighbourhood of the image of TX in $\mathbb{R}^{2n}$. We now obtain a composition

$$\mathrm{K}^*_{\mathcal{G}}(TX) \to \mathrm{K}^*_{\mathcal{G}}(|\nu|) \to \mathrm{K}^*_{\mathcal{G}}(\mathbb{R}^{2n}) \to \mathrm{K}^*_{\mathcal{G}}(\star) \cong \mathrm{Rep}(\mathcal{G}),$$

where the first map is the Thom isomorphism for the equivariantly K-oriented $\mathcal{G}$-vector bundle ν, the second is the map on equivariant K-theory induced by the open inclusion $|\nu| \hookrightarrow \mathbb{R}^{2n}$ and the third is equivariant Bott Periodicity ($\mathbb{R}^{2n} \cong \mathbb{C}^n$) – another instance of the $\mathcal{G}$-equivariant Thom isomorphism. The point is that this composition is defined in essentially topological, or geometric terms (the only constructions used are open embeddings and the Thom isomorphism). The index theorem asserts of course that the 'topological index' so defined agrees with the Fredholm index (a virtual representation of $\mathcal{G}$) of the equivariant elliptic operator with the given symbol.

For our purposes, it is the issue of embedding a *bundle* of smooth manifolds into a *bundle* of Euclidean spaces, equivariantly, that becomes problematic when passing from the group-equivariant situation, to the groupoid-equvariant one. We now explain this.

The embedding result in the group-equivariant case seems to be due to Mostow (see [27]). Very similar arguments also prove Lemma 3.2. We sketch the proof because it involves important ideas.

Let then $\mathcal{G}$ be a compact group acting smoothly on a manifold X. We will prove that X equivariantly embeds in a finite-dimensional representation of $\mathcal{G}$.

We may assume (or average using the Haar measure on $\mathcal{G}$) that X has a $\mathcal{G}$-invariant Riemannian metric. For any $x \in X$ the orbit $\mathcal{G}x$ is a smooth embedded submanifold of X isomorphic to $\mathcal{G}/\mathcal{G}^x_x$. The tangent space of X at x splits into the orthogonal sum of the tangent space to the orbit and its orthogonal complement $N_x := T_x(\mathcal{G}x)^\perp$. The latter is a finite-dimensional representation of $\mathcal{G}^x_x$, and inducing it results yields a $\mathcal{G}$-equivariant vector bundle

$$N := \mathcal{G} \times_{\mathcal{G}^x_x} N_x := \mathcal{G} \times N_x \,/\, (g, \xi) \in \mathcal{G} \times N_x \sim (gh, h^{-1}\xi) \text{ for } h \in \mathcal{G}^x_x,$$

on the orbit. This $\mathcal{G}$-equivariant vector bundle is obviously isomorphic to the the normal bundle to the embedded submanifold $\mathcal{G}x$; more explicitly, using the exponential map we get an equivariant diffeomorphism between the total space of N and an invariant open neighbourhood U of the orbit, fixing the zero section.

We now construct a smooth $\mathcal{G}$-equivariant embedding $U \to \mathbb{R}^n$ for some representation of $\mathcal{G}$ on $\mathbb{R}^n$ as follows.

By Lemma 3.1, the representation of $\mathcal{G}^x_x$ on N_x is contained in the restriction of some representation of $\mathcal{G}$ on some finite-dimensional vector space $\tilde{N}_x$. Inducing this as before yields a $\mathcal{G}$-equivariant vector bundle $\tilde{N}$ on the orbit $\mathcal{G}x$ containing N as a summand. But since $\tilde{N}_x$ is the restriction of a $\mathcal{G}$-representation, by a standard trick, $\tilde{N}$ is isomorphic to a product bundle $\mathcal{G}x \times \tilde{N}_x$. This provides a $\mathcal{G}$-equivariant map $U \to \tilde{N}_x$. To be explicit, this map sends $[(g, n)] \in U \cong N \subset \tilde{N}$ to the point $gn \in \tilde{N}_x$. It is of course not necessarily an embedding; to improve it to an embedding, fix a vector $v \in \tilde{N}_x$ whose isotropy in $\mathcal{G}$ is *exactly* $\mathcal{G}^x_x$ (for this see also [27]) and set

$$\varphi\colon U \cong \mathcal{G} \times_{\mathcal{G}^x_x} N_x \to \tilde{N}_x \oplus \tilde{N}_x \cong \mathbb{R}^n, \qquad \varphi([g, n]) := (gn, gv).$$

The map φ is an equivariant embedding as required.

If X is compact, then we can (carefully) paste together the local embeddings to get an embedding of X; see [15], or the source [27]. The reader should notice that the assumption that X has enough vector bundles is used implicitly to show that the representation of $\mathcal{G}_x^x$ on N_x can be extended to a $\mathcal{G}$-equivariant vector bundle on the orbit of x (the vector bundle $\tilde{N}$ induced from $\tilde{N}_x$). This was the statement of Lemma 3.1, and is just an explicit way of saying that $\mathcal{G}$ has enough vector bundles on its one-point base space.

If X is not compact, the argument fails.

3.4 Embedding theorems from [15]. More generally, in [15] the following is proved. Let X be a $\mathcal{G}$-space, where $\mathcal{G}$ is a proper groupoid.

We say that X is a *smooth $\mathcal{G}$-manifold* if we can cover X by charts of the form $U \times \mathbb{R}^n$ where $U \subset Z$ is open, so that with respect to this product structure the anchor map $\varrho_X \colon X \to Z$ identifies with the first coordinate projection, and such that groupoid elements and change of coordinates are smooth in the vertical direction.

A *smooth equivariant open embedding* between $\mathcal{G}$-manifolds is a smooth equivariant map which is a fibrewise diffeomorphism onto an open subset of its codomain.

Theorem 3.1 *Let $\mathcal{G}$ be a (proper) groupoid and X and Y be smooth $\mathcal{G}$-manifolds. Suppose that either*

A. *The $\mathcal{G}$-space Z has enough vector bundles and $\mathcal{G}\backslash X$ is compact, or*
B. *Z has a full vector bundle and $\mathcal{G}\backslash X$ has finite covering dimension.*

Then, given a smooth, $\mathcal{G}$-equivariant map $f \colon X \to Y$, there exists

- *A smooth $\mathcal{G}$-equivariant vector bundle V over X,*
- *A smooth $\mathcal{G}$-equivariant vector bundle E over Z,*
- *An smooth, equivariant open embedding $\varphi \colon V \to E^Y$,*

such that

$$f = \pi_{E^Y} \circ \varphi \circ \zeta_V. \tag{3.3}$$

Furthermore, under the union of hypotheses **A**, **B**, *any $\mathcal{G}$-equivariant vector bundle over X (or Y) is subtrivial.*

Recall that the notation E^Y means the pullback of E to Y using the anchor map $\varrho_Y \colon Y \to Z$.

We call a factorisation of a map $f \colon X \to Y$ of the form (3.3) a *normal factorisation.*

3.5 Normally non-singular maps. As in the previous section, $\mathcal{G}$ is a proper groupoid. The constructions of the previous section motivate the following definition.

Definition 3.3 A *$\mathcal{G}$-equivariant normally non-singular map* Φ from X to Y is a triple $(V, E, \hat{f})$ where V is a $\mathcal{G}$-equivariant subtrivial vector bundle over X, E is a $\mathcal{G}$-equivariant vector bundle over Z and $\hat{f} \colon V \to E^Y$ is a $\mathcal{G}$-equivariant open embedding.

- The *trace* of Φ is the composition $\pi_{E^Y} \circ \hat{f} \circ \zeta_V$.
- The *stable normal bundle* of Φ is the class $[V] - [E^X] \in \mathrm{VK}_{\mathcal{G}}(X)$.
- The *degree* of Φ is $\dim(V) - \dim(E)$.
- The normally non-singular map Φ is K-*oriented* if the V and E are equivariantly K-oriented.

- The normally non-singular map Φ is *smooth* if X and Y are smooth $\mathcal{G}$-manifolds and V and E are smooth equivariant vector bundles on which $\mathcal{G}$ acts smoothly, and if $\hat{f}$ is a smooth embedding.

Of course the trace of a $\mathcal{G}$-equivariant smooth normally non-singular map is itself a smooth equivariant map, and the content of Theorem 3.1 is that, conversely, any smooth equivariant map between smooth $\mathcal{G}$-manifolds is the trace of some smooth normally non-singular map , under some hypotheses about the availability of equivariant vector bundles. (This statement is improved in Theorem 3.2.)

Example 3.6 The simplest example of a normally non-singular map is the zero section and bundle projection

$$\zeta_V : X \to V, \qquad \pi_V : |V| \to X,$$

of a $\mathcal{G}$-equivariant vector bundle V over a *compact* $\mathcal{G}$-space X, where $\mathcal{G}$ is a compact group. The zero section is the trace of the normally non-singular map $(V, 0_{|V|}, \mathrm{id})$. The stable normal bundle is the class $[V] \in \mathrm{VK}_{\mathcal{G}}(X)$ of the vector bundle itself. Since X is compact, V is subtrivial. If X is not compact, this can fail, *cf.* Example 3.2.

With the same (compact) X, V *etc*, the bundle projection $\pi_V : |V| \to X$ is the trace of a normally non-singular map $(\pi_V^*(V'), E, \varphi)$, where $V' \in \mathrm{Vect}_{\mathcal{G}}(X)$ is a choice of $\mathcal{G}$-vector bundle on X such that $V \oplus V'$ is a trivial bundle E^X, and $\varphi : |V \oplus V'| \cong |\pi_V^*(V')| \xrightarrow{\cong} E^X$ is a trivialisation. The stable normal bundle is $\pi_V^*([V']) - [E^X] \in \mathrm{VK}_{\mathcal{G}}(|V|)$ respectively.

The normally non-singular map just described seems to depend on the choice of trivialisation of V, but it can be checked that any two choices yield equivalent normally non-singular maps in the sense explained below.

Example 3.7 Let $\mathcal{G}$ be a compact group acting smoothly on manifolds X, Y with X compact. By the discussion in §3.3 we can fix a smooth, equivariant embedding $i : X \to E$ in a linear representation of $\mathcal{G}$. Define V to be the normal bundle to the embedding $x \mapsto \big(f(x), i(x)\big)$ of X in $E^Y := Y \times E$. Let $\varphi : V \to E^Y$ be the corresponding tubular neighbourhood embedding. Then the trace of the composition $\pi_{E^Y} \circ \varphi \circ \zeta_V$ of the normally non-singular map $(V, E, \hat{f})$ is f. Since $TX \oplus V \cong f^*(TY) \oplus E^X$, the stable normal bundle is $f^*([TY]) - [TX] \in \mathrm{VK}_{\mathcal{G}}(X)$.

Example 3.8 If G is a discrete group with a G-compact model for $\mathcal{E}G$, then Lück and Oliver have shown in [24] that there is a full $\mathcal{G}$-equivariant vector bundle on $\mathcal{E}G$, where $\mathcal{G} := G \ltimes \mathcal{E}G$, (so that the base of $\mathcal{G}$ is $Z := \mathcal{E}G$.) Let X and Y be smooth manifolds equipped with smooth actions of G and $f : X \to Y$ be a smooth, G-equivariant map. As above let $\mathcal{G}$ be the proper groupoid $G \ltimes \mathcal{E}G$. Applying the Baum-Connes procedure of §3.1 to this situation we get smooth $\mathcal{G}$-manifolds $X \times \mathcal{E}G$ and $Y \times \mathcal{E}G$ and a smooth $\mathcal{G}$-map $f \times \mathrm{id}_{\mathcal{E}G} : X \times \mathcal{E}G \to Y \times \mathcal{E}G$. It is the trace of a normally non-singular map because of Theorem 3.1 and the result of Lück and Oliver.

Two normally non-singular maps are *isomorphic* if there are vector bundle isomorphisms $V_0 \cong V_1$ and $E_0 \cong E_1$ that intertwine the open embeddings f_0 and f_1. The *lifting* of a normally non-singular map $\Phi = \Psi = (V, \varphi, E)$ along an equivariant vector bundle E^+ over Z is the normally non-singular map $\Phi \oplus E^+ := (V \oplus (E^+)^X, E \oplus E^+, \hat{f} \times_Z \mathrm{id}_{E^+})$. Two normally non-singular map s are *stably isomorphic* if there are $\mathcal{G}$-equivariant vector bundles E_0^+ and E_1^+ such that $\Phi_0 \oplus E_0^+$

is isomorphic to $\Phi_1 \oplus E_1^+$. Finally, two normally non-singular map s Φ_0 and Φ_1 are *isotopic* if there is a continuous 1-parameter family of normally non-singular maps whose values at the endpoints are stably isomorphic to Φ_0 and Φ_1 respectively (see [15] for the exact definition), and are *equivalent* if they have isotopic liftings. There is an obvious notion of *smooth equivalence* of smooth normally non-singular maps.

There are obvious K-oriented analogues of the above relations. For example, lifting must only use K-oriented trivial bundles, and isomorphism must preserve the given K-orientations. Referring to this kind of equivalence we will speak of K-*oriented equivalence* of K-oriented normally non-singular maps.

3.6 Manifolds with smooth normally non-singular maps to Z. A useful hypothesis covering a number of geometric situations is that a given smooth $\mathcal{G}$-manifold X admits a smooth normally non-singular map to the object space Z of $\mathcal{G}$. By the theorem above this is the case if **A** or **B** holds. It means explicitly that we have a triple $(N_X, \hat{g}, E)$ where N_X is a smooth subtrivial vector bundle over X, E is an equivariant vector bundle over Z and $\hat{g}$ is a smooth open equivariant embedding $N_X \to E$. Note that $N_X \oplus TX \cong E^X$. If we like, we can replace E by $E \oplus E$, which can be equivariantly K-oriented by the usual $\mathcal{G}$-equivariant complex structure on a doubled vector space, and replace N_X by $N_X \oplus E^X$. We get a new $\mathcal{G}$-equivariant normally non-singular map which is stably isomorphic to the old one, but for which the trivial bundle component is equivariantly K-oriented.

For a normally non-singular $\mathcal{G}$ map $(N_X, \hat{g}, E)$ from X to Z for which a fixed equivariant K-orientation on E has been fixed, equivariant K-orientations on TX are in 1-1 correspondence with ones on N_X, because of the 2-out-of-3 lemma.

Theorem 3.2 *Let X and Y be smooth $\mathcal{G}$-manifolds, and assume that X admits a smooth, normal $\mathcal{G}$-map to Z and that $f^*(TY)$ is subtrivial.*

Then any smooth $\mathcal{G}$-map from X to Y is the trace of a smooth normal $\mathcal{G}$-map, and two smooth normally non-singular maps from X to Y are smoothly equivalent if and only if their traces are smoothly homotopic.

Furthermore, smooth equivalence classes of smooth K*-oriented normally non-singular maps from X to Y are in 1-1 correspondence with pairs (f, τ) where f is a smooth homotopy class of equivariant smooth map $X \to Y$ and τ is an equivariant* K*-orientation on $N_X \oplus f^*(TX)$.*

We sketch the existence part of this proof. Fix a smooth equivariant normally non-singular map $(N_X, \hat{g}, E)$ from X to Z. We can assume by replacing E by $E \oplus E$ and N_X by $N_X \oplus E^X$ if needed that E is equivariantly K-oriented. Let $g\colon X \xrightarrow{\zeta_{N_X}} N_X \xrightarrow{\hat{g}} E$ be the composite smooth embedding.

Let Y be another smooth $\mathcal{G}$-manifold and $f\colon X \to Y$ be a smooth map. Then one obtains a a smooth embedding $X \to Y \times_Z E = E^Y$, $x \mapsto \big(f(x), g(x)\big)$. It has a (smooth) normal bundle V with a smooth open embedding in E^Y. Since $V \cong N_X \oplus f^*(TY)$, V is subtrivial and $(V, E, \hat{f})$ is a smooth normally non-singular map with trace f and stable normal bundle $[V] - [E^X] \in \mathrm{VK}_{\mathcal{G}}(X)$. Note that $V \oplus TX \cong f^*(TY) \oplus E^X$.

The stable normal bundle is $[V] - [E^X] = f^*([TY]) - [TX] \in \mathrm{VK}_{\mathcal{G}}(X)$. Equivariant K-orientations on V are in 1-1 correspondence with equivariant K-orientations on $N_X \oplus f^*(TY)$.

3.7 Correspondences. Let $\mathcal{G}$ continue to denote a proper groupoid.

Definition 3.4 Let X and Y be $\mathcal{G}$-spaces. A *$\mathcal{G}$-equivariant correspondence from X to Y* is a quadruple (M, b, f, ξ) where M is a $\mathcal{G}$-space, $f\colon M \to Y$ is a $\mathcal{G}$-equivariantly K-oriented normally non-singular map , $b\colon M \to X$ is a $\mathcal{G}$-equivariant map, and $\xi \in \mathrm{RK}^*_{\mathcal{G},X}(M)$ is a $\mathcal{G}$-equivariant K-theory class with X-compact support (see §1.4) where the $\mathcal{G} \ltimes X$-structure on M is that determined by the $\mathcal{G}$-equivariant map $b\colon M \to X$.

The *degree* of the correspondence (M, b, f, ξ) is the sum of the degrees of f and ξ.

Remark 3.1 Thus a significant difference from the set-up of Connes and Skandalis in [8] is that the map $b\colon M \to X$ is not required to be proper; we have replaced this by a support condition on ξ.

Several equivalence relations on correspondences are imposed. The first is to consider two correspondences (M, b_0, f_0, ξ) and (M, b_1, f_1, ξ) to be equivalent if their normally non-singular maps are equivalent. The second is to consider *bordant* correspondences equivalent (we will not discuss this at all in this survey.) The third is most interesting, and is called *Thom modification.* The Thom modification of a correspondence (M, b, f, ξ) using a subtrivial K-oriented vector bundle V over M is the correspondence

$$\big(V, b \circ \pi_V, f \circ \pi_V, \tau_V(\xi)\big),$$

where $\tau_V\colon \mathrm{RK}^*_{\mathcal{G},X}(M) \xrightarrow{\cong} \mathrm{RK}^{*+\dim(V)}_{\mathcal{G},X}(|V|)$ is the Thom isomorphism, *i.e.* $\tau_V(\xi) := \pi_V^*(\xi) \cdot \xi_V$, where $\xi_V \in \mathrm{RK}^{\dim(V)}_{\mathcal{G},M}(|V|)$ is the Thom class. We declare a correspondence and its Thom modification to be Thom equivalent. Note that applying Thom modification to a correspondence does not change its degree.

The equivalence relation on correspondences is that generated by equivalence of normally non-singular maps, bordism and Thom equivalence.

Definition 3.5 Let $\mathcal{G}$ be a proper groupoid and X and Y be $\mathcal{G}$-spaces. We let $\widehat{KK}^*_{\mathcal{G}}(X, Y)$ denote the $\mathbb{Z}/2$-graded set of equivalence classes of $\mathcal{G}$-equivariant correspondences from X to Y, graded by degree.

A correspondence (M, b, f, ξ) from X to Y is *smooth* if X, Y and M are smooth manifolds, f is a smooth normally non-singular map (see §3.3), and b is a smooth map. There is a rather obvious notion of *smooth equivalence* of smooth correspondences. This gives rise to a parallel theory using only smooth equivalence classes of smooth correspondences; we do not use notation for this.

3.8 $\widehat{KK}^{\mathcal{G}}$ as a category. Classes of correspondences form a category with analogous properties to Kasparov's equivariant KK. The composition of correspondences is called the *composition product.* For composition we use similar notation to Kasparov's: if $\Psi \in \widehat{KK}^{\mathcal{G}}_i(X, Y)$ is a topological morphism, *i.e.* an equivalence class of equivariant correspondence from X to Y, and if $\Phi \in \widehat{KK}^{\mathcal{G}}_j(Y, W)$ is another, then we write $\Psi \otimes_Y \Phi \in \widehat{KK}^{\mathcal{G}}_{i+j}(X, W)$ for their composition.

The composition product is different from the product operation, due to Connes and Skandalis in [8], which uses transversality. To distinguish them, we call the Connes-Skandalis construction the *intersection* product (we will review it below). It only works well in the setting of smooth correspondences between smooth manifolds; even then, it may be problematic in the equivariant situation. The difficulty is that in the equivariant setting, two smooth equivariant maps may *not* in general be deformable to be transverse to each other (see Example 3.9.) This means that if we attempt to define composition of correspondences using transversality, we will not get a category. Instead, our strategy has been to use a more restrictive class of wrong-way maps – the *normally non-singular maps* – and a trick of Baum and Block to ensure that composition of correspondences is well-defined in general.

The intersection product of Connes and Skandalis can still be used in specific *examples* but one has to check a few things.

In any case, although the existence of the general composition product justifies the whole theory of normally non-singular maps, it is less geometrically satisfying in some ways than the intersection product, and so in this survey we will restrict ourselves to discussing the latter – which, in any case, gives more of the flavour of the subject of correspondences.

Recall that two smooth $\mathcal{G}$-maps $f_1\colon M_1 \to Y$ and $b_2\colon M_2 \to Y$ are *transverse* if for every $(p_1, p_2) \in M_1 \times M_2$ such that $f_1(p_1) = b_2(p_2)$, we have $D_{p_1} f_1(T_{p_1} M_1) + D_{p_2} b_2(TM_2) = T_{f_1(p_1)}(X)$. Transversality ensures that the *coincidence space*

$$M_1 \times_X M_2 := \{(p_1, p_2) \in M_1 \times M_2 \mid f_1(p_1) = b_2(p_2)\}$$

has the structure of a smooth $\mathcal{G}$-manifold.

Theorem 3.3 *Let $\Phi_1 = (M_1, b_1, f_1, \xi_1)$ and $\Phi_2 = (M_2, b_2, f_2, \xi_2)$ be smooth correspondences from X to Y and from Y to U, respectively. Assume that both M_1 and M_2 admit smooth normally non-singular maps to Z (see §3.5), so that we lose nothing if we view f_1 and f_2 as* K*-oriented smooth maps (see Theorem 3.2).*

Assume also that f_1 and b_2 are transverse, so that $M_1 \times_Y M_2$ is a smooth $\mathcal{G}$-manifold; it has a smooth normally non-singular map to Z as well.

Then the intersection product of Φ_1 and Φ_2 is the class of the correspondence

$$\big(M_1 \times_Y M_2, b_1 \circ \pi_1, f_2 \circ \pi_2, \pi_1^*(\xi_1) \cdot \pi_2^*(\xi_2)\big),$$

where $\pi_j\colon M_1 \times_Y M_2 \to M_j$ for $j = 1, 2$ are the canonical projections.

In the non-equivariant situation, any two smooth maps can be perturbed to be transverse, and in [8] this is shown to give rise to a bordism of correspondences. As a result, one can compose bordism classes of correspondence by the recipe described in Theorem 3.3.

However, this fails in the equivariant situation because pairs of smooth maps cannot in general be perturbed equivariantly to be transverse; this happsn in even some of the simplest situations.

Example 3.9 Let μ be the non-trivial character of $\mathbb{Z}/2$. The corresponding one-dimensional representation is denoted by $\mathbb{C}_\mu$. We regard this as an equivariant vector bundle $\mathbb{C}_\mu$ over a point. Its total space is $|\mathbb{C}_\mu|$. The equivariant vector bundle $\mathbb{C}_\mu$ is equivariantly K-oriented, since the $\mathbb{Z}/2$-action preserves the complex structure on $\mathbb{C}_\mu$. We therefore obtain a smooth normal equivariant map $\star \to |\mathbb{C}_\mu|$ of degree 2. But since the origin is the only fixed point of the $\mathbb{Z}/2$-action, this map cannot be perturbed to be transverse to itself (transverse for two points embedded in $\mathbb{C}$ meaning that they are not the same point.)

With the given architecture of equivariant correspondences, we compose them roughly as follows. Let $X \xleftarrow{b} (M, \xi) \xrightarrow{f} Y$ be such an equivariant correspondence from X to Y. Let $f = (V, E, \hat{f})$. The equivariant vector bundles V and E are K-oriented by assumption. Thom modification using V results in the correspondence

$X \xleftarrow{b} (|V|, \xi_V \cdot \xi) \xrightarrow{f \circ \pi_V} Y$. An obvious smooth isotopy of normally non-singular maps replaces this by $X \xleftarrow{b} (|V|, \xi_V \cdot \xi) \xrightarrow{\pi_{E^Y} \circ \hat{f}} Y$. Since $\pi_{E^Y} \circ \hat{f}$ is a submersion, it is transverse to any other smooth map to Y. Hence this correspondence can be composed using the transversality recipe of Theorem 3.3 with any other one (on the right).

3.9 Further properties of topological KK**-theory.** We have said that $\widehat{KK}^{\mathcal{G}}$ is a category under composition product. It is also additive, with the sum operation on correspondences defined by disjoint union. The other important property is the existence of external products. This means that there exists a map

$$\widehat{KK}^{\mathcal{G}}_i(X,Y) \times \widehat{KK}^{\mathcal{G}}_j(U,V) \to \widehat{KK}^{\mathcal{G}}_{i+j}(X \times_Z U, Y \times_Z V).$$

This endows $\widehat{KK}^{\mathcal{G}}$ with the structure of a symmetric monoidal category.

Finally, there is a natural map $\widehat{KK}^{\mathcal{G}} \to \mathrm{KK}^{\mathcal{G}}$. This is defined not using the pseudodifferential calculus, as in [8], but by purely topological considerations. Indeed, by definition, a normal K-oriented $\mathcal{G}$-map from X to Y factors, by definition, as a composite of a zero section of an equivariantly K-oriented vector bundle, an equivariant open embedding, and the projection map for another equivariantly K-oriented vector bundle. Zero sections and bundle projections yield wrong-way morphisms in $\mathrm{KK}^{\mathcal{G}}$ because of the Thom isomorphism of [23] – these wrong-way morphisms *are* the Thom isomorphism classes in the sense that composition with them gives the Thom isomorphism in equivariant K-theory.

Open embeddings clearly determine elements morphisms in $\mathrm{KK}^{\mathcal{G}}$ because they determine equivariant *-homomorphisms.

The various naturality properties of the Thom isomorphism imply corresponding facts about the map $\widehat{KK}^{\mathcal{G}} \to \mathrm{KK}^{\mathcal{G}}$. Other functorial properties of $\widehat{KK}^{\mathcal{G}}$, e.g., with respect to homomorphisms $\mathcal{G}' \to \mathcal{G}$ of groupoids, are explained in detail in [15]. It should be emphasised that proving functoriality for the map $\widehat{KK}^{\mathcal{G}} \to \mathrm{KK}^{\mathcal{G}}$ with this set-up is not difficult (it should not be, since all computations are purely topological). This is in contrast with the situation of Connes and Skandalis, where functoriality is difficult because they prove it analytically.

In any case, the Atiyah-Singer Index Theorem is essentially the statement that the two frameworks (analytic and topological) agree.

4 Topological duality and the topological Lefschetz map

We have organized this survey around the goal of computing the Lefschetz map for smooth $\mathcal{G}$-manifolds. This problem is intertwined with that of computing equivariant KK-groups topologically and we will solve both problems at once in this section. The first step is to describe a class of $\mathcal{G}$-spaces to which the general theory of duality described in §1 applies. Subject to the resulting constraints on X we will obtain a *topological model* of the Lefschetz map and simultaneously a proof that the map $\widehat{KK}^{\mathcal{G}} \to \mathrm{KK}^{\mathcal{G}}$ is an isomorphism on both the domain and range of the Lefschetz map. This will complete our computation of Lef for a fairly wide spectrum of $\mathcal{G}$-spaces X.

4.1 Normally non-singular $\mathcal{G}$-spaces.

Definition 4.1 A *normally non-singular* $\mathcal{G}$*-space* X is a $\mathcal{G}$-space equipped with a $\mathcal{G}$-equivariant normally non-singular map $(N_X, E, \hat{g})$ from X to Z.

We also require of a normally non-singular $\mathcal{G}$-space X that every $\mathcal{G}$-equivariant vector bundle on X is subtrivial.

The vector bundle N_X is called the *stable normal bundle of* X.

We denote its total space by P for brevity.

We may assume, after replacing the initial normally non-singular map with a stably isomorphic one, that E is equivariantly K-orientable; fix an equivaraint K-orientation. Since the zero bundle is always uniquely K-oriented, we get an equivariant K-oriented normally non-singular map $(0_P, E, \hat{g})$ from P to Z. Let $D \in \widehat{KK}^{\mathcal{G}}(P, Z)$ be the corresponding class.

Since $\widehat{KK}^{\mathcal{G}\ltimes X}$ has external products, we can define a map

$$\widehat{KK}^{\mathcal{G}\ltimes X}(X \times_Z U, Y \times_Z V) \to \widehat{KK}^{\mathcal{G}\ltimes X}(P \times_Z U, P \times_Z V)$$

If every $\mathcal{G}$-equivariant vector bundle over X is subtrivial, then there is a forgetful functor $\widehat{KK}^{\mathcal{G}\ltimes X} \to \widehat{KK}^{\mathcal{G}}$ and this results in a map

$$\widehat{KK}^{\mathcal{G}\ltimes X}(P \times_Z U, P \times_Z V) \to \widehat{KK}^{\mathcal{G}}((P \times_Z U, P \times_Z V).$$

Composing with the previous one yields a topological analogue

$$\widehat{KK}^{\mathcal{G}\ltimes X}(X \times_Z U, X \times_Z V) \to \widehat{KK}^{\mathcal{G}}(P \times_Z U, P \times_Z V)$$

of the functor denoted T_P in the discussion of duality in §1, and, composing further with the morphism D gives a topological analogue of the Kasparov duality map

$$\mathrm{PD}^*\colon \widehat{KK}^{\mathcal{G}\ltimes X}(X \times_Z U, X \times_Z V) \to \widehat{KK}^{\mathcal{G}}(P \times_Z U, V)$$

of Theorem 1.1. However, it is of course *not* the case in general that every equivariant vector bundle over X is subtrivial, which is why we have added this as a hypothesis.

Example 4.1 Let X be the integers with the trivial action of the circle group $\mathcal{G} := \mathbb{T}$. Then there are (*cf.* Example 3.2) equivariant vector bundles on X which are not subtrivial. This is despite the fact that X admits a normally non-singular map to a point, since it smoothly embeds in the trivial representation of $\mathcal{G}$ on $\mathbb{R}$, with trivial normal bundle.

In any case, X is not normally non-singular.

Any smooth $\mathcal{G}$-manifold satisfying one of the hypotheses of Theorem 3.1 is normally non-singular.

To define a topological analogue of the map denoted PD we need a class $\Theta \in \widehat{KK}^{\mathcal{G}\ltimes X}(X, X \times_Z P)$. Combining composition with this class and the map

$$\mathrm{KK}^{\mathcal{G}}(P \times_Z U, V) \to \widehat{KK}^{\mathcal{G}\ltimes X}(X \times_Z P \times_Z U, X \times_Z V),$$

which is defined in the topological category in the same way as in the analytic one, we will obtain the required topological map

$$\mathrm{PD}\colon\ \mathrm{KK}^{\mathcal{G}}(P \times_Z U, V) \to \widehat{KK}^{\mathcal{G}\ltimes X}(X \times_Z U, X \times_Z V).$$

We just describe Θ in a heuristic fashion. Assume for simplicity that the base Z of $\mathcal{G}$ is a point. So E is just a Euclidean space. Choose a point $x \in X$. Using the zero section of N_X and the map $\hat{g}$, we see x as a point in the open subset $P := |N_X|$ of E, and hence by rescaling E (fibrewise) into a sufficiently small open ball around x

we obtain an open embedding of E into P. Explicitly, we use an open embedding of the form

$$\hat{\varrho}_x \colon |E| \xrightarrow{\beta_x} B_\epsilon\big(\hat{g}(\zeta_{N_X}(x))\big) \subset P := |N_X|$$

where the first map is the re-scaling. This yields an obvious normally nonsingular map $(E, 0_P, \hat{\varrho}_x)$ from a point to P. It is easily checked that this can be carried out continuously in the parameter $x \in X$, and we obtain an X-equivariant normal K-oriented map δ from X to $X \times_Z P$ with trace the graph of the zero-section $\Xi \colon X \to X \times_Z P$, $\Xi(x) := (x, (x, 0))$. This yields an element of $\widehat{KK}^X(X, X \times_Z P)$. The same can be checked to work equivariantly, and a fibrewise version works for groupoids with nontrivial base. We let $\Theta \in \widehat{KK}^{\mathcal{G} \ltimes X}(X, X \times_Z P)$ be the corresponding class.

Theorem 4.1 *In the notation above, let X be a normally non-singular $\mathcal{G}$-space, let N_X be the stable normal bundle, $P := |N_X|$ its total space, and $D \in \mathrm{KK}^{\mathcal{G}}(P, Z)$ and $\Theta \in \mathrm{KK}^{\mathcal{G} \ltimes X}(X, X \times_Z P)$ the classes constructed above.*

Then (P, D, Θ) is a Kasparov dual for X in $\widehat{KK}^{\mathcal{G}}$, and the maps PD *and* PD* *are isomorphisms.*

The same formal computations as in §1 then imply that the maps PD and PD* are isomorphisms.

Again, the stronger result is proved in [15] that one gets a symmetric Kasparov dual; this, remember, is designed to give, as well, an isomorphism of the form

$$\widehat{KK}^*_{\mathcal{G}}(X \times_Z U, V) \cong \widehat{KK}^*_{\mathcal{G} \ltimes X}(X \times_Z U, P \times_Z V)$$

for any pair of $\mathcal{G}$-spaces U and V. We do not give the details. As a consequence one deduces the following theorem by using duality to reduce from the bivariant to the monovariant case.

Theorem 4.2 *Let X be a normally non-singular $\mathcal{G}$-space and Y be an arbitrary $\mathcal{G}$-space. Then the natural transformation $\widehat{KK}^*_{\mathcal{G}}(X, Y) \to \mathrm{KK}^{\mathcal{G}}_*\big(\mathcal{C}_0(X), \mathcal{C}_0(Y)\big)$ is invertible.*

Example 4.2 Consider the space $X := \mathbb{Z}$ with the trivial action of $\mathcal{G} := \mathbb{T}$. This space is not normally non-singular, though it is a smooth $\mathcal{G}$-manifold admitting a normally non-singular map to a point. The map

$$\widehat{KK}^{\mathcal{G}}(X, \star) \to \mathrm{KK}^{\mathcal{G}}(\mathcal{C}_0(X), \mathbb{C}) \tag{4.1}$$

is *not* an isomorphism in this case. By duality in analytic Kasparov theory, the elements of $\mathrm{KK}^{\mathcal{G}}(\mathcal{C}_0(X), \mathbb{C})$ are parameterised by $\mathbb{Z}/2$-graded $\mathcal{G}$-equivariant complex vector bundles on X. One can check that the elements of $\widehat{KK}^{\mathcal{G}}(X, \star)$ are by contrast parameterised by $\mathcal{G}$-equivariant complex vector bundles which only involve a finite number of representations of $\mathcal{G}$. In other words, (4.1) is equivalent to the embedding

$$\oplus_{n \in \mathbb{Z}} \mathrm{Rep}(\mathbb{T}) \to \prod_{n \in \mathbb{Z}} \mathrm{Rep}(\mathbb{T})$$

of the direct sum into the direct product of representation rings.

If we dropped the subtriviality requirement on vector bundles that we imposed on cycles for $\widehat{KK}$ then (4.1) would be an isomorphism, but then we would not be able to define the intersection product of correspondences in general.

Remark 4.1 Restricting to smooth correspondences and smooth equivalence classes of correspondences yields a parallel 'smooth' theory. If X is a smooth normally non-singular $\mathcal{G}$-space and Y is smooth, then it follows that the smooth and non-smooth versions of $\widehat{KK}^*_{\mathcal{G}}(X, Y)$ agree.

We are now in a position to solve the problem we have been working towards: a topological computation of the Lefschetz map for a normally non-singular $\mathcal{G}$-space. Firstly, we define the *topological Lefschetz map*

$$\mathrm{Lef}\colon \widehat{KK}^{\mathcal{G}\ltimes X}_*(X \times_Z X, X) \to \widehat{KK}^*_{\mathcal{G}}(X, Z)$$

for any normally non-singular $\mathcal{G}$-space of finite type as in Definition 1.5, using the topological Kasparov dual constructed above out of the stable normal bundle. Since a topological dual maps to an analytic dual, the diagram

$$\begin{array}{ccc} \widehat{KK}^{\mathcal{G}\ltimes X}_*(X \times_Z X, X) & \xrightarrow{\mathrm{Lef}} & \widehat{KK}^*_{\mathcal{G}}(X, Z) \\ \downarrow & & \downarrow \\ \mathrm{KK}^{\mathcal{G}\ltimes X}_*\big(\mathcal{C}_0(X \times_Z X), \mathcal{C}_0(X)\big) & \xrightarrow{\mathrm{Lef}} & \mathrm{KK}^{\mathcal{G}}_*\big(\mathcal{C}_0(X), \mathcal{C}_0(Z)\big) \end{array} \quad . \tag{4.2}$$

commutes. The finite-type hypothesis implies (Theorem 4.2) that the vertical maps are both isomorphisms. Therefore we have obtained a complete description of Lef in purely topological terms. We now proceed to describe it explicitly for smooth $\mathcal{G}$-manifolds.

4.2 Explicit computation of Lef. Let X be a smooth $\mathcal{G}$-manifold satisfying one of the hypotheses **A** or **B** of Theorem 3.1. From that Theorem, and by Remark 4.1, the map $X \to Z$ is the trace of an essentially unique smooth normally non-singular $\mathcal{G}$-map; moreover any $\mathcal{G}$-equivariant vector bundle on X is subtrivial. Thus X is a smooth normally non-singular $\mathcal{G}$-space and so has duality in topological equivariant KK-theory. The Lefschetz map exists for such a space.

We can represent the general morphism in $\widehat{KK}^{\mathcal{G}\ltimes X}(X \times_Z X, X)$ by a smooth, $\mathcal{G} \ltimes X$-equivariant correspondence

$$\Psi := (M, b, f, \xi), \text{ or in diagram form } X \times_Z X \xleftarrow{b} (M, \xi) \xrightarrow{f} X$$

from $X \times_Z X$ to X. Let Ψ denote its class. Recall that the X-structure on $X \times_Z X$ is on the *first* coordinate. Hence the map b, since it is $\mathcal{G} \ltimes X$-equivariant, has the form $b(m) = \big(\varrho^{\mathcal{G}\ltimes X}_M(m), b'(m)\big)$ for some $\mathcal{G}$-map $b'\colon M \to X$. Here $\varrho^{\mathcal{G}\ltimes X}_M\colon M \to X$ is the anchor map for the $\mathcal{G} \ltimes X$-space M; it is in fact the same map as f (or, more precisely, as the trace of f) since the latter is $\mathcal{G} \ltimes X$-equivariant.

By the definitions, the hypothesis that M is a smooth $\mathcal{G} \ltimes X$*-manifold* (by the anchor map $\varrho^{\mathcal{G}\ltimes X}_M\colon M \to X$) means that the anchor map $\varrho^{\mathcal{G}\ltimes X}_M$ (= the trace of f) gives M the structure of a bundle of smooth manifolds over X. Since X is assumed throughout to be itself a smooth $\mathcal{G}$-manifold, and all maps appearing here to be (fibrewise) smooth, it means that $\varrho^{\mathcal{G}\ltimes X}_M$ must be a smooth fibrewise submersion.

The following is our general position assumption.

Assumption 4.1 *Assume that for all $m \in M$ for which $\varrho^{\mathcal{G}\ltimes X}_G(m) = b'(m)$, the linear map*

$$T_m M \to T_{\varrho_M(m)} X, \qquad \zeta \mapsto D_m b'(\zeta) - D_m \varrho^{\mathcal{G}\ltimes X}_M(\zeta) \tag{4.3}$$

is non-singular.

This implies that the *coincidence space*

$$F_\Psi := \{m \in M \mid \rho_M^{\mathcal{G}\ltimes X}(m) = b'(m)\},$$

is a *smooth, equivariantly* K*-oriented* $\mathcal{G}$*-manifold*; in fact, more, by Theorem 3.3 it implies that the projection $F_\Psi \to Z$ is a smooth, K-oriented normally non-singular map . Finally, it is easily checked that the restriction of ξ to F_Ψ has compact vertical support with respect to the map $(\varrho_M)^{\mathcal{G}\ltimes X}_{|F_\Psi}\colon F_\Psi \to X$, (this is already implied by Theorem 3.3), so we get a $\mathcal{G}$-equivariant correspondence from X to Z:

$$(F_\Psi, (\varrho_M^{\mathcal{G}\ltimes X})_{|F_\Psi}, \varrho^{\mathcal{G}}_{F_\Psi}, \xi_{|F_\Psi}),$$

in the usual notation for anchor maps. We call this the *coincidence cycle* of Ψ.

Theorem 4.3 *Let* $\Psi \in \widehat{KK}{}^{\mathcal{G}\ltimes X}_*(X \times_Z X, X)$*; let*

$$\mathrm{Lef}\colon \widehat{KK}{}^*_{\mathcal{G}}(X \times_Z X, X) \to \widehat{KK}{}^*_{\mathcal{G}}(X, Z)$$

be the Lefschetz map in topological equivariant KK*-theory. Then the topological Lefschetz invariant of the class of a correspondence* Ψ *in general position in the sense explained above, is the class of the coincidence cycle of* Ψ*,*

$$\mathrm{Lef}\big([(M,b,f,\xi)]\big) = [(F_\Psi, (\varrho_M)^{\mathcal{G}\ltimes X}_{|F_\Psi}, \varrho^{\mathcal{G}}_{F_\Psi}, \xi_{|F_\Psi})] \in \widehat{KK}{}^*_{\mathcal{G}}(X, Z).$$

Similar statements follow in analytic KK.

Namely, the Lefschetz invariant of the $\mathrm{KK}^{\mathcal{G}\ltimes X}(X \times_Z X, X)$-morphism $\mathrm{KK}^{\mathcal{G}\ltimes X}(\Psi)$ determined by Ψ is the class $\mathrm{KK}^{\mathcal{G}}(\mathrm{Lef}(\Psi))$. Furthermore, this class is the pushforward under the map $(\varrho_M)^{\mathcal{G}\ltimes X}_{|F_\Psi}\colon F_\Psi \to X$ of the class of the Dirac operator on the K-oriented coincidence manifold F_Ψ, twisted by ξ (by an appeal to the Index Theorem.)

Even if the general position assumption fails, the Lefschetz map still exists, and is topologically computable, but in the interests of brevity, we omit the discussion.

A variant of the above theorem applies to self-morphisms of X. We compute the composition

$$\widehat{KK}{}^{\mathcal{G}}(X, X) \to \widehat{KK}{}^{\mathcal{G}\ltimes X}(X \times_Z X, X) \xrightarrow{\mathrm{Lef}} \mathrm{KK}^{\mathcal{G}}(X, Z),$$

where the first map is the composition of the canonical inflation map

$$p_X^*\colon \widehat{KK}{}^{\mathcal{G}}(X, X) \to \widehat{KK}{}^{\mathcal{G}\ltimes X}(X \times_Z X, X \times_Z X)$$

and the map

$$\widehat{KK}{}^{\mathcal{G}\ltimes X}(X \times_Z X, X \times_Z X) \to \widehat{KK}{}^{\mathcal{G}\ltimes X}(X \times_Z X, X)$$

of composition with the diagonal restriction class $\Delta_X \in \mathrm{KK}^{\mathcal{G}\ltimes X}(X \times_Z X, X)$.

Recall that the latter is the class of the $\mathcal{G} \ltimes X$-equivariant correspondence $X \times_Z X \xleftarrow{\delta_X} X \xrightarrow{\mathrm{id}} X$ where δ_X is the diagonal embedding. Let $\Psi = (M, b, f, \xi)$ be a smooth, $\mathcal{G}$-equivariant correspondence from X to X. Of course this implies that there is a smooth normally non-singular map $M \to Z$, by composing $f\colon M \to X$ and $\varrho_X\colon X \to Z$. The inflation map replaces Ψ by the $\mathcal{G}\ltimes X$-equivariant correspondence

$$X \times_Z X \xleftarrow{\mathrm{id}_X \times_Z b} (X \times_Z M, \xi) \xrightarrow{\mathrm{id} \times_Z f} X \times_Z X.$$

The X-structures are all on the first variable. In order to compose (on the right) with the diagonal restriction class using transversality, we check that Theorem 3.3) applies as long as the smooth maps $\mathrm{id}_X \times_Z f\colon X \times_X M \to X \times_X X$ and

$\Delta_X\colon X \to X \times_Z X$ are transverse in the sense of Theorem 3.3, in the category of $\mathcal{G} \ltimes X$-equivariant smooth maps. *They are transverse if and only the smooth $\mathcal{G}$-map $f\colon M \to X$ is a submersion.* If this condition is met, then $f\colon M \to X$ gives M not just the structure of a smooth $\mathcal{G}$-manifold, but the structure of a smooth $\mathcal{G} \ltimes X$-manifold, *i.e.* a bundle of smooth manifolds over X with morphisms in $\mathcal{G}$ acting by diffeomorphisms between the fibres. Composing with Δ_X using transversality then yields the $\mathcal{G} \ltimes X$-equivariant correspondence $X \times_Z X \xleftarrow{(f,b)} (M,\xi) \xrightarrow{f} X$ where $(f,b)(m) := (f(m), b(m))$. Theorem 4.3 then implies the following.

Theorem 4.4 *Let X be a smooth normally non-singular $\mathcal{G}$-manifold, let $\Psi = (M, b, f, \xi)$ be a smooth, $\mathcal{G}$-equivariant correspondence from X to X such that f is a fibrewise submersion and such that for every $m \in M$ such that $b(m) = f(m)$ the linear map $T_m M \to T_{b(m)} X$,*

$$\zeta \mapsto D_m b(\zeta) - D_m f(\zeta)$$

is non-singular. Then the Lefschetz invariant of Ψ is the class of the smooth, $\mathcal{G}$-equivariant correspondence

$$X \xleftarrow{b} (F'_\Psi, \xi_{|_{F'_\Psi}}) \to Z$$

from X to Z, where $F'_\Psi := \{m \in M \mid b(m) = f(m)\}$ with its induced K*-orientation.*

Example 4.3 If $b\colon X \to X$ is a smooth $\mathcal{G}$-equivariant map, then $X \xleftarrow{b} (X,\xi) \xrightarrow{\mathrm{id}} X$ is a smooth, zero-dimensional correspondence from X to X. Since $\mathrm{id}\colon X \to X$ is obviously a submersion, the transversality condition amounts to saying that for every $x \in X$ which is fixed by b, the linear map $\mathrm{id} - D_x b\colon T_x X \to T_x X$ is non-singular. This is the classical condition. The coincidence space is of course the fixed point set of b, suitably K-oriented. This means that a sign is attached to each point, which can be checked to agree with the usual assignment. Thus the Lefschetz invariant is the algebraic fixed-point set.

Thus, when interpreted topologically, the Lefschetz map amounts to a simple and rather natural geometric procedure. Of course we could have simply *defined* the Lefschetz map this way. However, proving its functoriality using the direct definition would be difficult. In our framework, the functoriality properties of Lefschetz invariants follow from the functoriality properties of duals.

One can go further, and look for a *homological* interpretation of Lefschetz invariants, as in the classical situation. If $\mathcal{G}$ is a compact group, X a compact $\mathcal{G}$-space, and $\Psi \in \mathrm{KK}^{\mathcal{G}}_*(\mathcal{C}(X), \mathcal{C}(X))$ is a morphism, then $\mathrm{Lef}(\Psi) \in \mathrm{KK}^{\mathcal{G}}_*(\mathcal{C}(X), \mathbb{C})$ has an equivariant index in $\mathrm{KK}^{\mathcal{G}}_*(\mathbb{C}, \mathbb{C}) = \mathrm{Rep}(\mathcal{G})$. If $\mathcal{G}$ is trivial and Ψ is represented by a smooth self-map in general position, this is exactly the usual Lefschetz number $l(f)$ of f (= the number of algebraic fixed points). Since, by the Lefschetz fixed-point theorem, this integer equals the trace of the induced map on cohomology, one might hazard the conjecture that, for compact groups $\mathcal{G}$, a $\mathrm{Rep}(\mathcal{G})$-valued module trace on $\mathrm{Rep}(\mathcal{G})$-linear endomorphisms of $\mathrm{K}^*_{\mathcal{G}}(X)$ can be defined so as to fit into an equivariant Lefschetz fixed-point formula, thus equating the (graded) module trace of the module map $\Lambda_*\colon \mathrm{K}^*_{\mathcal{G}}(X) \to \mathrm{K}^*_{\mathcal{G}}(X)$ of a morphism $\Lambda \in \mathrm{KK}^{\mathcal{G}}_*(\mathcal{C}(X), \mathcal{C}(X))$, with the equivariant index of the Lefschetz invariant, in our sense.

The formulation and proof of such a statement is more difficult than it may first appear, however. The first issue is to make sense of the notion of *trace* in this context: the trace of an endomorphism of a module is only defined in the obvious way if the module is *free*, and $\mathrm{K}^*_{\mathcal{G}}(X)$ is a free module over $\mathrm{Rep}(\mathcal{G})$ only in special circumstances. The Hattori-Stallings trace is useful for these purposes, although one still has to restrict to compact *connected* groups, (with torsion-free fundamental group) whose representation rings have finite homological dimension in a suitable sense (contrast finite groups).

But the proof of the proposed 'Lefschetz theorem' becomes quite technical. The exact statement and proof will appear in a forthcoming article.

References

[1] M. F. Atiyah and I. M. Singer, *The index of elliptic operators. I*, Ann. of Math. (2) **87** (1968), 484–530.

[2] ———, *The index of elliptic operators. III*, Ann. of Math. (2) **87** (1968), 546–604. MR 0236952 (38 #5245)

[3] Paul Baum and Jonathan Block, *Equivariant bicycles on singular spaces*, C. R. Acad. Sci. Paris Sér. I Math. **311** (1990), no. 2, 115–120 (English, with French summary).

[4] ———, *Excess intersection in equivariant bivariant K-theory*, C. R. Acad. Sci. Paris Sér. I Math. **314** (1992), no. 5, 387–392 (English, with French summary).

[5] Paul Baum, Alain Connes, and Nigel Higson, *Classifying space for proper actions and K-theory of group C^*-algebras*, C^*-Algebras: 1943–1993 (San Antonio, TX, 1993), Contemp. Math., vol. 167, Amer. Math. Soc., Providence, RI, 1994, pp. 240–291.

[6] Paul Baum, Nigel Higson, and Thomas Schick, *On the equivalence of geometric and analytic K-homology*, Pure and Applied Mathematics Quarterly **3**, no. 1, 1–24. Special issue in honor of Robert MacPherson, Part 3 of 3.

[7] Peter Abramenko and Kenneth S. Brown, *Buildings*, Graduate Texts in Mathematics, vol. 248, Springer, New York, 2008. Theory and applications.

[8] A. Connes and G. Skandalis, *The longitudinal index theorem for foliations*, Publ. Res. Inst. Math. Sci. **20** (1984), no. 6, 1139–1183.

[9] Siegfried Echterhoff, Heath Emerson, and Hyun Jeong Kim, *KK-theoretic duality for proper twisted actions*, Math. Ann. **340** (2008), no. 4, 839–873.

[10] ———, *A Lefschetz fixed-point formula for certain orbifold C^*-algebras*, J. Noncomm.Geom. **4** (2009), no. 1, 125–155.

[11] Heath Emerson and Ralf Meyer, *Equivariant representable K-theory*, J.Topology (2008), no. 2, 123–156.

[12] ———, *Euler characteristics and Gysin sequences for group actions on boundaries*, Math. Ann. **334** (2006), no. 4, 853–904.

[13] ———, *Equivariant Lefschetz maps for simplicial complexes and smooth manifolds*, Math. Ann. **345** (2009), no. 3, 599–630.

[14] ———, *Equivariant embedding theorems and topological index maps*, Adv. Math. **225** (2010), 2840–2882.

[15] ———, *Bivariant K-theory via correspondences*, Adv. Math. **225** (2010), 2883–2919.

[16] ———, *Dualities in equivariant Kasparov theory*, New York J. Math. **16** (2010), 245–313.

[17] Heath Emerson and Robert Yuncken, *Equivariant correspondences and the Borel-Bott-Weil theorem* (2009), eprint.

[18] Nigel Higson and John Roe, *Analytic K-homology*, Oxford Mathematical Monographs, Oxford University Press, Oxford, 2000. Oxford Science Publications.

[19] Nigel Higson and Gennadi Kasparov, *E-theory and KK-theory for groups which act properly and isometrically on Hilbert space*, Invent. Math. **144** (2001), no. 1, 23–74.

[20] Gennadi G. Kasparov, *The operator K-functor and extensions of C^*-algebras*, Izv. Akad. Nauk SSSR Ser. Mat. **44** (1980), no. 3, 571–636, 719 (Russian); English transl., Math. USSR-Izv. **16** (1981), no. 3, 513–572 (1981).

[21] ———, *Equivariant KK-theory and the Novikov conjecture*, Invent. Math. **91** (1988), no. 1, 147–201.

[22] G. G. Kasparov and G. Skandalis, *Groups acting on buildings, operator K-theory, and Novikov's conjecture*, K-Theory **4** (1991), no. 4, 303–337.

[23] Pierre-Yves Le Gall, *Théorie de Kasparov équivariante et groupoïdes. I*, K-Theory **16** (1999), no. 4, 361–390 (French, with English and French summaries).

[24] Wolfgang Lück and Bob Oliver, *The completion theorem in K-theory for proper actions of a discrete group*, Topology **40** (2001), no. 3, 585–616.
[25] Wolfgang Lück and Jonathan Rosenberg, *The equivariant Lefschetz fixed point theorem for proper cocompact G-manifolds*, High-dimensional manifold topology, World Sci. Publ., River Edge, NJ, 2003, pp. 322–361.
[26] ———, *Equivariant Euler characteristics and K-homology Euler classes for proper cocompact G-manifolds*, Geom. Topol. **7** (2003), 569–613.
[27] George Daniel Mostow, *Equivariant embeddings in Euclidean space*, Ann. of Math. (2) **65** (1957), 432–446.
[28] Jeff Raven, *An equivariant bivariant Chern character*, PhD Thesis, the Pennsylvannia State University (2004). http://etda.libraries.psu.edu/theses/approved/WorldWideIndex/ETD-723/index.html.
[29] John Roe, *Elliptic operators, topology and asymptotic methods*, 2nd ed., Pitman Research Notes in Mathematics Series, vol. 395, Longman, Harlow, 1998.
[30] Juliane Sauer, *K-Theory for proper smooth actions of totally disconnected groups*, High-dimensional manifold topology, World Sci. Publ., River Edge, NJ, 2003, pp. 427–448.
[31] Georges Skandalis, *Kasparov's bivariant K-theory and applications*, Exposition. Math. **9** (1991), no. 3, 193–250.
[32] Jean-Louis Tu, *La conjecture de Baum-Connes pour les feuilletages moyennables*, K-Theory **17** (1999), no. 3, 215–264 (French, with English and French summaries).
[33] Jacques Tits, *Buildings of spherical type and finite BN-pairs*, Lecture Notes in Mathematics, Vol. 386, Springer-Verlag, Berlin, 1974.
[34] Jean-Louis Tu, *La conjecture de Novikov pour les feuilletages hyperboliques*, K-Theory **16** (1999), no. 2, 129–184 (French, with English and French summaries).
[35] Julia Weber, *The universal functorial equivariant Lefschetz invariant*, K-Theory **36** (2006), no. 1–2, 169–207.

Fields Institute Communications
Volume **61**, 2011

Twisted Spectral Triples and Connes' Character Formula

Farzad Fathizadeh
Department of Mathematics and Statistics
York University
Toronto, ON Canada M3J 1P3
ffathiza@mathstat.yorku.ca

Masoud Khalkhali
Department of Mathematics
The University of Western Ontario
London, ON Canada N6A 5B7
masoud@uwo.ca

Abstract. We give a proof of an analogue of Connes' Hochschild character theorem for twisted spectral triples obtained from twisting a spectral triple by scaling automorphisms, under some suitable conditions. We also survey some of the properties of twisted spectral triples that are known so far.

1 Introduction

In [8] Connes and Moscovici introduced a refinement of the notion of spectral triple called *twisted spectral triple*. As is well known by now, the metric aspects of a geometric space in noncommutative geometry are encoded by a *spectral triple* [5]. This concept however puts rather severe restrictions on the underlying noncommutative space, as it forces the existence of a tracial state. In particular it is inapplicable in *type III situations* according to Murray-von Neumann's classification of von Neumann algebras into types. It was to deal with these type III examples that twisted spectral triples were introduced.

In a twisted spectral triple, the twist is afforded by an *automorphism*. More precisely, there is an automorphism σ of the underlying noncommutative space $\mathcal{A}$ such that all twisted commutators $[D, a]_\sigma = Da - \sigma(a)D$ are bounded operators (D is a selfadjoint operator playing the role of the Dirac operator). While there are many natural examples of twisted spectral triples, proving general results about them, parallel to the untwisted case, has proved to be a challenging task. The most celebrated among such questions is perhaps the extension of Connes' character formula for the Hochschild class of the Connes-Chern character of a spectral triple [5], as well as an extension of Connes-Moscovici's local index formula, [7] to twisted

2010 *Mathematics Subject Classification.* Primary 58J42; Secondary 58J22.

spectral triples. In this paper we give a proof of a character formula for twisted spectral triples obtained from twisting a spectral triple by scaling automorphisms, under suitable conditions. We shall also briefly outline Moscovici's local index formula for twisted spectral triples obtained by conformal perturbation of a spectral triple [17].

This paper is organized as follows. In Section 2 we recall the basic notions of spectral triples, noncommutative integration and residue functionals for a spectral triple. We also recall Connes and Moscovici's local index formula [7]. We then recall the notion of twisted spectral triples [8]. In Section 3 we recall several general methods to construct twisted spectral triples from [8], [17] and [11]. An important idea here is twisting a spectral triple by scaling automorphisms and the corresponding algebra of twisted pseudodifferential operators introduced by Moscovici [17]. In Section 4 we recall some non-obvious properties of twisted spectral triples from [8], among them the fact that one can define a Connes-Chern character for twisted spectral triples with values in ordinary non-twisted cyclic cohomology. Equivalently one can define a pairing between a finitely summable twisted spectral triple and K-theory. In section 5 we give a proof of an analogue of Connes' Hochschild character theorem for twisted spectral triples obtained from twisting a spectral triple by scaling automorphisms. Finally, in the last section we recall Moscovici's Ansatz and proof of the local index formula for twisted spectral triples obtained by scaling automorphisms [17].

2 From spectral triples to twisted spectral triples

The notion of a geometric space in noncommutative geometry is encoded by a *spectral triple.* This concept however puts rather severe restrictions on the underlying noncommutative space that renders it inapplicable in *type III situations* according to the Murray-von Neumann classification of von Neumann algebras into types. We shall first try to explain this point.

We start with a quick review of the *Dixmier trace* and the *noncommutative integral*, following closely [5]. Let $\mathcal{H}$ be a Hilbert space and let $\mathcal{K}(\mathcal{H})$ denote the two-sided ideal of compact operators on $\mathcal{H}$. For a compact operator $T : \mathcal{H} \to \mathcal{H}$, let

$$\mu_1(T) \geq \mu_2(T) \geq \cdots \geq 0$$

denote the sequence of eigenvalues of its absolute value $|T| := (T^*T)^{\frac{1}{2}}$, written in decreasing order. Thus, by the minimax principle, $\mu_1(T) = ||T||$, and in general

$$\mu_n(T) = \inf ||T|_V||, \quad n \geq 1,$$

where the infimum is over the set of subspaces of codimension $n-1$, and $T|_V$ denotes the restriction of T to the subspace V. A compact operator $T \in \mathcal{K}(\mathcal{H})$ is called trace class if $\sum \mu_n(T) < \infty$, and in that case the trace of T is defined by

$$\text{Trace}(T) = \sum_n (Te_n, e_n)$$

where $\{e_n\}_{n=1}^{\infty}$ is an orthonormal basis for $\mathcal{H}$. It is known that this is the unique normal trace on the C^*-algebra $\mathcal{L}(\mathcal{H})$ of bounded operators on $\mathcal{H}$. The question of existence of non-normal traces on $\mathcal{L}(\mathcal{H})$ was left open until it was settled affirmatively by Jacques Dixmier. In [9] Dixmier shows that there are uncountably many non-normal traces on $\mathcal{L}(\mathcal{H})$. Many years later, Alain Connes discovered that these non-normal traces can in fact be used to define a process of noncommutative integration in noncommutative geometry, as we describe next.

The natural domain of a Dixmier trace is the set of operators

$$\mathcal{L}^{1,\infty}(\mathcal{H}) = \{T \in \mathcal{K}(\mathcal{H}); \quad \sum_1^N \mu_n(T) = O\,(\log N)\}.$$

Notice that trace class operators are automatically in $\mathcal{L}^{1,\infty}(\mathcal{H})$. The Dixmier trace of an operator $T \in \mathcal{L}^{1,\infty}(\mathcal{H})$ measures the *logarithmic divergence* of its ordinary trace. More precisely, we are interested in the *limit* of the bounded sequence

$$\sigma_N(T) = \frac{\sum_1^N \mu_n(T)}{\log N}, \quad N = 1, 2, \ldots$$

as $N \to \infty$. The first problem of course is that, while by our assumption the sequence is bounded, the usual limit may not exist and must be replaced by a *generalized limit*, similar to *Banach limits* of non-convergent bounded sequences. A more challenging task is to make sure that our generalized limit still defines a trace.

To this end, let $\text{Trace}_\Lambda(T), \Lambda \in [1, \infty)$, be the piecewise affine interpolation of the partial trace function $\text{Trace}_N(T) = \sum_1^N \mu_n(T)$. Recall that a state on a C^*-algebra is a non-zero positive linear functional on the algebra. Let $\omega : C_b[e, \infty) \to \mathbb{C}$ be a normalized state on the algebra of bounded continuous functions on $[e, \infty)$ such that $\omega(f) = 0$ for all f vanishing at ∞. Now, using ω, the Dixmier trace of a positive operator $T \in \mathcal{L}^{1,\infty}(\mathcal{H})$ is defined as

$$\text{Tr}_\omega(T) := \omega(\tau_\Lambda(T)),$$

where

$$\tau_\Lambda(T) = \frac{1}{\log \Lambda} \int_e^\Lambda \frac{\text{Trace}_r(T)}{\log r} \frac{dr}{r}$$

is the Cesáro mean of the function $\frac{\text{Trace}_r(T)}{\log r}$ over the multiplicative group $\mathbb{R}_+^*$. One then extends Tr_ω to all of $\mathcal{L}^{1,\infty}(\mathcal{H})$ by linearity.

The resulting linear functional Tr_ω is a positive trace on $\mathcal{L}^{1,\infty}(\mathcal{H})$. It is easy to see from its definition that if T actually happens to be a trace class operator then $\text{Tr}_\omega(T) = 0$ for all ω, i.e., the Dixmier trace is invariant under perturbations by trace class operators. This is a very useful property and makes Tr_ω a flexible tool in computations.

The Dixmier trace, Tr_ω, in general depends on the limiting procedure ω; however, for the class of operators T for which $\text{Lim}_{\Lambda \to \infty} \tau_\Lambda(T)$ exists, it is independent of the choice of ω and is equal to $\text{Lim}_{\Lambda \to \infty} \tau_\Lambda(T)$. One of the main results proved in [4] is that if M is a closed n-dimensional manifold, E is a smooth vector bundle on M, P is a pseudodifferential operator of order $-n$ acting between L^2-sections of E, and $\mathcal{H} = L^2(M, E)$, then $P \in \mathcal{L}^{1,\infty}(\mathcal{H})$ and, for any choice of ω, $\text{Tr}_\omega(P) = n^{-1}\text{Res}(P)$. Here Res denotes Wodzicki's noncommutative residue [19, 15]. For example, if D is an elliptic first order differential operator, $|D|^{-n}$ is a pseudodifferential operator of order $-n$, and the Dixmier trace $\text{Tr}_\omega(|D|^{-n})$ is independent of the choice of ω.

Next, we would like to explain the notion of spectral triple. This concept has its roots in K-homology and Riemannian geometry simultaneously. We start by explaining the notion of a Fredholm module, which is the conformal counterpart of a spectral triple.

An *odd Fredholm module* over a unital algebra $\mathcal{A}$ is a pair $(\mathcal{H}, F)$, where $\mathcal{H}$ is a Hilbert space on which the algebra $\mathcal{A}$ acts by bounded operators and $F \in \mathcal{L}(\mathcal{H})$ is a selfadjoint operator such that $F^2 = \text{id}$, and such that the commutators $[F, \pi(a)]$ are compact operators for all $a \in \mathcal{A}$. Here $\pi : \mathcal{A} \to \mathcal{L}(\mathcal{H})$ denotes a unital action of $\mathcal{A}$ on $\mathcal{H}$. A Fredholm module is called *p-summable* $(1 \leq p < \infty)$, if $[F, \pi(a)] \in \mathcal{L}^p(\mathcal{H})$ for all $a \in \mathcal{A}$, where $\mathcal{L}^p(\mathcal{H})$ is the *Schatten ideal* of p-summable compact operators [3]. Fredholm modules should be thought of as representing K-homology classes defined by abstract elliptic partial differential operators on the noncommutative space $\mathcal{A}$.

Spectral triples provide a refinement of Fredholm modules. Going from Fredholm modules to spectral triples is akin to going from the *conformal class* of a Riemannian metric to the metric itself. Spectral triples simultaneously provide a notion of *Dirac operator* in noncommutative geometry, as well as a Riemannian type *distance function* for noncommutative spaces, as we shall explain next.

To motivate the definition of a spectral triple, we recall that the Dirac operator D on a compact Riemannian Spin^c manifold acts as an unbounded selfadjoint operator on the Hilbert space $L^2(M, S)$ of L^2-spinors on the manifold M. If we let $C^\infty(M)$ act on $L^2(M, S)$ by multiplication operators, then one can check that for any smooth function f, the commutator $[D, f] = Df - fD$ extends to a bounded operator on $L^2(M, S)$. Now the *geodesic distance* d on M can be recovered from the following beautiful *distance formula* of Connes [5]:

$$d(p, q) = \text{Sup}\{|f(p) - f(q)|; \quad \|[D, f]\| \leq 1\}, \qquad \forall p, q \in M.$$

The triple $(C^\infty(M), L^2(M, S), D)$ is a commutative example of a spectral triple. Its general definition, in the odd case, is as follows.

Definition 2.1 Let $\mathcal{A}$ be a unital algebra. An odd spectral triple on $\mathcal{A}$ is a triple $(\mathcal{A}, \mathcal{H}, D)$ consisting of a Hilbert space $\mathcal{H}$, a selfadjoint unbounded operator $D : \text{Dom}(D) \subset \mathcal{H} \to \mathcal{H}$ with compact resolvent, i.e., $(D+\lambda)^{-1} \in \mathcal{K}(\mathcal{H})$ for all $\lambda \notin \mathbb{R}$, and a unital representation $\pi : \mathcal{A} \to \mathcal{L}(\mathcal{H})$ of $\mathcal{A}$ such that for all $a \in \mathcal{A}$, the commutator $[D, \pi(a)]$ is defined on $\text{Dom}(D)$ and extends to a bounded operator on $\mathcal{H}$.

The finite summability assumption for Fredholm modules has a finer analogue for spectral triples. For simplicity we shall assume that D is invertible (in general, since $\text{Ker}(D)$ is finite dimensional, by restricting to its orthogonal complement we can always reduce to this case). A spectral triple is called *finitely summable* if for some $n \geq 1$,

$$|D|^{-n} \in \mathcal{L}^{1,\infty}(\mathcal{H}). \tag{2.1}$$

A simple example of an odd spectral triple is $(C^\infty(S^1), L^2(S^1), D)$, where $S^1 = \mathbb{R}/2\pi\mathbb{Z}$ is the circle and D is the unique selfadjoint extension of the operator $\frac{1}{i}\frac{d}{dx}$. The eigenvalues of $|D|$ are $|n|; n \in \mathbb{Z}$ which shows that if we restrict D to the orthogonal compliment of constant functions then $|D|^{-1} \in \mathcal{L}^{1,\infty}(L^2(S^1))$.

Given a spectral triple $(\mathcal{A}, \mathcal{H}, D)$, one can obtain a Fredholm module $(\mathcal{A}, \mathcal{H}, F)$ by choosing $F = \text{Sign}\,(D) = D|D|^{-1}$. Connes' Hochschild character formula gives a local expression for the Hochschild class of the Connes-Chern character of $(\mathcal{A}, \mathcal{H}, F)$ in terms of D itself. For this one has to assume that the spectral triple $(\mathcal{A}, \mathcal{H}, D)$ is *regular* in the sense that for all $a \in \mathcal{A}$,

$$a \text{ and } [D, a] \in \cap\, \text{Dom}(\delta^k)$$

where the derivation δ is given by $\delta(x) = [|D|, x]$.

Now, assuming (2.1) holds, Connes defines an $(n+1)$-linear functional φ on $\mathcal{A}$ by

$$\varphi(a^0, a^1, \ldots, a^n) = \mathrm{Tr}_\omega(a^0[D, a^1] \cdots [D, a^n]|D|^{-n}).$$

It can be shown that φ is a Hochschild n-cocycle on $\mathcal{A}$. We recall that a Hochschild n-cycle $c \in Z_n(A, A)$ is an element $c = \sum a^0 \otimes a^1 \otimes \cdots \otimes a^n \in A^{\otimes(n+1)}$ such that its Hochschild boundary $b(c) = 0$, where b is defined by (5.4). The following result, known as *Connes' Hochschild character formula*, computes the Hochschild class of the Chern character by a local formula, i.e., in terms of φ:

Theorem 2.1 [5] *Let $(\mathcal{A}, \mathcal{H}, D)$ be a regular spectral triple. Let $F = Sign(D)$ denote the sign of D and $\tau_n \in HC^n(\mathcal{A})$ denote the Connes-Chern character of $(\mathcal{H}, F)$. For every n-dimensional Hochschild cycle $c = \sum a^0 \otimes a^1 \otimes \cdots \otimes a^n \in Z_n(\mathcal{A}, \mathcal{A})$, one has*

$$\langle \tau_n, c \rangle = \sum \varphi(a^0, a^1, \ldots, a^n).$$

Identifying the full cyclic cohomology class of the Connes-Chern character of $(\mathcal{A}, \mathcal{H}, D)$ by a local formula is the content of Connes-Moscovici's local index formula. For this we have to assume the spectral triple satisfies another technical condition. Let $\mathcal{B}$ denote the subalgebra of $\mathcal{L}(\mathcal{H})$ generated by operators $\delta^k(a)$ and $\delta^k([D, a])$, $k \geq 1$. A spectral triple is said to have a discrete *dimension spectrum* Σ if $\Sigma \subset \mathbb{C}$ is discrete and for any $b \in \mathcal{B}$ the function

$$\zeta_b(z) = \mathrm{Trace}(b|D|^{-z}), \qquad \mathrm{Re}\, z > n,$$

extends to a holomorphic function on $\mathbb{C} \setminus \Sigma$. It is further assumed that Σ is *simple* in the sense that $\zeta_b(z)$ has only simple poles in Σ.

The local index formula of Connes and Moscovici [7] is given by the following theorem (we have used the formulation in [6]):

Theorem 2.2 [6] *1. The equality*

$$⨍ P = \mathrm{Res}_{z=0}\, \mathrm{Trace}(P|D|^{-z})$$

defines a trace on the algebra generated by $\mathcal{A}$, $[D, \mathcal{A}]$, and $|D|^z$, $z \in \mathbb{C}$.
2. There are only a finite number of non-zero terms in the following formula which defines the odd components $(\varphi_n)_{n=1,3,\ldots}$ of an odd cyclic cocycle in the (b, B) bicomplex of $\mathcal{A}$: For each odd integer n let

$$\varphi_n(a^0, \ldots, a^n) := \sum_k c_{n,k} ⨍ a^0[D, a^1]^{(k_1)} \cdots [D, a^n]^{(k_n)}|D|^{-n-2|k|}$$

where $T^{(k)} := \nabla^k$ and $\nabla(T) = D^2T - TD^2$, k is a multi-index, $|k| = k_1 + \cdots + k_n$ and

$$c_{n,k} := (-1)^{|k|}\sqrt{2i}(k_1! \ldots k_n!)^{-1}((k_1 + 1) \cdots (k_1 + k_2 + \cdots k_n))^{-1}\Gamma(|k| + \frac{n}{2}).$$

3. The pairing of the cyclic cohomology class $(\varphi_n) \in HC^*(\mathcal{A})$ with $K_1(\mathcal{A})$ gives the Fredholm index of D with coefficients in $K_1(\mathcal{A})$.

Given an n-summable regular spectral triple $(\mathcal{A}, \mathcal{H}, D)$, the linear functional

$$a \mapsto \mathrm{Tr}_\omega(a|D|^{-n})$$

defines a trace on the algebra $\mathcal{A}$ (*cf.* Proposition 4.1). Thus to deal with type III algebras which carry no non-trivial traces, the notion of spectral triple must be modified. In [8] Connes and Moscovici define a notion of *twisted spectral triple*, where the twist is afforded by an algebra automorphism (related to the modular automorphism group). More precisely, one postulates that there exists an automorphism σ of $\mathcal{A}$ such that the twisted commutators $Da - \sigma(a)D$ are bounded operators for all $a \in \mathcal{A}$. Here is the full definition:

Definition 2.1 Let $\mathcal{A}$ be an algebra which is represented by bounded operators in a Hilbert space $\mathcal{H}$, and D be an unbounded selfadjoint operator in $\mathcal{H}$ with compact resolvent. With σ being an automorphism of $\mathcal{A}$, the triple $(\mathcal{A}, \mathcal{H}, D)$ is called a twisted spectral triple or a σ-spectral triple if for any $a \in \mathcal{A}$, the twisted commutator

$$[D, a]_\sigma := Da - \sigma(a)D$$

is defined on the domain of D, and extends to a bounded operator in $\mathcal{H}$.
A twisted spectral triple is said to be Lipschitz-regular if the twisted commutators $|D|a - \sigma(a)|D|$ are bounded as well for all $a \in \mathcal{A}$.
A graded twisted spectral triple is one that is endowed with a grading operator $\gamma = \gamma^* \in \mathcal{L}(\mathcal{H})$ such that $\gamma^2 = \text{id}$, γ commutes with the action of $\mathcal{A}$, and anticommutes with D.

When the algebra $\mathcal{A}$ is involutive, the representation is assumed to be involutive as well. In this case, it is natural to impose the following compatibility condition between the automorphism and the involution:

$$\sigma(a)^* = \sigma^{-1}(a^*), \ \ \forall a \in \mathcal{A}.$$

It is shown in [8] that in the twisted case, the Dixmier trace induces a twisted trace on the algebra $\mathcal{A}$, but surprisingly, under some regularity conditions, the Connes-Chern character of the phase space lands in ordinary cyclic cohomology. Thus its pairing with ordinary K-theory makes sense, and it can be recovered as the index of Fredholm operators. This suggests the significance of developing a local index formula for twisted spectral triples, *i.e.* finding a formula for a cocycle, cohomologous to the Connes-Chern character in the (b, B)-bicomplex, which is given in terms of twisted commutators and residue functionals. We believe that this new theme of twisted spectral triples, and type III noncommutative geometry in general, shall dominate the subject in the near future.

For example, very recently a local index formula has been proved for a class of twisted spectral triples by Henri Moscovici [17]. We will discuss this result in detail in Section 6 of this paper. This class is obtained by twisting an ordinary spectral triple $(\mathcal{A}, \mathcal{H}, D)$ by a subgroup G of *scaling automorphisms* of the triple, *i.e.* the set of all unitary operators $U \in \mathcal{U}(\mathcal{H})$ such that $U\mathcal{A}U^* = \mathcal{A}$, and $UDU^* = \mu(U)D$, with $\mu(U) > 0$. It is shown that the crossed product algebra $\mathcal{A} \rtimes G$ admits an automorphism σ, given by the formula $\sigma(aU) = \mu(U)^{-1}aU$, for all $a \in \mathcal{A}, U \in G$; and $(\mathcal{A} \rtimes G, \mathcal{H}, D)$ is a twisted spectral triple.

3 Examples of twisted spectral triples

In this section, we recall general methods to construct twisted spectral triples.

3.1 Perturbing spectral triples. In [8], it is shown that starting from an ordinary spectral triple $(\mathcal{A}, \mathcal{H}, D)$ and a selfadjoint element $h = h^* \in \mathcal{A}$, the perturbed triple

$$(\mathcal{A}, \mathcal{H}, D'), \ \ D' = e^h D e^h$$

is a σ-spectral triple, where the automorphism σ is given by

$$\sigma(a) = e^{2h} a e^{-2h}, \quad a \in \mathcal{A}.$$

In fact, for any $a \in \mathcal{A}$, one has

$$D'a - \sigma(a)D' = e^h D e^h a - e^{2h} a e^{-2h} e^h D e^h = e^h [D, e^h a e^{-h}] e^h.$$

Therefore the twisted commutators $[D', a]_\sigma$ are bounded since the commutators $[D, b]$ are bounded operators for all $b \in \mathcal{A}$. A concrete example of this construction is obtained when one compares the Dirac operators of conformally equivalent Riemannian metrics as follows:

Example 3.1 Let (M, g) be a compact Riemannian Spin manifold and $D = D^g$ be the associated Dirac operator acting on the Hilbert space of L^2-spinors $\mathcal{H} = L^2(M, S^g)$. Let $g' = e^{-4h} g$ be a conformally equivalent metric, where $h \in C^\infty(M)$ is a selfadjoint element. It can be shown that the gauge transform operator ${}^g D^{g'} = \beta_g^{g'} \circ D^{g'} \circ \beta_{g'}^g$ has the form

$${}^g D^{g'} = e^{(n+1)h} \circ D^g \circ e^{-(n-1)h}.$$

After the canonical identification of the space of g-spinors with g'-spinors, it can be seen [8] that the gauge transformed spectral triple is obtained from the original one by replacing D with

$$D' = e^h D e^h.$$

More generally, one can start from a σ-spectral triple $(\mathcal{A}, \mathcal{H}, D)$ and a selfadjoint element $h = h^* \in \mathcal{A}$, and investigate the properties of the perturbed triple

$$(\mathcal{A}, \mathcal{H}, D'), \quad D' = e^h D e^h.$$

Lemma 3.2 *Let $(\mathcal{A}, \mathcal{H}, D)$ be a σ-spectral triple, and $h = h^* \in \mathcal{A}$. Then the perturbed triple*

$$(\mathcal{A}, \mathcal{H}, D'), \quad D' = e^h D e^h$$

is a σ'-spectral triple where $\sigma' \in Aut(\mathcal{A})$ is given by

$$\sigma'(a) = e^h \sigma(e^h a e^{-h}) e^{-h}, \quad a \in \mathcal{A}.$$

Proof For any $a \in \mathcal{A}$, one has

$$D'a - \sigma'(a)D' = e^h D e^h a - e^h \sigma(e^h a e^{-h}) e^{-h} e^h D e^h = e^h [D, e^h a e^{-h}]_\sigma e^h.$$

Therefore the twisted commutators $[D', a]_{\sigma'}$ are bounded since the twisted commutators $[D, b]_\sigma$ are bounded operators for all $b \in \mathcal{A}$. □

3.2 Twisted connections on crossed product algebras. In [11], a method for constructing twisted spectral triples on crossed product algebras is suggested. This method uses 1-cocycles to construct automorphisms for crossed product algebras, and explains how one can obtain twisted traces, twisted derivatives, and twisted connections on these algebras. We shall explain this method and show that it can reconstruct an example of twisted spectral triples, first given by Connes and Moscovici in [8]. First we recall some definitions:

Definition 3.3 Let $\sigma : \mathcal{A} \to \mathcal{A}$ be an automorphism of an algebra $\mathcal{A}$.

1. A σ-derivation on $\mathcal{A}$ is a linear map $\delta : \mathcal{A} \to \mathcal{A}$ such that

$$\delta(ab) = \delta(a)b + \sigma(a)\delta(b), \qquad \forall a, b \in \mathcal{A}.$$

2. A σ-trace on $\mathcal{A}$ is a linear map $\tau : \mathcal{A} \to \mathbb{C}$ such that

$$\tau(ab) = \tau(\sigma(b)a), \qquad \forall a, b \in \mathcal{A}.$$

Now, let A be an algebra with a right action of a group Γ by automorphisms:

$$A \times \Gamma \to A, \qquad (a, \gamma) \mapsto a \cdot \gamma.$$

We consider the algebraic crossed product $A \rtimes \Gamma$ with the standard multiplication:

$$(a \otimes \gamma)(b \otimes \mu) = (a \cdot \mu)b \otimes \gamma\mu, \qquad a, b \in A, \ \gamma, \mu \in \Gamma.$$

Let $Z(A)$ denote the center of the algebra A, and A^* its group of invertible elements.

Definition 3.4 1. A map $j : \Gamma \to Z(A) \cap A^*$ is a 1-cocycle if

$$j(\gamma\mu) = ((j\gamma) \cdot \mu) j\mu, \qquad \forall \gamma, \mu \in \Gamma.$$

2. Given a map $j : \Gamma \to A$, a linear functional $\tau : A \to \mathbb{C}$ is said to have the change of variable property with respect to j if

$$\tau((a \cdot \gamma) j\gamma) = \tau(a), \qquad \forall a \in A, \ \gamma \in \Gamma.$$

Notice that (3.2) amounts to saying that j is a (multiplicative) group 1−cocycle for $H^1(\Gamma, Z(A) \cap A^*)$.

Proposition 3.5 [11]*Let A be an algebra with a right action of a group Γ by automorphisms, and $j : \Gamma \to Z(A) \cap A^*$ be a 1-cocycle.*

1. *The map $\sigma : A \rtimes \Gamma \to A \rtimes \Gamma$ given by*

$$\sigma(a \otimes \gamma) = \left((j\gamma^{-1}) \cdot \gamma\right) a \otimes \gamma$$

is an automorphism.

2. *Let $\delta : A \to A$ be a derivation such that $\delta(a \cdot \gamma) = (\delta(a) \cdot \gamma) j\gamma$ for all $a \in A, \gamma \in \Gamma$. Then for any $s = 1, 2, \ldots,$ the map $\delta'_s : A \rtimes \Gamma \to A \rtimes \Gamma$ defined by*

$$\delta'_s(a \otimes \gamma) = \left(\delta\left((a \cdot \gamma^{-1})(j\gamma^{-1})^s\right)(j\gamma^{-1})^{-s} \otimes 1\right)\left(1 \otimes \gamma\right)$$

is a σ-derivation on $A \rtimes \Gamma$. Also, $\delta'_s \circ \sigma = \sigma \circ \delta'_s$ if $\delta \circ j = 0$.

3. *If $\tau : A \to \mathbb{C}$ is a trace such that $\tau \circ \delta = 0$, then $\tau' \circ \delta'_s = 0$, where the linear functional $\tau' : A \rtimes \Gamma \to \mathbb{C}$ is defined by*

$$\tau'(a \otimes \gamma) = 0 \ if \ \gamma \neq 1, \ \ and \ \ \tau'(a \otimes 1) = \tau(a).$$

Also, if τ has the change of variable property with respect to j, then τ' is a σ-trace on $A \rtimes \Gamma$.

Example 3.6 1. Let M be a smooth oriented manifold and ω a volume form on M. Let $\Gamma = Diff(M)$ be the group of diffeomorphisms of M. The map $j : \Gamma \to C^\infty(M)$ defined by

$$\gamma^*(\omega) = j(\gamma)\omega$$

is easily seen to be a 1-cocycle.

2. Let $\chi : \Gamma \to \mathbb{C}^*$ be a 1-dimensional character of a group Γ which acts on an algebra A by automorphisms, and let $j(\gamma) = \chi(\gamma)1_A$. Then j is a 1-cocycle, and a derivation $\delta : A \to A$ is compatible with j if and only if $\delta(a \cdot \gamma) = \chi(\gamma)\, \delta(a) \cdot \gamma$ for any $a \in A$, and $\gamma \in \Gamma$.

Definition 3.7 Let $\mathcal{A}$ be an algebra, $\delta : \mathcal{A} \to \mathcal{A}$ a derivation, and E a left $\mathcal{A}$-module. A linear map $\nabla : E \to E$ is said to be a connection if it satisfies the Leibniz rule, i.e.

$$\nabla(a\xi) = \delta(a)\xi + a\nabla(\xi), \ \ \forall a \in \mathcal{A}, \ \ \xi \in E.$$

If $\sigma : \mathcal{A} \to \mathcal{A}$ is an automorphism and $\delta : \mathcal{A} \to \mathcal{A}$ a σ-derivation, then a linear map $\nabla : E \to E$ is said to be a twisted connection if it satisfies the twisted Leibniz rule:

$$\nabla(a\xi) = \delta(a)\xi + \sigma(a)\nabla(\xi), \ \ \forall a \in \mathcal{A}, \ \ \xi \in E.$$

This notion of twisted connection comes from [18].

Given a twisted connection ∇, one can try to define a twisted spectral triple by letting $D = \nabla$. Then $\nabla a - \sigma(a)\nabla = \delta(a)$ shows that the boundedness condition is satisfied provided $\delta(a)$ acts by a bounded operator.

In the following proposition, A is an algebra endowed with a right action of a group Γ by automorphisms, with representations $\pi : A \to \mathrm{End}\,(E)$, and $\rho : \Gamma \to \mathrm{GL}(E)$, defining a covariant system, i.e.

$$\pi(a \cdot \gamma) = \rho(\gamma^{-1})\pi(a)\rho(\gamma), \ \ \forall a \in A, \ \ \gamma \in \Gamma.$$

Then we obtain a representation $\pi' : A \rtimes \Gamma \to \mathrm{End}(E)$ given by

$$\pi'(a \otimes \gamma) = \rho(\gamma)\pi(a), \ \ \forall a \in A, \ \ \gamma \in \Gamma.$$

Also, let $j : \Gamma \to Z(A) \cap A^*$ be a 1-cocycle and $\delta : A \to A$ a derivation as in Proposition 3.5. Therefore we have an automorphism $\sigma : A \rtimes \Gamma \to A \rtimes \Gamma$ and we fix a σ-derivation $\delta'_s : A \rtimes \Gamma \to A \rtimes \Gamma$ for some $s \in \mathbb{N}$.

Proposition 3.8 [11]*A connection $\nabla : E \to E$ for A is a twisted connection for $A \rtimes \Gamma$ if and only if*

$$\nabla\rho(\gamma) = \rho(\gamma)\Big((s-1)\,\pi\big(\delta(j\gamma^{-1} \cdot \gamma)\big) + \nabla\pi(j\gamma^{-1} \cdot \gamma)\Big), \ \ \forall\gamma \in \Gamma. \tag{3.2}$$

Remark 3.9 Let $s > 0$ be a real number and assume $\big(j(\gamma)\big)^s \in A$ is defined for all $\gamma \in \Gamma$. Propositions 3.5 and 3.8 continue to hold for these values of s as well. For this we need the extra condition $\delta(x^s) = sx^{s-1}\delta(x)$ to hold for all $x = j(\gamma)$, $\gamma \in \Gamma$.

Example 3.10 Let $C^\infty(S^1)$ be the algebra of smooth functions on the circle $S^1 = \mathbb{R}/\mathbb{Z}$, and $\Gamma \subset Diff^+(S^1)$ a group of orientation preserving diffeomorphisms of the circle as in [8]. We represent the algebra $C^\infty(S^1)$ by bounded operators in the Hilbert space $L^2(S^1)$ by

$$(\pi(g)\,\xi)(x) = g(x)\,\xi(x), \ \ \forall g \in C^\infty(S^1), \ \ \xi \in L^2(S^1), \ \ x \in \mathbb{R}/\mathbb{Z}.$$

Define a representation of Γ by bounded operators in $L^2(S^1)$ by

$$(\rho(\phi^{-1})\,\xi)(x) = \phi'(x)^{\frac{1}{2}}\,\xi(\phi(x)), \ \ \forall \phi \in \Gamma, \ \ \xi \in L^2(S^1), \ \ x \in \mathbb{R}/\mathbb{Z}.$$

The group Γ acts on $C^\infty(S^1)$ from the right by composition and one can easily check that the above representations give a covariant system which yields the representation of $C^\infty(S^1) \rtimes \Gamma$ as in [8]. The map $j : \Gamma \to C^\infty(S^1)$ defined by $j(\phi) = \phi'$ is a 1-cocycle and the derivation $\delta : C^\infty(S^1) \to C^\infty(S^1), \delta(f) = \frac{1}{i}f'$ is compatible with j. Now, by using Proposition 3.5, we obtain an automorphism σ of $C^\infty(S^1) \rtimes \Gamma$ which agrees with the automorphism in [8], and a twisted derivation $\delta'_{\frac{1}{2}}$. Note that since it is possible to take the square root of the elements in the image of j in this example, we can let $s = \frac{1}{2}$ to obtain a twisted derivation. Now if we let $\nabla = \frac{1}{i}\frac{d}{dx}$, one can see that the equality (3.2) holds, therefore ∇ is a twisted connection for $C^\infty(S^1) \rtimes \Gamma$ by Proposition 3.8.

Also, the linear map $\tau : C^\infty(S^1) \to \mathbb{C}$ defined by

$$\tau(g) = \int_{\mathbb{R}/\mathbb{Z}} g(x)\, dx, \quad \forall g \in C^\infty(S^1),$$

is a trace which has the change of variable property with respect to j, and $\tau \circ \delta = 0$. Therefore, by Proposition 3.5, one obtains a twisted trace $\tau' : C^\infty(S^1) \rtimes \Gamma \to \mathbb{C}$ such that $\tau' \circ \delta'_{\frac{1}{2}} = 0$.

3.3 Twisting spectral triples by scaling automorphisms. The last example of the preceding subsection gives a special case of a class of twisted spectral triples that arise naturally in conformal geometry [17]. Let (M, g) be a connected compact Riemannian Spin manifold of dimension n and $D = D_g$ be the associated Dirac operator acting on the Hilbert space of L^2-spinors $\mathcal{H} = L^2(M, S^g)$. Let $SCO(M, [g])$ denote the Lie group of diffeomorphisms of M that preserve the conformal structure $[g]$ consisting of all Riemannian metrics that are conformally equivalent to g, the orientation, and the Spin structure, and let $G = SCO(M, [g])_0$ denote the connected component of the identity. In [17], using a suitable automorphism of the crossed product algebra $C^\infty(M) \rtimes G$, a twisted spectral triple of the form $(C^\infty(M) \rtimes G, L^2(M, S^g), D)$ is constructed. Similarly, by endowing $\mathbb{R}^n$ with the Euclidean metric, and considering the group G of conformal transformations of $\mathbb{R}^n$, a twisted spectral triple is constructed over the crossed product algebra $C_c^\infty(\mathbb{R}^n) \rtimes G$. An abstract formulation of this class of twisted spectral triples leads in [17] to the idea of twisting an ordinary spectral triple by its *scaling automorphisms* or *conformal similarities*, which we explain in this subsection.

Using scaling automorphisms of a spectral triple $(\mathcal{A}, \mathcal{H}, D)$, one can construct a twisted spectral triple [17]. The set of scaling automorphisms, also called conformal similarities, of a spectral triple $(\mathcal{A}, \mathcal{H}, D)$, denoted by $\mathrm{Sim}(\mathcal{A}, \mathcal{H}, D)$, consists of all unitary operators U on $\mathcal{H}$ such that

$$U\mathcal{A}U^* = \mathcal{A}, \text{ and } UDU^* = \mu(U)D, \text{ for some } \mu(U) > 0.$$

It is easy to see that $\mathrm{Sim}(\mathcal{A}, \mathcal{H}, D)$ is a group and the map $\mu : \mathrm{Sim}(\mathcal{A}, \mathcal{H}, D) \to (0, \infty)$ is a character. We fix a subgroup $G \subset \mathrm{Sim}(\mathcal{A}, \mathcal{H}, D)$ and let $\mathcal{A}_G$ be the crossed product algebra $\mathcal{A} \rtimes G$. It is shown in [17] that the formula

$$\sigma(aU) = \mu(U)^{-1} aU, \quad \forall a \in \mathcal{A}, U \in G,$$

defines an automorphism of $\mathcal{A}_G$, and $(\mathcal{A}_G, \mathcal{H}, D)$ is a twisted spectral triple. In fact the twisted commutators

$$[D, aU]_\sigma := DaU - \sigma(aU)D = [D, a]U$$

are bounded operators for all $a \in \mathcal{A}, U \in G$.

For this class of twisted spectral triples, one can form the crossed product algebra $\Psi(\mathcal{A} \rtimes G, \mathcal{H}, D) := \Psi(\mathcal{A}, \mathcal{H}, D) \rtimes G$, where $\Psi(\mathcal{A}, \mathcal{H}, D)$ is the algebra of pseudodifferential operators associated to the base spectral triple [7, 13, 14], and under the *extended simple dimension spectrum hypothesis*, the residue functional

$\fint_D : \Psi(\mathcal{A}_G, \mathcal{H}, D) \to \mathbb{C}$ given by

$$\fint_D P := \mathrm{Res}_{z=0}\mathrm{Trace}(P|D|^{-2z})$$

is a trace [17]. This seems to be in agreement with the twisted analogue of the Adler-Manin trace [1, 16, 11]: For a triple (A, σ, δ) consisting of an algebra A, an algebra automorphism $\sigma : A \to A$, and a σ-derivation $\delta : A \to A$, we define an algebra of formal twisted pseudodifferential symbols $\Psi(A, \sigma, \delta)$ whose elements are formal series of the form

$$\sum_{n=-\infty}^{N} a_n \xi^n, \quad N \in \mathbb{Z},\, a_n \in A,\, \forall n \leq N.$$

The multiplication in this algebra is essentially derived from these relations

$$\xi a = \sigma(a)\xi + \delta(a), \forall a \in A, \;\; \xi\xi^{-1} = \xi^{-1}\xi = 1.$$

In [11], we prove that starting from a δ-invariant twisted trace on A, the noncommutative residue is a trace on $\Psi(A, \sigma, \delta)$:

Theorem 3.11 [11]*Let $\sigma : A \to A$ be an automorphism of an algebra A. If $\tau : A \to \mathbb{C}$ is a σ-trace and $\delta : A \to A$ is a σ-derivation such that $\tau \circ \delta = 0$, then the linear functional* $\mathrm{Res} : \Psi(A, \sigma, \delta) \to \mathbb{C}$ *defined by*

$$\mathrm{Res}\left(\sum_{i=-\infty}^{n} a_i \xi^i \right) = \tau(a_{-1})$$

is a trace on $\Psi(A, \sigma, \delta)$.

4 Properties of twisted spectral triples

In this section, we recall the basic properties of twisted spectral triples [8].

4.1 Twisted trace. Given a σ-spectral triple $(\mathcal{A}, \mathcal{H}, D)$ with $D^{-1} \in \mathcal{L}^{n,\infty}$, it is observed in [8] that

$$D^{-n}a - \sigma^{-n}(a)D^{-n} \in \mathcal{L}_0^{1,\infty}(\mathcal{H}), \;\; \forall a \in \mathcal{A}.$$

Here $\mathcal{L}_0^{1,\infty}(\mathcal{H})$ is the ideal

$$\{T \in \mathcal{K}(\mathcal{H}); \quad \sum_{i=0}^{N} \mu_i(T) = o(\log N)\},$$

on which the Dixmier trace Tr_ω vanishes. Then it follows that the Dixmier trace induces a twisted hypertrace on $\mathcal{A}$:

Proposition 4.1 [8]*Let $(\mathcal{A}, \mathcal{H}, D)$ be a σ-spectral triple with $D^{-1} \in \mathcal{L}^{n,\infty}$. Then the linear functional $\varphi : \mathcal{A} \to \mathbb{C}$ defined by*

$$\varphi(a) = \mathrm{Tr}_\omega(aD^{-n}), \;\; a \in \mathcal{A}$$

is a σ^n-hypertrace, i.e.

$$\mathrm{Tr}_\omega(TaD^{-n}) = \mathrm{Tr}_\omega(\sigma^n(a)TD^{-n})$$

for any $a \in \mathcal{A}, T \in \mathcal{L}(\mathcal{H})$. In particular φ is a twisted trace on $\mathcal{A}$. Note that for Lipschitz-regular twisted spectral triples, the same holds when D^{-n} is replaced by $|D|^{-n}$.

4.2 Connes-Chern character. Let $(\mathcal{A}, \mathcal{H}, D)$ be a Lipschitz-regular σ-spectral triple such that $D^{-1} \in \mathcal{L}^{n,\infty}(\mathcal{H})$ for some $n \in \mathbb{N}$. Here

$$\mathcal{L}^{n,\infty}(\mathcal{H}) = \{T \in \mathcal{K}(\mathcal{H}); \quad \sum_{i=0}^{N} \mu_i(T) = O(N^{1-\frac{1}{n}})\}, \text{ if } n > 1,$$

$$\mathcal{L}^{1,\infty}(\mathcal{H}) = \{T \in \mathcal{K}(\mathcal{H}); \quad \sum_{i=0}^{N} \mu_i(T) = O(\log N)\},$$

where, for any compact operator $T \in \mathcal{K}(\mathcal{H})$, its singular values are written in decreasing order: $\mu_1(T) \geq \mu_2(T) \geq \cdots \geq 0$.

We recall from [8] that passage to the phase $F = D|D|^{-1}$ of such a twisted spectral triple gives a finitely summable Fredholm module which has a well-defined Connes-Chern character in cyclic cohomology given by

$$\Phi_F(a_0, a_1, \ldots, a_n) = \mathrm{Trace}(\gamma F[F, a_0][F, a_1] \cdots [F, a_n]), \; a_0, a_1, \ldots, a_n \in \mathcal{A},$$

where $\gamma = \mathrm{id}$ in the ungraded case. Since, for any $a \in \mathcal{A}$

$$[F, a] = |D|^{-1}([D, a]_\sigma - [|D|, a]_\sigma F),$$

the commutators $[F, a_i]$ are differentials of the same order as D^{-1} and the operator

$$\gamma F[F, a_0][F, a_1] \cdots [F, a_n]$$

is a trace class operator. One can see that Φ_F is a cyclic cocycle, *i.e.*

$$\Phi_F(a_n, a_0, \ldots, a_{n-1}) = (-1)^n \Phi_F(a_0, a_1, \ldots, a_n), \; \forall a_0, a_1, \ldots, a_n \in \mathcal{A},$$

$$b\Phi_F = 0,$$

where b is the Hochschild coboundary operator:

$$b\Phi_F(a_0, a_1, \ldots, a_{n+1}) = \sum_{i=0}^{n} (-1)^i \Phi_F(a_0, \ldots, a_i a_{i+1}, \ldots, a_{n+1}) + (-1)^{n+1} \Phi_F(a_{n+1} a_0, a_1, \ldots, a_n),$$

for all $a_0, a_1, \ldots, a_{n+1} \in \mathcal{A}$.

Moreover, if $D^{-1} \in \mathcal{L}^{n,\infty}(\mathcal{H})$ for an even $n \in \mathbb{N}$, then the Connes-Chern character can be defined without the Lipschitz-regularity assumption:

Proposition 4.2 [8]*Let $(\mathcal{A}, \mathcal{H}, D)$ be a graded σ-spectral triple with $D^{-1} \in \mathcal{L}^{n,\infty}(\mathcal{H})$ for some even $n \in \mathbb{N}$. Then the multilinear functional $\Phi_{D,\sigma}$ defined by*

$$\Phi_{D,\sigma}(a_0, a_1, \ldots, a_n) = \mathrm{Trace}(\gamma D^{-1}[D, a_0]_\sigma D^{-1}[D, a_1]_\sigma \cdots D^{-1}[D, a_n]_\sigma) \quad (4.3)$$

for $a_0, \ldots, a_n \in \mathcal{A}$, is a cyclic cocycle.

4.3 Index pairing and K-theory. In [8], it is shown that in the twisted case, the index pairing with ordinary K-theory makes sense and is given by the pairing of the Connes-Chern character with K-theory.

Given a graded σ-spectral triple $(\mathcal{A}, \mathcal{H}, D)$ with $D^{-1} \in \mathcal{L}^{n,\infty}(\mathcal{H})$ for some even $n \in \mathbb{N}$, one can write an orthogonal decomposition for the Hilbert space $\mathcal{H}$ using the grading γ:

$$\mathcal{H} = \mathcal{H}_+ \oplus \mathcal{H}_-,$$

where

$$\mathcal{H}_+ = \{x \in \mathcal{H}; \quad \gamma(x) = x\}, \; \mathcal{H}_- = \{x \in \mathcal{H}; \quad \gamma(x) = -x\}.$$

Now a close look at the Connes-Chern character $\Phi_{D,\sigma}$ defined by (4.3) shows the existence of a pair of Fredholm modules over $\mathcal{A}$ and two cocycles as follows. With respect to the decomposition $\mathcal{H} = \mathcal{H}_+ \oplus \mathcal{H}_-$, we can write

$$D = \begin{bmatrix} 0 & D_- \\ D_+ & 0 \end{bmatrix}, \qquad a = \begin{bmatrix} a_+ & 0 \\ 0 & a_- \end{bmatrix}, \ \forall a \in \mathcal{A},$$

since we have $D\gamma = -\gamma D$, and $a\gamma = \gamma a$, for all $a \in \mathcal{A}$. Then define two Hilbert spaces

$$\widetilde{\mathcal{H}}^+ = \mathcal{H}_+ \oplus \mathcal{H}_+, \ \ \widetilde{\mathcal{H}}^- = \mathcal{H}_- \oplus \mathcal{H}_-.$$

There are representations π^+ and π^- of the algebra $\mathcal{A}$ on $\mathcal{L}(\widetilde{\mathcal{H}}^+)$ and $\mathcal{L}(\widetilde{\mathcal{H}}^-)$, respectively, given by

$$\pi^{\pm}(a) = \begin{bmatrix} a_{\pm} & 0 \\ 0 & D_{\pm}^{-1}\sigma(a)_{\mp}D_{\pm} \end{bmatrix}.$$

In [8], letting

$$F = \begin{bmatrix} 0 & I_{\pm} \\ I_{\pm} & 0 \end{bmatrix},$$

where I_+ and I_- are identity operators on $\mathcal{H}_+$ and $\mathcal{H}_-$ respectively, it is shown that the commutators $[F^{\pm}, \pi^{\pm}(a)]$ are compact operators, hence they obtain a pair of Fredholm modules over the algebra $\mathcal{A}$. It is also shown that for any idempotent $e \in \mathcal{A}$, the bounded closure of operators the $D_{\pm}^{-1}\sigma(e)_{\mp}D_{\pm}$, denoted by $f_{\pm}$, are idempotents, and $f_{\pm}e_{\pm} : e_{\pm}\mathcal{H}_{\pm} \to f_{\pm}\mathcal{H}_{\pm}$ are Fredholm operators. Since the index depends only on the K-theory class of the idempotent, they define a pair of index maps by

$$\mathrm{Index}^{\pm}[e] = \mathrm{Index}(f_{\pm}e_{\pm}),$$

for all idempotents $e \in M_N(\mathcal{A})$.

On the other hand, the cyclic cocycle $\Phi_{D,\sigma}$ defined by (4.3) consists of a pair of cocycles $\Phi^{\pm}_{D,\sigma}$ given by

$$\begin{aligned} &\Phi^{\pm}_{D,\sigma}(a^0, \dots, a^n) \\ &\quad = \mathrm{Trace}(D_{\pm}^{-1}(D_{\pm}a^0_{\pm} - \sigma(a^0)_{\mp}D_{\pm}) \cdots D_{\pm}^{-1}(D_{\pm}a^n_{\pm} - \sigma(a^n)_{\mp}D_{\pm})), \end{aligned}$$

for $a^0, \dots, a^n \in \mathcal{A}$.

The following proposition states that the index pairing can be expressed as the pairing of these cocycles with K-theory:

Proposition 4.3 [8]*Given a graded σ-spectral triple $(\mathcal{A}, \mathcal{H}, D)$ with $D^{-1} \in \mathcal{L}^{n,\infty}(\mathcal{H})$ for some even $n \in \mathbb{N}$, and an idempotent $e \in M_N(\mathcal{A})$, one has*

$$\mathrm{Index}^{\pm}[e] = \Phi^{\pm}_{D,\sigma}(e, \dots, e).$$

If $e^* = \sigma(e)$, then

$$\mathrm{Index}^{+}[e] = -\mathrm{Index}^{-}[e].$$

If there exists a strongly continuous 1-parameter group of isometric automorphisms $\{\sigma_t\}_{t\in\mathbb{R}}$ with an analytic extension coinciding with σ at $t = -i$, then the above index maps coincide. This assumption is denoted by (1PG). Such an analytic extension defined on a dense subalgebra $\mathcal{O}$ of $\mathcal{A}$ is assured to exist by a theorem of Bost in [2].

Proposition 4.4 [8]*If $\mathcal{A}$ satisfies (1PG), then*

$$\mathrm{Index}^{+}[e] = -\mathrm{Index}^{-}[e], \ \ \forall e \in M_N(\mathcal{A}).$$

Accordingly, in [8], the relation between the cyclic cocycles $\Phi^{\pm}_{D,\sigma}$ has been studied under the (1PG) assumption.

Theorem 4.5 [8]*Let $(\mathcal{A}, \mathcal{H}, D)$ be a graded σ-spectral triple as above and assume that $\mathcal{A}$ satisfies (1PG). Then*

$$[\Phi^{-}_{D,\sigma}] = -[(\Phi^{+}_{D,\sigma})^*] \in HP^{ev}(\mathcal{O}),$$

where

$$(\Phi^{+}_{D,\sigma})^*(a_0, \ldots, a_n) := \overline{\Phi^{+}_{D,\sigma}(a_n^*, \ldots, a_0^*)}, \quad \forall a_0, \ldots, a_n \in \mathcal{O}.$$

We note that in the proof of the latter [8], the homotopy invariance of the Connes-Chern character of a finitely summable Fredholm module, established in Lemma 1, in Part I, section 5 of [3] plays a crucial role.

4.4 Local Hochschild cocycle. In [8], as a first step to extend the local index formula [7, 13, 14] to twisted spectral triples, using the Dixmier trace, a local Hochschild cocycle is constructed for twisted spectral triples:

Proposition 4.6 [8]*Let $(\mathcal{A}, \mathcal{H}, D)$ be a graded σ-spectral triple such that $D^{-1} \in \mathcal{L}^{n,\infty}(\mathcal{H})$ for some even $n \in \mathbb{N}$. Then the $(n+1)$-linear form on $\mathcal{A}$ defined by*

$$\begin{aligned} &\Psi_{D,\sigma}(a_0, a_1, \ldots, a_n) \\ = \; &\mathrm{Tr}_\omega(\gamma a_0 [D, \sigma^{-1}(a_1)]_\sigma \cdots [D, \sigma^{-n}(a_n)]_\sigma |D|^{-n}) \end{aligned}$$

for $a_0, \ldots, a_n \in \mathcal{A}$, is a Hochschild cocycle.

In the ungraded case, for a Lipschitz-regular σ-spectral triple of odd summability degree, the Hochschild cocycle is given by

$$\begin{aligned} &\Psi_{D,\sigma}(a_0, a_1, \ldots, a_n) \\ = \; &\mathrm{Tr}_\omega(a_0 [D, \sigma^{-1}(a_1)]_\sigma \cdots [D, \sigma^{-n}(a_n)]_\sigma |D|^{-n}), \end{aligned}$$

for any $a_0, \ldots, a_n \in \mathcal{A}$.

The above cocycle identities are proved in [8] using Proposition 4.1 and the fact that for any $a, b \in \mathcal{A}$:

$$[D, ab]_\sigma = [D, a]_\sigma b + \sigma(a)[D, b]_\sigma.$$

In [8, 17], the above local Hochschild cocycle is obtained in a heuristic manner as follows. For an ordinary spectral triple $(\mathcal{A}, \mathcal{H}, D)$, consider the local Hochschild cocycle

$$\Psi_D(a_0, \ldots, a_n) = \mathrm{Tr}_\omega(\gamma a_0 [D, a_1] \cdots [D, a_n] D^{-n}),$$

for any $a_0, \ldots, a_n \in \mathcal{A}$. One can move D^{-n} to the left and distribute it among the factors to write this cocycle in the form

$$\Psi_D(a_0, \ldots, a_n) = \mathrm{Tr}_\omega(\gamma a_0 (D a_1 D^{-1} - a_1) \cdots (D^n a_n D^{-n} - D^{n-1} a_n D^{-n+1}) D^{-n}).$$

In order to construct a local Hochschild cocycle for twisted spectral triples, they replace each $D^k a D^{-k}$ in the latter by $D^k \sigma^{-k}(a) D^{-k}$ and reverse the process of moving the D^{-n} to the left, which leads to the above local Hochschild cocycle for twisted spectral triples. We note that the Connes-Chern character introduced in Proposition 4.2 can be obtained in a similar manner.

5 Hochschild class of the Connes-Chern character

For any algebra $\mathcal{A}$, there is an obvious pairing between the space of Hochschild n-cochains $C^n(\mathcal{A}, \mathcal{A}^*)$, *i.e.* the space of $(n+1)$-linear functionals on $\mathcal{A}$, and the space of Hochschild n-chains $\mathcal{A}^{\otimes(n+1)}$, given by

$$\langle \varphi, a_0 \otimes a_1 \otimes \cdots \otimes a_n \rangle := \varphi(a_0, \dots, a_n).$$

This pairing satisfies

$$\langle b\varphi, c \rangle = \langle \varphi, b(c) \rangle$$

where the Hochschild operators for the cochains and chains are given by

$$\begin{aligned} b\varphi(a_0, a_1, \dots, a_{n+1}) &= \sum_{i=0}^{n} (-1)^i \varphi(a_0, \dots, a_i a_{i+1}, \dots, a_{n+1}) + \\ &\quad (-1)^{n+1} \varphi(a_{n+1} a_0, a_1, \dots, a_n), \\ b(a_0 \otimes a_1 \otimes \cdots \otimes a_{n+1}) &= \sum_{i=0}^{n} (-1)^i a_0 \otimes \cdots \otimes a_i a_{i+1} \otimes \cdots \otimes a_{n+1} + \\ &\quad (-1)^{n+1} a_{n+1} a_0 \otimes a_1 \otimes \cdots \otimes a_n, \end{aligned} \tag{5.4}$$

for $a_0, \dots, a_{n+1} \in \mathcal{A}$. It follows that, if two cocycles φ_1 and φ_2 are cohomologous with $\varphi_1 - \varphi_2 = b\psi$, then they yield the same value on any Hochschild cycle, since we have

$$\langle \varphi_1, c \rangle - \langle \varphi_2, c \rangle = \langle b\psi, c \rangle = \langle \psi, b(c) \rangle = 0 \text{ if } b(c) = 0.$$

As we saw in Section 4, to any twisted spectral triple, in particular to an ordinary spectral triple, with certain conditions one can associate the Connes-Chern character and a local Hochschild cocycle. Connes' character formula (or Connes' Hochschild character theorem) states that in the case of ordinary spectral triples, these two cocycles yield the same value on any Hochschild cycle [5, 12, 14].

In this section the analogue of Connes' character formula for the class of twisted spectral triples introduced in Subsection 3.3 is investigated, which we explain next following [10].

5.1 Connes' character formula and twisting by scaling automorphisms. In Subsection 3.3, we saw that using conformal similarities of a spectral triple $(\mathcal{A}, \mathcal{H}, D)$, one can construct a twisted spectral triple (*cf.* [17]). We recall that the set of conformal similarities of a spectral triple $(\mathcal{A}, \mathcal{H}, D)$, denoted by $\mathrm{Sim}(\mathcal{A}, \mathcal{H}, D)$, consists of all unitary operators U on $\mathcal{H}$, such that

$$U\mathcal{A}U^* = \mathcal{A}, \text{ and } UDU^* = \mu(U)D, \text{ for some } \mu(U) > 0.$$

Recall that $\mathrm{Sim}(\mathcal{A}, \mathcal{H}, D)$ is a group and the map $\mu : \mathrm{Sim}(\mathcal{A}, \mathcal{H}, D) \to (0, \infty)$ is a character. We fix a subgroup $G \subset \mathrm{Sim}(\mathcal{A}, \mathcal{H}, D)$ and let $\mathcal{A}_G$ be the crossed product algebra $\mathcal{A} \rtimes G$. The formula

$$\sigma(aU) = \mu(U)^{-1} aU, \ \forall a \in \mathcal{A}, U \in G,$$

defines an automorphism of $\mathcal{A}_G$, and $(\mathcal{A}_G, \mathcal{H}, D)$ is a σ-spectral triple [17]. We will assume that the base spectral triple $(\mathcal{A}, \mathcal{H}, D)$ is regular, *i.e.* the operators $\mathcal{A}, [D, \mathcal{A}]$ are in the domain of all powers of the derivation $\delta(\cdot) = [|D|, \cdot]$. We also assume that $D^{-1} \in \mathcal{L}^{n,\infty}(\mathcal{H})$ for some fixed even number $n \in 2\mathbb{N}$, and that the twisted spectral triple $(\mathcal{A}_G, \mathcal{H}, D)$ is graded.

From the regularity of the base spectral triple, it easily follows that the twisted spectral triple $(\mathcal{A}_G, \mathcal{H}, D)$ is Lipschitz-regular. Therefore, by passage to the phase, one can associate the Connes-Chern character to the Fredholm module $(\mathcal{A}_G, \mathcal{H}, F = D|D|^{-1})$. To recall, this is a cyclic n-cocycle given by

$$\Phi_F(a_0U_0, a_1U_1, \ldots, a_nU_n) = \mathrm{Trace}(\gamma F[F, a_0U_0][F, a_1U_1] \cdots [F, a_nU_n])$$

for all $a_i \in \mathcal{A}, U_i \in G, i = 0, \ldots, n$. Here γ denotes the grading. Also, there is a Hochschild n-cocycle given by

$$\begin{aligned}\Psi_{D,\sigma}&(a_0U_0, a_1U_1, \ldots, a_nU_n) \\ &= \mathrm{Tr}_\omega(\gamma a_0U_0[D, \sigma^{-1}(a_1U_1)]_\sigma \cdots [D, \sigma^{-n}(a_nU_n)]_\sigma |D|^{-n})\end{aligned}$$

for all $a_i \in \mathcal{A}, U_i \in G, i = 0, \ldots, n$.

We recall from [12] how

$$\Phi_F(a_0U_0, a_1U_1, \ldots, a_nU_n) = \mathrm{Trace}(\gamma F[F, a_0U_0][F, a_1U_1] \cdots [F, a_nU_n])$$

can be approximated by the trace of finite rank operators using a cutoff. Let $g : \mathbb{R} \to \mathbb{R}$ be a smooth function which takes the value 1 on $[0, \frac{1}{2}]$, decreases to 0 on $[\frac{1}{2}, 1]$, is 0 for $t > 1$, $g(-t) = g(t)$ for $t < 0$, and $\int_0^\infty g'(u)du = -1$. Define the operators $A_t = g(t|D|)$ for $t > 0$. The operators A_t, $t > 0$, are positive, finite rank, and satisfy $P_{1/2t} \leq A_t \leq P_{1/t}$, where P_N is the spectral projector of $|D|$ on the interval $[0, N]$. Given a Hochschild n-cycle

$$c = \sum_{j=1}^{k} a_{0j}U_{0j} \otimes a_{1j}U_{1j} \otimes \cdots \otimes a_{nj}U_{nj},$$

it is proved in [12] that

$$\Phi_F(c) = 2 \lim_{t \downarrow 0} \Psi_t(c), \tag{5.5}$$

where

$$\begin{aligned}\Psi_t&(a_0U_0, a_1U_1, \ldots, a_nU_n) \\ :=& -\mathrm{Trace}(\gamma a_0U_0[F, a_1U_1] \cdots [F, a_{n-1}U_{n-1}]F[A_t, a_nU_n]).\end{aligned}$$

The proof of this is essentially based on two facts (*cf.* [12]). The first is that the operator trace is normal, and the second is that, since $b(c) = 0$, one has

$$\begin{aligned}&\sum_{j=1}^{k} a_{0j}U_{0j}[F, a_{1j}U_{1j}] \cdots [F, a_{n-1,j}U_{n-1,j}]a_{nj}U_{nj} \\ &\quad = \sum_{j=1}^{k} a_{nj}U_{nj}a_{0j}U_{0j}[F, a_{1j}U_{1j}] \cdots [F, a_{n-1,j}U_{n-1,j}].\end{aligned}$$

For convenience, we drop the index j and the summation in the formula for the Hochschild cycle c, and simply write

$$c = a_0U_0 \otimes a_1U_1 \otimes \cdots \otimes a_nU_n.$$

Lemma 5.1 [10] *The operator* $a_0U_0[F, a_1U_1] \cdots [F, a_{n-1}U_{n-1}]F|D|^{n-1}$ *is bounded.*

Proof Using the identity $U[F,a] = [F, UaU^*]U$, for all $a \in A, U \in G$, one can see that

$$\begin{aligned} &a_0U_0[F,a_1U_1]\cdots[F,a_{n-1}U_{n-1}]F|D|^{n-1} \\ &= \mu(U_0U_1\cdots U_{n-1})^{n-1}a_0[F,U_0a_1U_0^*]\cdots \\ &\quad [F,U_0U_1\cdots U_{n-2}a_{n-1}U_{n-2}^*\cdots U_1^*U_0^*]F|D|^{n-1}U_0U_1\cdots U_{n-1}. \end{aligned}$$

The boundedness of the latter follows from the fact that the operator

$$a_0'[F,a_1']\cdots[F,a_{n-1}']F|D|^{n-1}$$

is bounded for any $a_0', \ldots, a_{n-1}' \in A$, which is proved in [12]. □

For convenience, let

$$R = -\gamma a_0U_0[F,a_1U_1]\cdots[F,a_{n-1}U_{n-1}]F|D|^{n-1} \in \mathcal{L}(H).$$

Then we have

$$\Psi_t(a_0U_0, a_1U_1, \ldots, a_nU_n) = \mathrm{Trace}(R|D|^{-(n-1)}[A_t, a_nU_n]).$$

In the sequel we will impose the following extra condition on the Hochschild cycle $c = a_0U_0 \otimes a_1U_1 \otimes \cdots \otimes a_nU_n$ (note that we have dropped the summation):

Condition 5.2 We shall assume that

$$\lim_{t\downarrow 0} \mathrm{Trace}(R|D|^{-(n-1)}[A_t, a_nU_n]) = \lim_{t\downarrow 0} \mathrm{Trace}(R|D|^{-(n-1)}[A_t, a_n]U_n).$$

For example, if $\mu(U_n) = 1$ then this condition is satisfied.

The function $\Psi_t'(a_0U_0, a_1U_1, \ldots, a_nU_n) := \mathrm{Trace}(R|D|^{-(n-1)}[A_t, a_n]U_n)$ is continuous on the interval $\frac{1}{t} \geq e$, and has a limit as $\frac{1}{t} \to \infty$, therefore it is bounded. Hence the evaluation of the state ω on the corresponding element in the C^*-algebra $C_b([e,\infty))/C_0([e,\infty))$ will yield this limit, which will be denoted by

$$\lim_{t^{-1}\to\omega} \Psi_t'(a_0U_0, a_1U_1, \ldots, a_nU_n).$$

To compute this limit, one can use the following proposition of Connes. For a detailed discussion, we refer the reader to [12].

Proposition 5.3 *Let $f : [0,\infty) \to [0,\infty)$ be a continuous function, $p > 1$, and $\sum_k m_k(f)e^{pk} < \infty$, where for each k*

$$m_k(f) = \sup\{f(u); \quad k \leq \log u \leq k+1\}.$$

Then $M_p = \int_0^\infty f(u)u^{p-1}du$ is finite and

$$\lim_{t^{-1}\to\omega} t^p \mathrm{Trace}(f(t|D|)S) = M_p \mathrm{Tr}_\omega(S|D|^{-p}),$$

provided $S \in \mathcal{L}(\mathcal{H})$, and $D^{-1} \in \mathcal{L}^{p,\infty}(\mathcal{H})$.

Also we use Lemma 10.29 of [12].

Lemma 5.4 *If $g(t) = h(t)^2$ where $h \in \mathcal{D}(\mathbb{R})$ is also a cutoff, and if $a \in \mathcal{A}$, then*

$$\| [g(t|D|), a] - \frac{1}{2}\{g'(t|D|), t\delta a\} \|_{n-} = o(t) \text{ as } t \downarrow 0.$$

We note that $\|\cdot\|_{n-}$ is defined in [12].

The argument following the above lemma in [12] applies verbatim to our case by simply replacing R by U_nR, and it follows that

$$\begin{aligned}
&\lim_{t^{-1}\to\omega} \Psi_t'(a_0U_0, a_1U_1, \dots, a_nU_n)\\
&\quad= n\mathrm{Tr}_\omega(\gamma a_0U_0[F, a_1U_1]\cdots[F, a_{n-1}U_{n-1}]D^{-1}\delta(a_n)U_n)\\
&\quad= n\mathrm{Tr}_\omega(\gamma a_0U_0[F, a_1U_1]\cdots[F, a_{n-1}U_{n-1}]\delta_\sigma\sigma^{-1}(a_nU_n)D^{-1}), \qquad (5.6)
\end{aligned}$$

where $\delta(x) = [|D|, x]$ and $\delta_\sigma(x) = [|D|, x]_\sigma$.

The last step is to prove that the local Hochschild cocycle $\Psi_{D,\sigma}$ is cohomologous to ζ_n^σ defined by

$$\begin{aligned}
&\zeta_n^\sigma(a_0U_0, a_1U_1, \dots, a_nU_n)\\
&\quad= n\mathrm{Tr}_\omega(\gamma a_0U_0[F, a_1U_1]\cdots[F, a_{n-1}U_{n-1}]\delta_\sigma\sigma^{-1}(a_nU_n)D^{-1})
\end{aligned}$$

for any $a_i \in \mathcal{A}$, and $U_i \in G$, $i = 0, \dots, n$. This will be achieved by defining the cochains $\zeta_1^\sigma, \dots, \zeta_{n-1}^\sigma$ by defining $\zeta_k^\sigma(a_0U_0, a_1U_1, \dots, a_nU_n)$ as

$$n\mathrm{Tr}_\omega(\gamma a_0U_0[F, a_1U_1]\cdots[F, a_{k-1}U_{k-1}]D^{-1}\delta_\sigma(a_kU_k)[F, a_{k+1}U_{k+1}]\cdots[F, a_nU_n])$$

for any $a_i \in \mathcal{A}, U_i \in G$, $i = 0, \dots, n$. Then, using the following two lemmas, one can see that $\Psi_{D,\sigma}$ is cohomologous to ζ_n^σ, hence they yield the same value on any Hochschild cycle.

Lemma 5.5 [10] *The cochains $\zeta_1^\sigma, \dots, \zeta_n^\sigma$ are mutually cohomologous Hochschild cocycles.*

Proof First we show that $b\zeta_n^\sigma = 0$. For any $a_i \in \mathcal{A}, U_i \in G$, we have

$$\begin{aligned}
&b\zeta_n^\sigma(a_0U_0, \dots, a_{n+1}U_{n+1})\\
&= \mathrm{Tr}_\omega(\gamma a_0U_0a_1U_1[F, a_2U_2]\cdots[F, a_nU_n][|D|, \sigma^{-1}(a_{n+1}U_{n+1})]_\sigma D^{-1}) -\\
&\quad \mathrm{Tr}_\omega(\gamma a_0U_0[F, a_1U_1a_2U_2]\cdots[F, a_nU_n][|D|, \sigma^{-1}(a_{n+1}U_{n+1})]_\sigma D^{-1}) -\\
&\quad \cdots +\\
&\quad \mathrm{Tr}_\omega(\gamma a_0U_0[F, a_1U_1]\cdots[F, a_{n-1}U_{n-1}][|D|, \sigma^{-1}(a_nU_na_{n+1}U_{n+1})]_\sigma D^{-1})\\
&\quad -\mathrm{Tr}_\omega(\gamma a_{n+1}U_{n+1}a_0U_0[F, a_1U_1]\cdots[F, a_{n-1}U_{n-1}][|D|, \sigma^{-1}(a_nU_n)]_\sigma D^{-1})\\
&= \mathrm{Tr}_\omega(\gamma a_0U_0a_1U_1[F, a_2U_2]\cdots[F, a_nU_n][|D|, \sigma^{-1}(a_{n+1}U_{n+1})]_\sigma D^{-1})\\
&\quad -\mathrm{Tr}_\omega(\gamma a_0U_0a_1U_1[F, a_2U_2]\cdots[F, a_nU_n][|D|, \sigma^{-1}(a_{n+1}U_{n+1})]_\sigma D^{-1})\\
&\quad -\mathrm{Tr}_\omega(\gamma a_0U_0[F, a_1U_1]a_2U_2\cdots[F, a_nU_n][|D|, \sigma^{-1}(a_{n+1}U_{n+1})]_\sigma D^{-1})\\
&\quad \cdots\\
&\quad +\mathrm{Tr}_\omega(\gamma a_0U_0[F, a_1U_1]\cdots[F, a_{n-1}U_{n-1}]a_nU_n[|D|, \sigma^{-1}(a_{n+1}U_{n+1})]_\sigma D^{-1})\\
&\quad +\mathrm{Tr}_\omega(\gamma a_0U_0[F, a_1U_1]\cdots[F, a_{n-1}U_{n-1}][|D|, \sigma^{-1}(a_nU_n)]_\sigma\\
&\qquad\qquad \sigma^{-1}(a_{n+1}U_{n+1})D^{-1})\\
&\quad -\mathrm{Tr}_\omega(\gamma a_{n+1}U_{n+1}a_0U_0[F, a_1U_1]\cdots[F, a_{n-1}U_{n-1}][|D|, \sigma^{-1}(a_nU_n)]_\sigma D^{-1}).
\end{aligned}$$

The latter vanishes because of successive cancellations, where the last two terms cancel each other since using the identity

$$D^{-1}a_{n+1}U_{n+1} - \sigma^{-1}(a_{n+1}U_{n+1})D^{-1} = -D^{-1}[D, \sigma^{-1}(a_{n+1}U_{n+1})]_\sigma D^{-1},$$

one obtains a trace class operator which vanishes under the Dixmier trace.

Now we introduce cochains $\eta_1^\sigma, \ldots, \eta_{n-1}^\sigma$ such that $\zeta_k^\sigma - \zeta_{k+1}^\sigma = b\eta_k^\sigma$ for all $k = 1, \ldots, n-1$, and this will finish the proof.

Using the identities

$$D^{-1}\delta_\sigma(a_kU_k) - \delta_\sigma\sigma^{-1}(a_kU_k)D^{-1} = -D^{-1}\delta([D,a_k])D^{-1}U_k,$$
$$|D|^{-1}[F,a_jU_j] - [F,\sigma^{-1}(a_jU_j)]|D|^{-1} = -|D|^{-1}[F,\delta_\sigma\sigma^{-1}(a_jU_j)]|D|^{-1},$$
$$F[F,a_jU_j] = -[F,a_jU_j]F,$$

we can move D^{-1} in the formula for ζ_k^σ to the right under the Dixmier trace and obtain the following expression:

$$\begin{aligned}&\zeta_k^\sigma(a_0U_0, a_1U_1, \ldots, a_nU_n)\\ &= n\mathrm{Tr}_\omega(\gamma a_0U_0[F,a_1U_1]\cdots[F,a_{k-1}U_{k-1}]\\ &\quad D^{-1}\delta_\sigma(a_kU_k)[F,a_{k+1}U_{k+1}]\cdots[F,a_nU_n])\\ &= (-1)^{n-k}n\mathrm{Tr}_\omega(\gamma a_0U_0[F,a_1U_1]\cdots[F,a_{k-1}U_{k-1}]\\ &\quad \delta_\sigma(\sigma^{-1}(a_kU_k))[F,\sigma^{-1}(a_{k+1}U_{k+1})]\cdots[F,\sigma^{-1}(a_nU_n)]D^{-1}).\end{aligned}$$

Therefore $(\zeta_k^\sigma - \zeta_{k+1}^\sigma)(a_0U_0, a_1U_1, \ldots, a_nU_n)$ is equal to

$$\begin{aligned}&(-1)^{n-k}n\mathrm{Tr}_\omega(\gamma a_0U_0[F,a_1U_1]\cdots[F,a_{k-1}U_{k-1}]R_k^\sigma[F,\sigma^{-1}(a_{k+2}U_{k+2})]\\ &\qquad\cdots[F,\sigma^{-1}(a_nU_n)]D^{-1}),\end{aligned}$$

where

$$R_k^\sigma = \delta_\sigma\sigma^{-1}(a_kU_k)[F,\sigma^{-1}(a_{k+1}U_{k+1})] + [F,a_kU_k]\delta_\sigma\sigma^{-1}(a_{k+1}U_{k+1}).$$

Now we define

$$\begin{aligned}&\eta_k^\sigma(a_0U_0, \ldots, a_{n-1}U_{n-1})\\ &= (-1)^k n\mathrm{Tr}_\omega(\gamma a_0U_0[F,a_1U_1]\cdots[F,a_{k-1}U_{k-1}]\\ &\quad [F,\delta_\sigma\sigma^{-1}(a_kU_k)][F,\sigma^{-1}(a_{k+1}U_{k+1})]\cdots[F,\sigma^{-1}(a_{n-1}U_{n-1})]D^{-1}).\end{aligned}$$

Finally, using the identity

$$\begin{aligned}&[F,\delta_\sigma\sigma^{-1}(a_kU_ka_{k+1}U_{k+1})]\\ &= R_k^\sigma + [F,\delta_\sigma\sigma^{-1}(a_kU_k)]\sigma^{-1}(a_{k+1}U_{k+1}) + a_kU_k[F,\delta_\sigma\sigma^{-1}(a_{k+1}U_{k+1})],\end{aligned}$$

one can see that $\zeta_k^\sigma - \zeta_{k+1}^\sigma = b\eta_k^\sigma$. □

Lemma 5.6 [10] *The cochain* $\Psi_{D,\sigma} - \frac{1}{n}(\zeta_1^\sigma + \cdots + \zeta_n^\sigma)$ *is a Hochschild coboundary.*

Proof Let $a_0U_0, a_1U_1, \ldots, a_nU_n \in \mathcal{A}_G$. Since

$$\begin{aligned}&|D|^{-1}[D,a_jU_j]_\sigma - [D,\sigma^{-1}(a_jU_j)]_\sigma|D|^{-1}\\ &= |D|^{-1}[D,a_j]U_j - \mu(U_j)[D,a_j]U_j|D|^{-1}\\ &= |D|^{-1}[D,a_j]U_j - [D,a_j]|D|^{-1}U_j\\ &= -|D|^{-1}\delta([D,a])|D|^{-1}U,\end{aligned}$$

in the expression for $\Psi_{D,\sigma}$ we can replace each $[D,\sigma^{-1}(a_jU_j)]_\sigma|D|^{-1}$ by $|D|^{-1}[D,a_jU_j]_\sigma$. Therefore

$$\begin{aligned}&\Psi_{D,\sigma}(a_0U_0, a_1U_1, \ldots, a_nU_n)\\ &= \mathrm{Tr}_\omega(\gamma a_0U_0[D,\sigma^{-1}(a_1U_1)]_\sigma\cdots[D,\sigma^{-n}(a_nU_n)]_\sigma|D|^{-n})\\ &= \mathrm{Tr}_\omega(\gamma a_0U_0[D,\sigma^{-1}(a_1U_1)]_\sigma|D|^{-1}\cdots[D,\sigma^{-1}(a_nU_n)]_\sigma|D|^{-1}). \quad (5.7)\end{aligned}$$

Also we have

$$[D,\sigma^{-1}(a_jU_j)]_\sigma|D|^{-1}=[F,a_jU_j]+\delta_\sigma\sigma^{-1}(a_jU_j)D^{-1}+[F,\delta_\sigma\sigma^{-1}(a_jU_j)]|D|^{-1}.$$

Since

$$\begin{aligned}[F,\delta_\sigma\sigma^{-1}(a_jU_j)] &= \mu(U_j)[F,\delta(a_j)U_j]\\ &= \mu(U_j)[F,\delta(a_j)]U_j\\ &= \mu(U_j)|D|^{-1}(\delta([D,a_j])-\delta^2(a_j)F)U_j,\end{aligned}$$

the terms containing $[F,\delta_\sigma\sigma^{-1}(a_jU_j)]|D|^{-1}$ yield trace class operators which vanish under the Dixmier trace. So we can replace each

$$[D,\sigma^{-1}(a_jU_j)]_\sigma|D|^{-1}$$

by

$$[F,a_jU_j]+\delta_\sigma\sigma^{-1}(a_jU_j)D^{-1}.$$

Therefore (5.7) is the sum of 2^n terms. The term

$$\mathrm{Tr}_\omega(\gamma a_0U_0[F,a_1U_1]\cdots[F,a_nU_n])$$

is zero because one can write $a_0U_0=F[F,a_0U_0]+Fa_0U_0F$, and

$$\gamma F[F,a_0U_0][F,a_1U_1]\cdots[F,a_nU_n]$$

is trace class. Therefore the term is equal to

$$\begin{aligned}&\mathrm{Tr}_\omega(\gamma Fa_0U_0F[F,a_1U_1]\cdots[F,a_nU_n])\\ &= (-1)^n\mathrm{Tr}_\omega(\gamma Fa_0U_0[F,a_1U_1]\cdots[F,a_nU_n]F)\\ &= -\mathrm{Tr}_\omega(F\gamma a_0U_0[F,a_1U_1]\cdots[F,a_nU_n]F)\\ &= -\mathrm{Tr}_\omega(\gamma a_0U_0[F,a_1U_1]\cdots[F,a_nU_n]).\end{aligned}$$

Hence it has to be zero.

The terms having exactly one factor of the form $\delta_\sigma\sigma^{-1}(a_jU_j)D^{-1}$ add up to $\frac{1}{n}(\zeta_1^\sigma+\cdots+\zeta_n^\sigma)$, and to finish the proof we show that the terms with more than one factor of the form $\delta_\sigma\sigma^{-1}(a_jU_j)D^{-1}$ are Hochschild coboundaries. For example, let us consider the case when two consecutive factors of the above form yield the term

$$\begin{aligned}&\mathrm{Tr}_\omega(\gamma a_0U_0[F,a_1U_1]\cdots[F,a_{j-1}U_{j-1}]\delta_\sigma\sigma^{-1}(a_jU_j)\\ &\quad D^{-1}\delta_\sigma\sigma^{-1}(a_{j+1}U_{j+1})D^{-1}[F,a_{j+2}U_{j+2}]\cdots[F,a_nU_n])\\ &= \mathrm{Tr}_\omega(\gamma a_0U_0[F,a_1U_1]\cdots[F,a_{j-1}U_{j-1}]\delta_\sigma\sigma^{-1}(a_jU_j)\\ &\quad \delta_\sigma\sigma^{-2}(a_{j+1}U_{j+1})D^{-2}[F,a_{j+2}U_{j+2}]\cdots[F,a_nU_n])\\ &= \mathrm{Tr}_\omega(\gamma a_0U_0[F,a_1U_1]\cdots[F,a_{j-1}U_{j-1}]\delta_\sigma\sigma^{-1}(a_jU_j)\\ &\quad \delta_\sigma\sigma^{-2}(a_{j+1}U_{j+1})[F,\sigma^{-2}(a_{j+2}U_{j+2})]\cdots[F,\sigma^{-2}(a_nU_n)]D^{-2}).\end{aligned}\tag{5.8}$$

Now using the identity

$$\delta_\sigma^2\sigma^{-2}(a_ja_{j+1})=\delta_\sigma^2\sigma^{-2}(a_j)\sigma^{-2}(a_{j+1})+2\delta_\sigma\sigma^{-1}(a_j)\delta_\sigma\sigma^{-2}(a_{j+1})+a_j\delta_\sigma^2\sigma^{-2}(a_{j+1}),$$

one can see that (5.8) is equal to $b\varphi_j^\sigma(a_0U_0,a_1U_1,\ldots,a_nU_n)$ where

$$\begin{aligned}&\varphi_j^\sigma(a_0U_0,a_1U_1,\ldots,a_{n-1}U_{n-1})\\ &= \frac{(-1)^j}{2}\mathrm{Tr}_\omega(\gamma a_0U_0[F,a_1U_1]\cdots[F,a_{j-1}U_{j-1}]\\ &\quad \delta_\sigma^2\sigma^{-2}(a_jU_j)[F,\sigma^{-2}(a_{j+1}U_{j+1})]\cdots[F,\sigma^{-2}(a_{n-1}U_{n-1})]D^{-2}).\end{aligned}$$

□

Hence, considering (5.5), Condition 5.2, and (5.6) we have proved the following:

Theorem 5.7 [10] *The cyclic cocycle Φ_F and the local Hochschild cocycle $2\Psi_{D,\sigma}$ yield the same value on any Hochschild n-cycle,*

$$c = \sum_{j=1}^{k} a_{0j}U_{0j} \otimes a_{1j}U_{1j} \otimes \cdots \otimes a_{nj}U_{nj}$$

that satisfies Condition 5.2.

6 A local index formula for twisted spectral triples

In [17], an Ansatz for a local index formula for twisted spectral triples is given and its validity for twisted spectral triples, obtained from scaling automorphisms of spectral triples, has been verified. This formula is given in terms of residue functionals and twisted commutators. In this section we sketch very briefly some of the ideas in [17].

6.1 Moscovici's Ansatz. In Subsection 4.4, we explained the heuristic manner for obtaining the local Hochschild cocycle for twisted spectral triples from the one for ordinary spectral triples. In a rather similar manner one can obtain Moscovici's Ansatz for a local index formula for twisted spectral triples from the original local index formula of Connes and Moscovici discussed in Section 2. The Ansatz suggests that for a σ-spectral triple $(\mathcal{A}, \mathcal{H}, D)$, the twisted version of the local character should be given by

$$\varphi_n^{\sigma}(a_0, \ldots, a_n) :=$$
$$\sum_k c_{n,k} \int\!\!\!\!\!- a_0[D, \sigma^{-2k_1-1}(a_1)]_{\sigma}^{(k_1)} \cdots [D, \sigma^{-2(k_1+\cdots+k_n)-n}(a_n)]_{\sigma}^{(k_n)} |D|^{-n-2|k|},$$

where the *iterated twisted commutators* $[D, a]_{\sigma}^{(k)}$ are defined in [17]. Here again

$$\int\!\!\!\!\!- P = \mathrm{Res}_{z=0}\mathrm{Trace}(P|D|^{-2z}).$$

Therefore an analogue of the simple dimension spectrum hypothesis is assumed. Namely, one needs to assume that there exists a discrete subset of the complex plane such that for any operator P in an algebra of twisted pseudodifferential operators associated to the twisted spectral triple $(\mathcal{A}, \mathcal{H}, D, \sigma)$, $P|D|^{-2z}$ is a trace class operator provided $\mathrm{Re}(z)$ is large enough, and $\zeta_P(z) := \mathrm{Trace}(P|D|^{-2z})$ has a meromorphic extension to the plane with at most simple poles in this discrete set. We note that, as is emphasized in [17], there is *no* canonical way of constructing an algebra of twisted pseudodifferential operators for a general twisted spectral triple. Also, in the twisted case, for formulating an analogue of the regularity condition, one needs to postulate that the automorphism σ has an extension to a larger algebra which contains the *twisted differential forms.* Another necessary condition for the validity of the Ansatz is the σ-invariance of the residue functionals which is related to the Selberg principle for reductive Lie groups [17].

6.2 A local index formula for spectral triples twisted by scaling automorphisms. Let $(\mathcal{A}_G, \mathcal{H}, D)$ denote a twisted spectral triple obtained by scaling automorphisms of an ordinary finitely summable regular spectral triple, introduced in Subsection 3.3 and let $\Psi(\mathcal{A}_G, \mathcal{H}, D) := \Psi(\mathcal{A}, \mathcal{H}, D) \rtimes G$ denote the associated algebra of twisted pseudodifferential operators. The automorphism σ extends to an automorphism of $\Psi(\mathcal{A}_G, \mathcal{H}, D)$ and it is required that for any P in this algebra [17]:

$$⨍ P = ⨍ \sigma(P). \tag{6.9}$$

Definition 6.1 [17] The twisted JLO bracket of order q as a $(q+1)$-linear form on $\Psi(\mathcal{A}_G, \mathcal{H}, D)$ is defined by

$$\langle a_0 U_0^*, \dots, a_q U_q^* \rangle_D = \int_{\Delta_q} \mathrm{Trace}\big(\gamma\, a_0 U_0^*\, e^{-s_0 \mu(U_0)^2 D^2}\, a_1 U_1^*\, e^{-s_1 \mu(U_0 U_1)^2 D^2} \dots$$
$$\cdots a_q U_q^*\, e^{-s_q \mu(U_0 \cdots U_q)^2 D^2}\big)$$

for all $a_0, \dots, a_q \in \mathcal{A}$ and $U_0, \dots, U_q \in G$, where

$$\Delta_q := \{s = (s_0, \dots, s_q) \in \mathbb{R}^{q+1}; \quad s_j \geq 0, \quad s_0 + \cdots + s_q = 1\}.$$

In [17], it is assumed that $a_0, \dots, a_q$ are polynomials in D and elements of $\mathcal{A}$ and $[D, \mathcal{A}]$ which are homogeneous in λ as D is replaced by λD. The expression obtained from replacing each D occurring in $a_0, \dots, a_q$ by $\epsilon^{1/2} D$ is denoted by $\langle a_0 U_0^*, \dots, a_q U_q^* \rangle_D(\epsilon)$. Equivalently, one has

$$\langle a_0 U_0^*, \dots, a_q U_q^* \rangle_D(\epsilon) \; = \epsilon^{\frac{m}{2}} \langle a_0 U_0^*, \dots, a_q U_q^* \rangle_{\epsilon^{1/2} D},$$

where m is the total degree of λ in $a_0 \cdots a_q$ after replacing each D by λD.

Proposition 6.2 [17]*If $a_0 \in \mathcal{A}$ and $a_1, \dots, a_q \in [D, \mathcal{A}]$, then there is an asymptotic expansion*

$$\langle a_0 U_0^*, \dots, a_q U_q^* \rangle_D(\epsilon) \sim \sum_{j \in J} (c_j + c_j' \log \epsilon)\, \epsilon^{\frac{q}{2} - \rho_j} + O(1) \quad \text{as } \epsilon \searrow 0$$

where $\rho_0, \dots, \rho_m$ are points in the half plane $Re(z) \geq \frac{q}{2}$.

The twisted version of the JLO cocycles is defined by

$$J^q(D)(A_0, \dots, A_q) \; = \; \langle A_0, [D, \sigma^{-1}(A_1)]_\sigma, \dots, [D, \sigma^{-q}(A_q)]_\sigma \rangle_D$$

for $A_0, \dots, A_q \in \mathcal{A}_G$. Since in the twisted case this does not define a cocycle, in [17], by considering

$$J^q(\epsilon^{1/2} D)(A_0, \dots, A_q) \; = \; \epsilon^{\frac{q}{2}} \langle A_0, [D, \sigma^{-1}(A_1)]_\sigma, \dots, [D, \sigma^{-q}(A_q)]_\sigma \rangle_{\epsilon^{1/2} D}$$

for $\epsilon > 0$, and passing to the constant term using the above proposition, φ_q^σ is defined by

$$\varphi_q^\sigma(A_0, \dots, A_q) := \langle A_0, [D, \sigma^{-1}(A_1)]_\sigma, \dots, [D, \sigma^{-q}(A_q)]_\sigma \rangle_D \,|_0 \,.$$

It follows from the following two results [17] that for any twisted spectral triple obtained by conformal perturbation of an ordinary finitely summable regular spectral triple and satisfying a Selberg type invariance condition (6.9), one can associate a local cyclic cocycle for its Connes-Chern character.

Theorem 6.3 [17] *The cochain $\{\varphi_q^\sigma\}$ satisfies the cocycle identity in the (b, B)-bicomplex:*

$$b\varphi_{q-1}^\sigma(a_0U_0^*, \dots, a_qU_q^*) + B\varphi_{q+1}^\sigma(a_0U_0^*, \dots, a_qU_q^*) = 0,$$

for all $a_0, \dots, a_q \in \mathcal{A}$ and $U_0, \dots, U_q \in G$.

Theorem 6.4 [17] *The cocycle $\{\varphi_q^\sigma\}$ is cohomologous to the Connes-Chern character associated to the twisted spectral triple $(\mathcal{A}_G, \mathcal{H}, D)$ in the periodic cyclic cohomology $HP^*(\mathcal{A}_G)$.*

References

[1] M. Adler, *On a trace functional for formal pseudo differential operators and the symplectic structure of the Korteweg-de Vries type equations.* Invent. Math. 50, no. 3, 219–248, 1978/79.

[2] J.-B. Bost, *Principe d'Oka, K-théorie et systèmes dynamiques non commutatifs*, Invent. math., **101**, 261-333, 1990.

[3] A. Connes, *Noncommutative differential geometry,* Inst. Hautes Études Sci. Publ. Math. No. **62**, 257-360, 1985.

[4] A. Connes, *The action functional in noncommutative geometry.* Comm. Math. Phys. 117, no. 4, 673-683, 1988.

[5] A. Connes, *Noncommutative Geometry.* Academic Press, 1994.

[6] A. Connes, *Noncommutative geometry year 2000*, Highlights of mathematical physics (London, 2000), 49110, Amer. Math. Soc., Providence, RI, 2002.

[7] A. Connes and H. Moscovici, *The local index formula in noncommutative geometry.* Geom. Funct. Anal. 5, no. 2, 174–243, 1995.

[8] A. Connes and H. Moscovici, *Type III and spectral triples.* Traces in number theory, geometry and quantum fields, 57-71, Aspects Math., E38, Friedr. Vieweg, Wiesbaden, 2008.

[9] J. Dixmier, *Existence de traces non normales.* C. R. Acad. Sci. Paris Sér. A-B 262, A1107-A1108, 1966.

[10] F. Fathi Zadeh, *Twisted Spectral Triples, Connes' Character Formula, and the Gauss-Bonnet Theorem for Noncommutative Two Tori*, Ph.D. Thesis, 2010.

[11] F. Fathi Zadeh and M. Khalkhali, *The Algebra of Formal Twisted Pseudodifferential Symbols and a Noncommutative Residue.* Lett. Math. Phys. 94, no. 1, 41-61, 2010.

[12] J. M. Gracia-Bondía, J. C. Várilly, H. Figueroa, *Elements of Noncommutative Geometry.* Birkhäuser Advanced Texts, Birkhäuser Boston Inc., Boston, MA, 2001.

[13] N. Higson, *The local index formula in noncommutative geometry.* Contemporary developments in algebraic K-theory, 443–536 (electronic), ICTP Lect. Notes, XV, Abdus Salam Int. Cent. Theoret. Phys., Trieste, 2004.

[14] N. Higson, *The residue index theorem of Connes and Moscovici.* Surveys in noncommutative geometry, 71–126, Clay Math. Proc., 6, Amer. Math. Soc., Providence, RI, 2006.

[15] Ch. Kassel, *Le résidu non commutatif (d'après M. Wodzicki).* (French) [The noncommutative residue (after M. Wodzicki)] Séminaire Bourbaki, Vol. 1988/89. Astérisque No. 177-178, Exp. No. 708, 199–229, 1989.

[16] Yu. I. Manin, *Algebraic aspects of nonlinear differential equations.* (Russian) Current problems in mathematics, Vol. 11 (Russian), pp. 5–152. (errata insert) Akad. Nauk SSSR Vsesojuz. Inst. Naučn. i Tehn. Informacii, Moscow, 1978.

[17] H. Moscovici, *Local index formula and twisted spectral triples.* Quanta of maths, 465-500, Clay Math. Proc., 11, Amer. Math. Soc., Providence, RI, 2010.

[18] A. Polishchuk, *Holomorphic bundles on 2-dimensional noncommutative toric orbifolds,* Noncommutative geometry and number theory, 341–359, Aspects Math., E37, Vieweg, Wiesbaden, 2006.

[19] M. Wodzicki, *Noncommutative residue. I. Fundamentals.* K-theory, arithmetic and geometry (Moscow, 1984–1986). 320–399, Lecture Notes in Math., 1289, Springer, Berlin, 1987.

Fields Institute Communications
Volume **61**, 2011

Spectral Morphisms, K-Theory, and Stable Ranks

Bogdan Nica
Department of Mathematics and Statistics
University of Victoria
Victoria, BC Canada V8W 3R4
bogdan.nica@gmail.com

To the memory of Israel M. Gelfand (1913–2009)

Abstract. We give a brief account of the interplay between spectral morphisms, K-theory, and stable ranks in the context of Banach algebras.

1 Introduction

This article revolves around the notion of spectral morphism—that is, a morphism which preserves spectra of elements—in the setting of Banach algebras. The importance of spectral morphisms was firmly established, in the commutative case, by the Gelfand transform. In the general, non-commutative case, spectral morphisms with dense image have been very useful for K-theoretic purposes: the Karoubi–Swan theorem asserts that topological K-theory is invariant across such morphisms. Since K-theory is controlled by stable ranks—which are, very loosely speaking, noncommutative notions of dimension—the following question arises: are stable ranks invariant under spectral morphisms with dense image? In the literature, this is known as Swan's problem. Schematically, the interconnections between spectral morphisms, K-theory, and stable ranks are summarized by the following diagram:

2010 *Mathematics Subject Classification.* Primary 46L80, 19B10; Secondary 58B34.
Supported by a Postdoctoral Fellowship from the Pacific Institute for the Mathematical Sciences (PIMS).

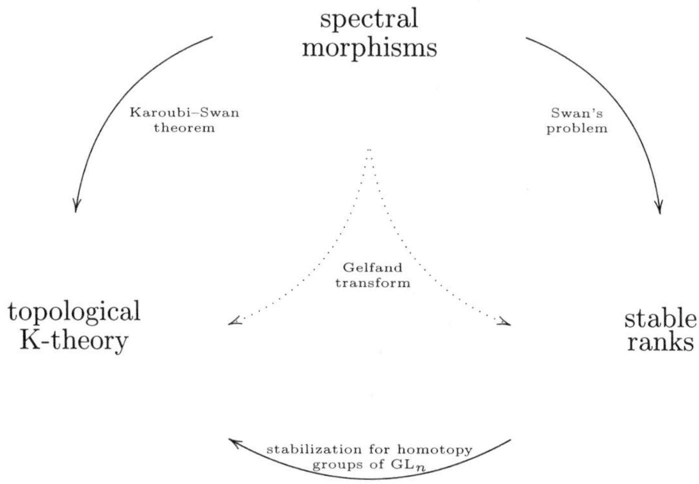

At the Noncommutative Geometry Workshop (May 2008), I gave a talk based on my paper [27]. A straightforward summary of the paper seemed uninteresting. So I decided to interpret the opportunity offered by these Proceedings as an occasion to describe the broader context captured in the above diagram. Results from [27], and from a more recent paper [28], are interwoven in this picture.

We follow the interplay between spectral morphisms and K-theory, on one hand, and spectral morphisms and stable ranks, on the other hand, in two different contexts. The first context is the Gelfand transform. The second context, typically non-commutative, deals with spectral morphisms with dense image; informally, the density assumption is supposed to make up for the loss of commutativity. We shall witness a nearly perfect parallelism between these two contexts.

The last section is devoted to stabilization phenomena for the homotopy groups of the general linear group over Banach algebras. The stable ranks play the key role in this discussion.

2 Spectral morphisms

In what follows, Banach algebras and their (continuous) morphisms are assumed to be unital.

A Banach algebra morphism $\phi : A \to B$ is *spectral* if it is spectrum-preserving, that is, for all $a \in A$ we have $\mathrm{sp}_B(\phi(a)) = \mathrm{sp}_A(a)$. Equivalently, the morphism ϕ is spectral if, for all $a \in A$, we have that a is invertible in A if and only if $\phi(a)$ is invertible in B.

In the case when ϕ is an inclusion—which is the typical case—A is spectral in B if and only if A is holomorphically closed in B; see, for instance, [34]. Recall that A is said to be holomorphically closed in B if, for each $a \in A$, we have that $f(a) \in A$ whenever f is holomorphic in a neighborhood of $\mathrm{sp}_B(a)$. The property that A is spectral in B appears under many other different names in the literature: Bourbaki ([8]) calls A full in B; Naimark ([26]) says that A and B form a Wiener pair; other sources say that A is inverse-closed in B.

The following example of spectral morphism is fundamental:

Example 2.1 (Gelfand transform) Let A be a unital, commutative Banach algebra. The *maximal ideal space* X_A is the set of characters of A, equipped with the topology of pointwise convergence; the name is due to the fact that we have a bijective correspondence, given by $\chi \mapsto \ker \chi$, between the characters of A and the maximal ideals of A. The maximal ideal space X_A is a compact Hausdorff space.

Let $\hat{a} \in C(X_A)$ denote the evaluation at $a \in A$. Then the evaluation map $a \mapsto \hat{a}$ defines a unital, continuous morphism $A \to C(X_A)$. This morphism—the *Gelfand transform* of A—is a spectral morphism.

Outside of the Gelfand context, the spectral morphisms we consider are assumed to have dense image. Henceforth, a morphism with dense image is called, simply, a *dense* morphism. The Gelfand transform may or may not be dense; an instance when it is dense is when the image is $*$-closed.

If A and B are commutative and $A \to B$ is a spectral morphism, then a simple argument involving the determinant shows that $\mathrm{M}_n(A) \to \mathrm{M}_n(B)$ is spectral for each $n \geq 1$. A noncommutative analogue is provided by the following useful fact, due to Swan [37] (see also Bost [7] and Schweitzer [34]):

Proposition 2.2 *If $A \to B$ is a dense and spectral morphism, then $\mathrm{M}_n(A) \to \mathrm{M}_n(B)$ is a dense and spectral morphism for each $n \geq 1$.*

Let us give some examples of morphisms which are dense and spectral. Especially motivating from the point of view of noncommutative geometry is the following commutative example:

Example 2.3 Let M be a compact smooth manifold. Then $C^k(M)$ is a Banach algebra under the norm $\|f\|_{(k)} := \max\{\|\partial^\alpha f\|_\infty : 0 \leq |\alpha| \leq k\}$ (defined using local charts on M), and the inclusion $C^k(M) \hookrightarrow C(M)$ is dense and spectral.

Passing from the smooth category to the metric category, we still have an analogue of the Banach algebra of C^1-functions on a compact manifold:

Example 2.4 Let X be a compact metric space. Then the algebra $\mathrm{Lip}(X)$ of Lipschitz functions on X is a Banach algebra under the norm $f \mapsto \max\{\|f\|_\infty, \|f\|_{\mathrm{Lip}}\}$, and the inclusion $\mathrm{Lip}(X) \hookrightarrow C(X)$ is dense and spectral.

Let us point out that the dense and spectral inclusions appearing in the previous examples are Gelfand transforms. This is due to the fact that a dense and spectral morphism $A \to B$ between commutative Banach algebras induces a homeomorphism $X_B \to X_A$; in particular, a dense and spectral morphism $A \to C(X)$, where X is a compact Hausdorff space, can be identified with the Gelfand transform of A.

The next two examples provide dense and spectral Banach subalgebras for the reduced C*-algebras of certain groups. Recall that the reduced C*-algebra $\mathrm{C}^*_\mathrm{r}\Gamma$ of a group Γ is the completion of the group algebra $\mathbb{C}\Gamma$ under the operator norm $\|\cdot\|$ coming from the regular representation of Γ on $\ell^2\Gamma$.

Example 2.5 Let Γ be a finitely generated group. If Γ has polynomial growth, then the dense inclusion $\ell^1\Gamma \hookrightarrow \mathrm{C}^*_\mathrm{r}\Gamma$ is spectral (Ludwig [25]).

Example 2.6 Let Γ be a finitely generated group, equippped with a word-length $|\cdot|$. Following Jolissaint [20], we define the s-Sobolev space $H^s\Gamma$ to be the

Banach space obtained by completing $\mathbb{C}\Gamma$ under the weighted ℓ^2-norm

$$\|\sum a_g g\|_{2,s} := \sqrt{\sum |a_g|^2(1+|g|)^{2s}}.$$

The group Γ is said to have *property RD (of order s)* if, for some $C, s \geq 0$, we have $\|a\| \leq C\|a\|_{2,s}$ for all $a \in \mathbb{C}\Gamma$. Free groups (Haagerup [16])—more generally, hyperbolic groups (de la Harpe [14])—have property RD; so do groups of polynomial growth (Jolissaint [20]), and many other groups.

If Γ has property RD of order s, then the following holds: for every $S > s$, the S-Sobolev space $H^S\Gamma$ is in fact a Banach algebra under $\|\cdot\|_{2,S}$, and the (continuous) dense inclusion $H^S\Gamma \hookrightarrow \mathrm{C}^*_{\mathrm{r}}\Gamma$ is spectral (Lafforgue [22]).

The following example, of a more general nature, is the noncommutative analogue of Example 2.3:

Example 2.7 Let B be a Banach algebra and $\delta : B \to B$ an unbounded derivation, i.e., δ is a linear map defined on a subalgebra $\mathrm{dom}(\delta)$ of B and satisfying the Leibniz rule $\delta(ab) = \delta(a)b + a\delta(b)$. Also, assume that δ is closed; then, for each $k \geq 1$, $B^{(k)} := \mathrm{dom}(\delta^k)$ is a Banach algebra under the norm $\|a\|_{(k)} := \max\{\|\delta^i(a)\| : 0 \leq i \leq k\}$.

If $B^{(k)}$ is dense in B, then the (continuous) inclusion $B^{(k)} \hookrightarrow B$ is spectral (Ji [19]).

As a non-example, we mention the following counterpoint to Example 2.5:

Example 2.8 Let Γ be a group containing a free subsemigroup on two generators. Then the dense inclusion $\ell^1\Gamma \hookrightarrow \mathrm{C}^*_{\mathrm{r}}\Gamma$ is not spectral (Jenkins [18]).

3 Spectral morphisms and K-theory

We recall the definition of the K-groups, the main purpose being that of introducing some notation. Following Taylor [39], we use the shorthand $[\,\cdot\,]$ for the set of path-connected components $\pi_0(\,\cdot\,)$.

Let A be a unital Banach algebra. Let $V(A) := \varinjlim\, [\mathrm{Idem}(\mathrm{M}_n(A))]$, where $\mathrm{Idem}(\cdot)$ denotes the set of idempotents and the direct limit is taken along the maps induced by $* \mapsto \left(\begin{smallmatrix} * & 0 \\ 0 & 0\end{smallmatrix}\right)$. Then $V(A)$ is an abelian monoid under $[a]+[a'] := \left[\left(\begin{smallmatrix} a & 0 \\ 0 & a'\end{smallmatrix}\right)\right]$. Also, let $\mathcal{P}(A)$ denote the set of isomorphism classes of finitely generated, projective right A-modules; equipped with the direct sum, $\mathcal{P}(A)$ is an abelian monoid. The monoids $V(A)$ and $\mathcal{P}(A)$ are naturally isomorphic, and the Grothendieck group of $V(A)$—or $\mathcal{P}(A)$—is the (abelian) group $K_0(A)$. On the other hand, we simply put $K_1(A) := \varinjlim\, [\mathrm{GL}_n(A)]$, where $\mathrm{GL}_n(A)$ denotes the invertible group of $\mathrm{M}_n(A)$ and the direct limit is taken along the maps induced by $* \mapsto \left(\begin{smallmatrix} * & 0 \\ 0 & 1\end{smallmatrix}\right)$. This time, the direct limit is already an abelian group under $[a][a'] := \left[\left(\begin{smallmatrix} a & 0 \\ 0 & a'\end{smallmatrix}\right)\right]$.

3.1 The Gelfand transform. Consider the Gelfand transform $A \to C(X_A)$ of a commutative Banach algebra A. Historically, the first result with a K-theoretic flavor is Shilov's idempotent theorem [35] saying that the induced map $\mathrm{Idem}(A) \to \mathrm{Idem}(C(X_A))$ is onto. It is easy to see, on the other hand, that the Gelfand transform is injective on idempotents: if p, q are idempotents with $\hat{p} = \hat{q}$, then $z := pq + (1-p)(1-q)$ is invertible since $\hat{z} = 1$, and from $pz = zq (= pq)$ we deduce—by the commutativity of A—that $p = q$. Thus, an equivalent form of Shilov's theorem is the statement that the Gelfand transform $A \to C(X_A)$ induces a bijection $\mathrm{Idem}(A) \to \mathrm{Idem}(C(X_A))$. With hindsight, this gives a strong hint that a K_0-isomorphism might be lurking in the background.

The next K-theoretic development, this time in the direction of K_1, is the theorem of Arens [1] and Royden [33]: the Gelfand transform $A \to C(X_A)$ induces an isomorphism $[A^\times] \to [C(X_A)^\times]$, where $(\,\cdot\,)^\times$ denotes the invertible group. Shortly after, Arens [2] extended the Arens–Royden theorem by showing that, for each $n \geq 1$, we have an isomorphism $[\mathrm{GL}_n(A)] \to [\mathrm{GL}_n(C(X_A))]$ induced by the Gelfand transform $A \to C(X_A)$.

Clearly, the Arens theorem implies that the Gelfand transform $A \to C(X_A)$ induces an isomorphism $K_1(A) \to K_1(C(X_A))$. This was first pointed out by Novodvorskii [30], who also handled the K_0-isomorphism for semisimple A. The semisimple hypothesis was later removed by Taylor; one of the results from [39] is that the Gelfand transform $A \to C(X_A)$ induces, for each $n \geq 1$, a bijection $[\mathrm{Idem}(\mathrm{M}_n(A))] \to [\mathrm{Idem}(\mathrm{M}_n(C(X_A)))]$. Summarizing, we have

Theorem 3.1 (Novodvorskii–Taylor) *The Gelfand transform $A \to C(X_A)$ induces an isomorphism $K_*(A) \to K_*(C(X_A))$.*

The Arens theorem has been greatly generalized by Davie [13]. As a particular case of Davie's theorem, we have the following fact: the Gelfand transform $A \to C(X_A)$ induces a homotopy equivalence $\mathrm{GL}_n(A) \to \mathrm{GL}_n(C(X_A))$ for each $n \geq 1$.

3.2 Dense and spectral morphisms. Versions of the following theorem are due to Karoubi [21] and Swan [37]; see also Connes [9]. An excellent account is given by Bost in [7, Appendix].

Theorem 3.2 (Karoubi–Swan) *A dense and spectral morphism $A \to B$ induces an isomorphism $K_*(A) \to K_*(B)$.*

The Karoubi–Swan theorem features, for instance, in the proof by Connes and Moscovici [10] that hyperbolic groups satisfy the Novikov conjecture, and in the remarkable work of Lafforgue [23]. In both cases, one needs to pass from the reduced C*-algebra $C^*_r\Gamma$ of a group Γ to a suitable group algebra of Γ in such a way that the K-theory is preserved.

The proof of the Karoubi–Swan theorem is fairly elementary. One shows, in fact, that a dense and spectral morphism $A \to B$ induces bijections $[\mathrm{GL}_n(A)] \to [\mathrm{GL}_n(B)]$ and $[\mathrm{Idem}(\mathrm{M}_n(A))] \to [\mathrm{Idem}(\mathrm{M}_n(B))]$ for all $n \geq 1$. In light of Proposition 2.2, it suffices to consider the case $n = 1$. Now, to prove that $[\mathrm{Idem}(A)] \to [\mathrm{Idem}(B)]$ is a bijection, one first enlarges the closed set of idempotents $\mathrm{Idem}(\,\cdot\,)$ to a suitable open set of "almost" idempotents; for instance, one could take the set $^\sim\mathrm{Idem}(\,\cdot\,)$ consisting of all elements whose spectrum does not meet the line $\mathrm{Re}\ z = \frac{1}{2}$. By holomorphic functional calculus, the set of idempotents $\mathrm{Idem}(\,\cdot\,)$ is a deformation retract of the set of almost idempotents $^\sim\mathrm{Idem}(\,\cdot\,)$; therefore, it suffices to prove that $[{}^\sim\mathrm{Idem}(A)] \to [{}^\sim\mathrm{Idem}(B)]$ is a bijection. This, and the fact that $[A^\times] \to [B^\times]$ is a bijection, are proved in a similar manner.

Actually, these two bijections are avatars of the following "spectral" picture (cf. [27]). Let Ω be an open subset of $\mathbb{C}$ containing the origin. For a unital Banach algebra A, we let A_Ω denote the set of elements of A whose spectrum is contained in Ω. For example, $A_\Omega = A^\times - 1$ when $\Omega = \mathbb{C} \setminus \{-1\}$, and $A_\Omega = {}^\sim\mathrm{Idem}(A)$ when $\Omega = \{z : \ \mathrm{Re}\ z \neq \frac{1}{2}\}$. Then a dense and spectral morphism $A \to B$ induces a bijection $[A_\Omega] \to [B_\Omega]$. As soon as we adopt this spectral picture, the K-theoretic pull is to define a "spectral" functor K_Ω in the same way as the usual K-functors are defined. Given a unital Banach algebra A, the direct limit

$$V_\Omega(A) := \varinjlim \left[\mathrm{M}_n(A)_\Omega\right]$$

is an abelian monoid under $[a] + [a'] = \left[\left(\begin{smallmatrix} a & 0 \\ 0 & a' \end{smallmatrix}\right)\right]$. Applying the Grothendieck construction to the abelian monoid $V_\Omega(A)$, we obtain an abelian group $K_\Omega(A)$. Thus, for each open set $\Omega \subseteq \mathbb{C}$ containing the origin, we have a *spectral K-functor* K_Ω. Of course, not every choice of Ω yields an interesting functor; but, at the very least, we can recover the classical functors K_0 and K_1. The Karoubi–Swan theorem for spectral K-functors reads as follows: a dense and spectral morphism $A \to B$ induces an isomorphism $K_\Omega(A) \to K_\Omega(B)$. The interesting part is not this generalized Karoubi–Swan theorem, but rather how the Karoubi–Swan theorem lead us to the generalized K-functors K_Ω.

The analogue, due to Bost ([7]), of Davie's result is the expected one: a dense and spectral morphism $A \to B$ induces, for each $n \geq 1$, a homotopy equivalence $\mathrm{GL}_n(A) \to \mathrm{GL}_n(B)$. In fact, with the previous notations we have that such a ϕ induces, for each $n \geq 1$, a homotopy equivalence $\mathrm{M}_n(A)_\Omega \to \mathrm{M}_n(B)_\Omega$.

4 Spectral morphisms and stable ranks

Bass [5] introduced the first notion of stable rank; more than a dozen stable ranks have appeared since then. Like the original stable rank introduced by Bass, some are purely ring-theoretic; other stable ranks take topology into account. Here, we consider the following four stable ranks: the Bass stable rank, the topological stable rank, the connected stable rank, and the general stable rank. The last three stable ranks are due to Rieffel [31].

Let A be a unital Banach algebra. For $n \geq 1$, consider the set of *unimodular n-tuples*

$$\mathrm{Lg}_n(A) = \{(a_1, \ldots, a_n) : \ Aa_1 + \cdots + Aa_n = A\} \subseteq A^n.$$

The last column of an invertible n-by-n matrix is an example of unimodular n-tuple. Left-multiplying the transpose of a unimodular n-tuple by an invertible n-by-n matrix yields another unimodular n-tuple; this describes an action of $\mathrm{GL}_n(A)$ on $\mathrm{Lg}_n(A)$. Under this action, the orbit of $(0, \ldots, 0, 1)$ consists of the last columns of invertible matrices.

The first two of the four stable ranks under consideration are defined as follows. The *Bass stable rank* of A, denoted bsr A, is the least $n \geq 1$ with the property that $\mathrm{Lg}_{n+1}(A)$ is "reducible" to $\mathrm{Lg}_n(A)$ in the following sense: if $(a_1, \ldots, a_{n+1}) \in \mathrm{Lg}_{n+1}(A)$, then $(a_1 + x_1 a_{n+1}, \ldots, a_n + x_n a_{n+1}) \in \mathrm{Lg}_n(A)$ for some $(x_1, \ldots, x_n) \in A^n$. The *topological stable rank* of A, denoted tsr A, is the least $n \geq 1$ with the property that $\mathrm{Lg}_n(A)$ is dense in A^n. Naturally, a stable rank is declared to be infinite if there is no integer n satisfying the required condition. The Bass and the topological stable ranks are related by the inequality bsr $A \leq$ tsr A ([31], [11]).

If X is a compact space then

$$\operatorname{bsr} C(X) = \operatorname{tsr} C(X) = \left\lfloor \frac{\dim X}{2} \right\rfloor + 1$$

where $\dim X$ denotes the covering dimension of X. In light of this formula—due to Vaserstein [43] for the Bass stable rank, and to Rieffel [31] for the topological stable rank—we think of the Bass and the topological stable ranks as being *dimensional* stable ranks.

The remaining two stable ranks are defined as follows. The *connected stable rank* of A, denoted $\operatorname{csr} A$, is the least $n \geq 1$ with the property that $\mathrm{GL}^0_m(A)$ acts transitively on $\mathrm{Lg}_m(A)$ for all $m \geq n$. By $\mathrm{GL}^0_m(A)$ we denote, as usual, the identity component of $\mathrm{GL}_m(A)$. It is not hard to see that the action of $\mathrm{GL}^0_m(A)$ on $\mathrm{Lg}_m(A)$ has open orbits. Hence, the *connected stable rank* of A can also be defined as the least $n \geq 1$ with the property that $\mathrm{Lg}_m(A)$ is connected for all $m \geq n$. Finally, the *general stable rank* of A, denoted $\operatorname{gsr} A$, is the least $n \geq 1$ with the property that $\mathrm{GL}_m(A)$ acts transitively on $\mathrm{Lg}_m(A)$ for all $m \geq n$. It turns out (Rieffel [31]) that an equivalent definition is the following: the *general stable rank* of A is the least $n \geq 1$ such that, for all $m \geq n$, if P is a right A-module satisfying $P \oplus A \simeq A^m$ then $P \simeq A^{m-1}$. Clearly, we have that $\operatorname{gsr} A \leq \operatorname{csr} A$.

The connected and the general stable ranks are invariant under homotopy equivalence ([29], [28]): if two Banach algebras A and B are homotopy equivalent (i.e., there are Banach algebra morphisms $\phi : A \to B$ and $\psi : B \to A$ such that $\psi\phi$ is homotopic to id_A and $\phi\psi$ is homotopic to id_B via paths of Banach algebra morphisms) then $\operatorname{csr} A = \operatorname{csr} B$ and $\operatorname{gsr} A = \operatorname{gsr} B$. For this reason, we think of the connected and the general stable ranks as being *homotopical* stable ranks (cf. [28]). The similarities between these two stable ranks, and the contrast between the homotopical and the dimensional stable ranks is the main theme of [28]. One such difference is that the homotopical stable ranks of $C(X)$ can be quite hard to compute, in general. Furthermore, here we encounter a difference amongst the homotopical stable ranks: $\operatorname{csr} C(X)$ can be computed whenever X satisfies a fairly mild cohomological hypothesis ([27], [28]), but no such general result is known for $\operatorname{gsr} C(X)$. In fact, even the computation of $\operatorname{gsr} C(T^d)$—where T^d denotes the familiar d-dimensional torus—is not known!

Stable ranks turn out to be relevant in K-theory; see Section 6. As we have seen in Section 3, K-theory is preserved by suitable spectral morphisms. The following question arises: are stable ranks also preserved by these spectral morphisms? This question, first suggested in [37], is called *Swan's problem.*

4.1 The Gelfand transform. We start with the behavior of the dimensional stable ranks under the Gelfand transform $A \to C(X_A)$ of a commutative Banach algebra A. For the Bass stable rank, we have $\operatorname{bsr} A \leq \operatorname{bsr} C(X_A)$ (Corach–Larotonda [11]); examples such as the disk algebra $A(D)$ or the Hardy algebra $H^\infty(D)$, where D denotes the open unit disk, show that the inequality can be strict. However, if the Gelfand transform $A \to C(X_A)$ is also dense then $\operatorname{bsr} A = \operatorname{bsr} C(X_A)$ (Vaserstein [43]). For the topological stable rank, no general comparison between $\operatorname{tsr} A$ and $\operatorname{tsr} C(X_A)$ is known.

On the other hand, the homotopical stable ranks are invariant under the Gelfand transform:

Theorem 4.1 (Novodvorskii–Taylor, Forster–Taylor) *Let* $A \to C(X_A)$ *be the Gelfand transform of a commutative Banach algebra* A*. Then* $\operatorname{csr} A = \operatorname{csr} C(X_A)$ *and* $\operatorname{gsr} A = \operatorname{gsr} C(X_A)$.

The invariance of the connected stable rank is due to the fact that the Gelfand transform $A \to C(X_A)$ induces a bijection $[\mathrm{Lg}_n(A)] \to [\mathrm{Lg}_n(C(X_A))]$ for all $n \geq 1$. When A is semisimple, this fact is a consequence of Novodvorskii's results from [30]; in general, it is proved by Taylor in [39].

The first result pointing towards the invariance of the general stable rank is due to Lin [24] (see also Taylor [39]) who showed that $(a_1, \ldots, a_n) \in \mathrm{Lg}_n(A)$ is the last column of a matrix in $\mathrm{GL}_n(A)$ if and only if $(\hat{a}_1, \ldots, \hat{a}_n) \in \mathrm{Lg}_n(C(X_A))$ is the last column of a matrix in $\mathrm{GL}_n(C(X_A))$; in particular, we have that $\mathrm{gsr}\, A \leq \mathrm{gsr}\, C(X_A)$. To obtain the desired invariance of the general stable rank, recall instead the argument which showed the invariance of K_0 under the Gelfand transform: $A \to C(X_A)$ induces a bijection $[\mathrm{Idem}(\mathrm{M}_n(A))] \to [\mathrm{Idem}(\mathrm{M}_n(C(X_A)))]$ for all $n \geq 1$, hence a monoid isomorphism $\mathcal{P}(A) \to \mathcal{P}(C(X_A))$ (Forster [15], Taylor [39]). Since the general stable rank can also be defined in terms of a certain cancellation property for projective modules, it trivially follows that the general stable rank is invariant under the Gelfand transform.

4.2 Dense and spectral morphisms. We start again with the dimensional stable ranks: the Bass stable rank and the topological stable rank. If $A \to B$ is a dense and spectral morphism, then

$$\mathrm{bsr}\, A \leq \mathrm{bsr}\, B \leq \mathrm{tsr}\, B \leq \mathrm{tsr}\, A.$$

The inequality $\mathrm{tsr} B \leq \mathrm{tsr} A$ is easily proved under the mere assumption that $A \to B$ has dense image. Since spectrality plays no role, this inequality cannot be taken as a hint for what should happen in the Gelfand context.

The inequality $\mathrm{bsr}\, A \leq \mathrm{bsr}\, B$, due to Swan [37], mirrors (and predates) the Corach–Larotonda result from the Gelfand context. An instance when equality holds is provided by the following result of Badea [3]: if A is a dense and spectral Banach $*$-subalgebra of a C*-algebra B, and A is closed under C^∞-functional calculus for self-adjoint elements, then $\mathrm{bsr}\, A = \mathrm{bsr}\, B$; this applies, for example, to the dense and spectral subalgebras constructed from derivations (Example 2.7).

Having $\mathrm{bsr}\, A = \mathrm{tsr}\, A$ would, of course, solve Swan's problem for both dimensional stable ranks. This is known to hold whenever A is a C*-algebra (Herman–Vaserstein [17]); unfortunately, the domain of a dense and spectral morphism is a C*-algebra only in trivial cases. Therefore, extensions of the Herman–Vaserstein theorem outside of the scope of C*-algebras would be very relevant here.

For the homotopical stable ranks, namely the connected stable rank and the general stable rank, we have a very satisfactory result:

Theorem 4.2 *If $A \to B$ is a dense and spectral morphism, then* $\mathrm{csr}\, A = \mathrm{csr}\, B$ *and* $\mathrm{gsr}\, A = \mathrm{gsr}\, B$.

The invariance of the connected stable rank was first proved in [27], as a consequence of the fact that a dense and spectral morphism $A \to B$ induces a bijection $[\mathrm{Lg}_n(A)] \to [\mathrm{Lg}_n(B)]$ for all $n \geq 1$; in [28], a different proof is given. The invariance of the general stable rank can be proved by a direct argument (see [28]), or by invoking the following analogue of the Forster–Taylor theorem from the Gelfand context: a dense and spectral morphism $A \to B$ induces a monoid isomorphism $\mathcal{P}(A) \to \mathcal{P}(B)$; this follows from the fact that $[\mathrm{Idem}(\mathrm{M}_n(A))] \to [\mathrm{Idem}(\mathrm{M}_n(B))]$ is a bijection for all $n \geq 1$. However, the direct argument has the added benefit of being adaptable to the ring-theoretic context of Swan [37].

5 Relatively spectral morphisms

Starting from the notion of spectral morphism, there are several ways of encountering into a weaker property. Firstly, consider a dense and spectral morphism of Banach algebras $A \to B$. Full surjectivity is not needed—on the B side, one approximates by elements from the image of A. This idea, that density suffices, could then be applied on the A side: the spectral condition could be required to hold on a dense subalgebra of A only. Secondly, instead of a dense inclusion consider a nesting of dense inclusions of Banach algebras $A \hookrightarrow B \hookrightarrow C$. As $\mathrm{sp}_A(a) \supseteq \mathrm{sp}_B(a) \supseteq \mathrm{sp}_C(a)$ for all $a \in A$, the inclusion $A \hookrightarrow C$ is spectral if and only if the inclusion $A \hookrightarrow B$ is spectral and the inclusion $B \hookrightarrow C$ is spectrum-preserving over the dense subalgebra A. Thirdly, let Γ be a group and consider the dense inclusion $\ell^1\Gamma \hookrightarrow \mathrm{C}^*_\mathrm{r}\Gamma$, or any other dense inclusion involving two Banach algebra completions of the group algebra $\mathbb{C}\Gamma$. Amidst the hazy halo of such a Banach algebra completion, the elements of $\mathbb{C}\Gamma$ have a reassuring concreteness. One is compelled to think of the possibility that the spectrum-invariance property for the dense inclusion $\ell^1\Gamma \hookrightarrow \mathrm{C}^*_\mathrm{r}\Gamma$ is only known for elements of $\mathbb{C}\Gamma$, instead of having it satisfied throughout $\ell^1\Gamma$.

These situations lead us to the following definition (cf. [27]). A Banach algebra morphism $\phi : A \to B$ is *relatively spectral* if it is spectrum-preserving over a dense subalgebra of A. That is, for some dense subalgebra X of A we have $\mathrm{sp}_B(\phi(x)) = \mathrm{sp}_A(x)$ for all $x \in X$; equivalently, for all $x \in X$ we have that x is invertible in A if and only if $\phi(x)$ is invertible in B.

One naturally wonders at this point:

(†) Let $A \to B$ be a relatively spectral morphism of Banach algebras. Is $A \to B$ spectral?

The answer to (†) is positive whenever A satisfies some form of spectral continuity (see [27]). It is unlikely, however, that the property in (†) is always true. As a particular aspect of (†), we do not know whether the relative analogue of Proposition 2.2 holds. So let us call a Banach algebra morphism $A \to B$ *completely relatively spectral* if each amplified morphism $\mathrm{M}_n(A) \to \mathrm{M}_n(B)$ is relatively spectral. This (slightly awkward) terminology follows the standard practice of calling a morphism $A \to B$ *completely* $\mathcal{P}$ whenever the amplified morphisms $\mathrm{M}_n(A) \to \mathrm{M}_n(B)$ satisfy a certain property $\mathcal{P}$.

Here are the relative analogues of Examples 2.5 and 2.7:

Example 5.1 Let Γ be a finitely generated group. If Γ has subexponential growth, then the dense inclusion $\ell^1\Gamma \hookrightarrow \mathrm{C}^*_\mathrm{r}\Gamma$ is completely relatively spectral.

Example 5.2 Let B be a Banach algebra and $\delta : B \to B$ an unbounded derivation. For $k \geq 1$, we let $B^{(k)}$ denote the Banach algebra obtained by completing the (unital) algebra $\mathrm{dom}(\delta^k)$ under the algebra norm

$$\|a\|_{(k)} := \max\{\|\delta^i(a)\| : 0 \leq i \leq k\}.$$

If $B^{(k)}$ is dense in B, then the (continuous) inclusion $B^{(k)} \hookrightarrow B$ is completely relatively spectral.

It is worth pointing out that Example 5.1 is much easier to prove than Example 2.5. Concerning Example 5.1, we also mention that the question whether $\ell^1\Gamma \hookrightarrow \mathrm{C}^*_\mathrm{r}\Gamma$ is spectral whenever Γ has subexponential growth—an instance of (†)—is equivalent to the question whether $\ell^1\Gamma$ is symmetric whenever Γ has subexponential growth. The latter question, still not settled, is a part of the following old problem: characterize the groups Γ for which the Banach algebra $\ell^1\Gamma$ is symmetric. Recall that a unital Banach $*$-algebra is *symmetric* if every self-adjoint element has real spectrum.

As for Example 5.2, it is a generalization of Example 2.7. Indeed, one shows that each inclusion $\mathrm{M}_n(B^{(k)}) \hookrightarrow \mathrm{M}_n(B)$ is spectral relative to the dense subalgebra $\mathrm{M}_n(\mathrm{dom}(\delta^k))$; if δ is assumed to be closed, then $\mathrm{dom}(\delta^k)$ coincides with $B^{(k)}$ and we recover the result of Example 2.7.

Why care about relatively spectral morphisms? The point is that we can reap the same benefits as in the case of spectral morphisms, but with less spectral information. Namely, a Banach algebra morphism which is dense and completely relatively spectral induces an isomorphism in K-theory and preserves the homotopical stable ranks ([27], [28]).

6 Stable ranks and K-theory

6.1 Stabilization in algebraic K-theory. Before we get to stabilization in topological K-theory, let us provide some motivation coming from the algebraic side. Bass [5] devised his "stable range" condition as a way to control the sequence

$$\cdots \to \mathrm{GL}_n(R)/\mathrm{E}_n(R) \to \mathrm{GL}_{n+1}(R)/\mathrm{E}_{n+1}(R) \to \cdots$$

which yields the algebraic K_1-group $K_1^{\mathrm{alg}}(R)$ in the limit. Recall that, for a unital ring R, the elementary group $\mathrm{E}_n(R)$ is the subgroup of $\mathrm{GL}_n(R)$ generated by the elementary matrices $\{1_n + e_{ij}\}_{1\leq i\neq j\leq n}$. We think of the quotients $\mathrm{GL}_n(R)/\mathrm{E}_n(R)$ as being the non-stable K_1-groups of R, despite the fact that $\mathrm{GL}_n(R)/\mathrm{E}_n(R)$ is not necessarily a group—that is, $\mathrm{E}_n(R)$ may not be normal in $\mathrm{GL}_n(R)$. The map $\mathrm{GL}_n(R) \to \mathrm{GL}_{n+1}(R)$, given by $* \mapsto \left(\begin{smallmatrix} * & 0 \\ 0 & 1 \end{smallmatrix}\right)$, sends $\mathrm{E}_n(R)$ to $\mathrm{E}_{n+1}(R)$, so it induces a well-defined map $\mathrm{GL}_n(R)/\mathrm{E}_n(R) \to \mathrm{GL}_{n+1}(R)/\mathrm{E}_{n+1}(R)$.

The following theorem was conjectured by Bass in [5]. Bass proved the surjectivity part, whereas the injectivity part, significantly more difficult, is due to Vaserstein [42].

Theorem 6.1 (Bass–Vaserstein) *The map*

$$\mathrm{GL}_n(R)/\mathrm{E}_n(R) \to \mathrm{GL}_{n+1}(R)/\mathrm{E}_{n+1}(R)$$

is surjective for $n \geq \mathrm{bsr}\, R$, *and injective for* $n \geq \mathrm{bsr}\, R + 1$.

In particular, $\mathrm{GL}_n(R)/\mathrm{E}_n(R)$ is a group isomorphic to $K_1^{\mathrm{alg}}(R)$ for $n \geq \mathrm{bsr} R+1$.

In [36], Suslin extended the Bass–Vaserstein theorem to higher algebraic K-theory. For a sequence of non-stable K_i-groups

$$K_{i,1}(R) \to K_{i,2}(R) \to \cdots \to K_{i,n}(R) \to K_{i,n+1}(R) \to \cdots$$

we expect stabilization whenever n is large enough compared to bsr R and i. As is usually the case with higher algebraic K-theory, a significant problem is how to define these non-stable K_i-groups. The two possibilities considered by Suslin are $K_{i,n}^V(R)$ in the sense of Volodin, and $K_{i,n}^Q(R)$ in the sense of Quillen; see [36] for the definitions. There is a canonical map $K_{i,n}^V(R) \to K_{i,n}^Q(R)$ for $n \geq 2i+1$. For these two interpretations of the non-stable K_i-groups, Suslin proved the following stabilization results:

(K_i^V) The map $K_{i,n}^V(R) \to K_{i,n+1}^V(R)$ is surjective for $n \geq \mathrm{bsr}\, R + i - 1$, and injective for $n \geq \mathrm{bsr}\, R + i$.

(K_i^Q) The map $K_{i,n}^Q(R) \to K_{i,n+1}^Q(R)$ is surjective for $n \geq \max\{i+1, \mathrm{bsr}\, R\}+i-1$, and injective for $n \geq \max\{i+1, \mathrm{bsr}\, R\} + i$.

A weaker version of (K_i^Q) was also obtained by Maazen and van der Kallen [41].

6.2 Stabilization in topological K-theory. Let A be a (unital) Banach algebra. For each $k \geq 0$, we have a direct sequence of identity-based homotopy groups:

$$(\pi_k) \qquad \pi_k(\mathrm{GL}_1(A)) \to \pi_k(\mathrm{GL}_2(A)) \to \cdots \to \pi_k(\mathrm{GL}_n(A)) \to \pi_k(\mathrm{GL}_{n+1}(A)) \to \cdots$$

The maps in this sequence are induced by the embeddings $\mathrm{GL}_n(A) \hookrightarrow \mathrm{GL}_{n+1}(A)$ given by $* \mapsto \left(\begin{smallmatrix} * & 0 \\ 0 & 1 \end{smallmatrix}\right)$. The direct limit of the homotopy sequence (π_k) is, by definition, the topological K-group $K_{k+1}(A)$. By Bott periodicity, $K_{k+1}(A)$ is isomorphic to $K_0(A)$ or $K_1(A)$ according to whether k is odd or even.

We are interested in the following stabilization problem: when do the maps in the homotopy sequence (π_k) become—and remain—injective/surjective/bijective? The first stabilization result in this direction appears in [31]. We formulate Rieffel's result in a way which makes it the topological analogue of the Bass–Vaserstein theorem:

Theorem 6.2 (Rieffel) *The map $\pi_0(\mathrm{GL}_n(A)) \to \pi_0(\mathrm{GL}_{n+1}(A))$ is surjective for $n \geq \mathrm{bsr}\, A$, and injective for $n \geq \mathrm{bsr}\, A + 1$.*

Actually, Rieffel shows that surjectivity holds for $n \geq \mathrm{csr}\, A - 1$; since $\mathrm{bsr}\, A \geq \mathrm{csr}\, A - 1$ ([31], [11]), surjectivity also holds for $n \geq \mathrm{bsr}\, A$. As for the fact that injectivity holds for $n \geq \mathrm{bsr}\, A + 1$, Rieffel proves this by invoking the injectivity part of the Bass–Vaserstein theorem.

Corach–Larotonda [12] then generalized Theorem 6.2 to higher homotopy groups as follows:

Theorem 6.3 (Corach–Larotonda) *The map $\pi_k(\mathrm{GL}_n(A)) \to \pi_k(\mathrm{GL}_{n+1}(A))$ is surjective for $n \geq \mathrm{bsr}\, A + k$, and injective for $n \geq \mathrm{bsr}\, A + k + 1$.*

The proof of this result holds the key to all stabilization results in the literature. It relies on the existence of a fibration $\mathrm{GL}_n(A) \to \mathrm{GL}_{n+1}(A) \to \mathrm{Lg}_{n+1}(A)$, generalizing the more familiar fibration $\mathrm{U}_n \to \mathrm{U}_{n+1} \to S^{2n+1}$. For $n \geq \mathrm{csr}\, A - 1$, $\mathrm{Lg}_{n+1}(A)$ is connected and we get a long exact homotopy sequence

$$\cdots \to \pi_{k+1}(\mathrm{Lg}_{n+1}(A)) \to \pi_k(\mathrm{GL}_n(A)) \to \pi_k(\mathrm{GL}_{n+1}(A)) \to \pi_k(\mathrm{Lg}_{n+1}(A)) \to \cdots$$

ending in

$$\cdots \to \pi_1(\mathrm{Lg}_{n+1}(A)) \to \pi_0(\mathrm{GL}_n(A)) \to \pi_0(\mathrm{GL}_{n+1}(A)) \to 0.$$

We identify $\pi_k(\mathrm{Lg}_{n+1}(A))$ with $\pi_0(\mathrm{Lg}_{n+1}(A(S^k)))$, where $A(S^k)$ denotes the Banach algebra of continuous maps from the k-sphere S^k to A. This brings the connected stable rank of $A(S^k)$ into play, for which—as one can easily show—we have $\mathrm{csr}\, A(S^k) \geq \mathrm{csr}\, A$. From the long exact homotopy sequence we conclude:

Theorem 6.4 *The map $\pi_k(\mathrm{GL}_n(A)) \to \pi_k(\mathrm{GL}_{n+1}(A))$ is surjective for $n \geq \mathrm{csr}\, A(S^k) - 1$, and injective for $n \geq \mathrm{csr}\, A(S^{k+1}) - 1$.*

To obtain Theorem 6.3 from Theorem 6.4, one needs to know that $\mathrm{bsr} A + k + 1 \geq \mathrm{csr}\, A(S^k)$. To that end, it suffices to know that $\mathrm{bsr}\, A + k \geq \mathrm{bsr}\, A(S^k)$, and this is precisely what Corach–Larotonda show in the remainder of their proof.

In the case $k = 0$, Theorem 6.4 together with the fact that $\mathrm{tsr} A + 1 \geq \mathrm{csr}\, A(S^1)$ (Rieffel [31]) yield the following significant result from [32]:

Theorem 6.5 (Rieffel) *The map $\pi_0(\mathrm{GL}_n(A)) \to \pi_0(\mathrm{GL}_{n+1}(A))$ is bijective for $n \geq \mathrm{tsr}\, A$.*

Comparing the two theorems of Rieffel, Theorem 6.2 and Theorem 6.5, we see that the latter strengthens the former when A is a C*-algebra—recall that the Bass and the topological stable ranks coincide in this case. I do not know whether this is still the case for general Banach algebras.

A variant of the Corach–Larotonda fibration, where the base space $\mathrm{Lg}_n(A)$ is replaced by the subspace $\mathrm{Lc}_n(A)$ consisting of last columns of invertible n-by-n matrices, is used by Rieffel in [32] and later on by Thomsen [40] and Badea [4], among others. This allows for the following sharpening of the injectivity bound from Theorem 6.4: the map $\pi_k(\mathrm{GL}_n(A)) \to \pi_k(\mathrm{GL}_{n+1}(A))$ is injective for $n \geq \mathrm{gsr}\, A(S^{k+1}) - 1$ (Badea [4]). A downside of this improvement is that the general stable rank is, in general, more difficult to compute than the connected stable rank.

Compared to the Corach–Larotonda theorem, Theorem 6.4 has a clear formal advantage. On the one hand, the injective/surjective stabilization of (π_k) has a three-fold invariance: with respect to homotopy equivalence, across Gelfand transforms, and across dense and spectral morphisms. On the other hand, the bounds from Theorem 6.4 enjoy the same three-fold invariance. A concrete example highlighting the power of Theorem 6.4 over the Corach–Larotonda theorem is provided by the irrational rotation C*-algebra A_θ (cf. [32]). It turns out that $\mathrm{csr}\, A_\theta(S^k) \leq 2$ for each $k \geq 0$; by Theorem 6.4, for each $k \geq 0$ we have that $\pi_k(\mathrm{GL}_n(A_\theta))$ is isomorphic to $K_{k+1}(A_\theta) \simeq \mathbb{Z}^2$ for all $n \geq 1$. Such a result would not be visible in its entirety through the Corach–Larotonda theorem.

All the results mentioned so far indicate an index n from which all maps in the homotopy sequence (π_k) are isomorphisms. But how to compute the smallest index n with the property that the sequence (π_k) stabilizes from n onwards? The non-triviality of such a computation is already suggested by the trivial case $A = \mathbb{C}$: it can be shown that the smallest such index is $n = \lfloor k/2 \rfloor + 1$, but one needs rather detailed information about the homotopy groups of unitary groups for this.

Let $\mathrm{bij}_k\, A$ denote the least $n \geq 1$ such that the map

$$\pi_k(\mathrm{GL}_m(A)) \to \pi_k(\mathrm{GL}_{m+1}(A))$$

is bijective for all $m \geq n$; equivalently, $\mathrm{bij}_k\, A$ is the least $n \geq 1$ such that the natural map $\pi_k(\mathrm{GL}_m(A)) \to K_{k+1}(A)$ is bijective for all $m \geq n$. By Theorem 6.4, we have the following upper bound:

$$\mathrm{bij}_k\, A \leq \max\{\mathrm{csr}\, A(S^k) - 1, \mathrm{csr}\, A(S^{k+1}) - 1\}$$

But the long exact homotopy sequence arising from the Corach–Larotonda fibration also yields the following inequality:

$$\mathrm{csr}\, A(S^{k+1}) - 1 \leq \max\{\mathrm{bij}_k\, A, \mathrm{bij}_{k+1}\, A, \mathrm{csr}\, A - 1\}$$

A variant of this inequality, formulated in terms of certain higher connected stable ranks, appears in [27]. These two inequalities allow us to compute $\max\{\mathrm{bij}_k\, A, \mathrm{bij}_{k+1}\, A\}$, which encodes the *joint* stabilization of (π_k) and (π_{k+1}). This stabilization for pairs of consecutive homotopy sequences has the following K-theoretic interpretation: if we think of stabilization in a single homotopy sequence (π_k) as stabilization for one of the two K-groups, then the joint stabilization of (π_k) and (π_{k+1}) should be understood as stabilization for the K-theory, i.e., for both K-groups, of A.

We first consider the case of commutative C*-algebras. Let $A = C(X^d)$, where X^d is a d-dimensional compact space, and assume that $H^d(X^d; \mathbb{Z}) \neq 0$. If $k \geq 0$

has the same parity as d, then $n = 1+\frac{1}{2}(d+k)$ is the least integer with the property that the natural maps

$$\pi_k(\mathrm{GL}_m(A)) \to K_{k+1}(A), \qquad \pi_{k+1}(\mathrm{GL}_m(A)) \to K_{k+2}(A)$$

are isomorphisms for all $m \geq n$.

As a non-commutative example, we consider tensor products of C*-extensions of compacts by commutative C*-algebras. For $1 \leq i \leq N$, let A_i be a unital C*-extension of $\mathcal{K}$ by $C(X_i)$, where X_i is a compact metrizable space. Put $A = A_1 \otimes \cdots \otimes A_N$, and $X = X_1 \times \cdots \times X_N$. Assume that $d = \dim X$ is non-zero, and $H^d(X^d; \mathbb{Z}) \neq 0$. If $k \geq 0$ has the same parity as d, then $n = 1 + \frac{1}{2}(d+k)$ is the least integer with the property that the natural maps

$$\pi_k(\mathrm{GL}_m(A)) \to K_{k+1}(A), \qquad \pi_{k+1}(\mathrm{GL}_m(A)) \to K_{k+2}(A)$$

are isomorphisms for all $m \geq n$.

Roughly speaking, this means that A has the same stabilization pattern as its "symbol space" $C(X)$. The key point in both examples is, of course, the computation of csr $A(S^k)$. This is immediate for the first example, but the computation for the second example is more involved. The ingredients of this computation, performed in [28], are estimates of stable ranks for quotients and extensions, as well as Nistor's result from [29] on the dimensional stable rank for such tensor products.

Acknowledgments I thank the referee for some useful suggestions.

References

[1] R. Arens: *The group of invertible elements of a commutative Banach algebra*, Studia Math. (Ser. Specjalna) Zeszyt 1 (1963), 21–23

[2] R. Arens: *To what extent does the space of maximal ideals determine the algebras?*, Proceedings of the International Symposium on Function Algebras (Tulane University, 1965), 164–168, Scott-Foresman 1966

[3] C. Badea: *The stable rank of topological algebras and a problem of R.G. Swan*, J. Funct. Anal. 160 (1998), no. 1, 42–78

[4] C. Badea: *Stable ranks, K-groups and Witt groups of some Banach and C*-algebras*, Function spaces (Edwardsville, IL, 1998), 55–64, Contemp. Math. 232, Amer. Math. Soc. 1999

[5] H. Bass: *K-theory and stable algebra*, Publ. IHES 22 (1964), 5–60

[6] B. Blackadar: *K-theory for operator algebras*, Second edition, MSRI Publications 5, Cambridge University Press 1998

[7] J.-B. Bost: *Principe d'Oka, K-théorie et systèmes dynamiques non commutatifs*, Invent. Math. 101 (1990), no. 2, 261–333

[8] N. Bourbaki: *Théories spectrales*, Éléments de mathématique Fasc. XXXII, Hermann 1967

[9] A. Connes: *An analogue of the Thom isomorphism for crossed products of a C*-algebra by an action of* $\mathbb{R}$, Adv. in Math. 39 (1981), no. 1, 31–55

[10] A. Connes, H. Moscovici: *Cyclic cohomology, the Novikov conjecture and hyperbolic groups*, Topology 29 (1990), no. 3, 345–388

[11] G. Corach, A.R. Larotonda: *Stable range in Banach algebras*, J. Pure Appl. Algebra 32 (1984), no. 3, 289–300

[12] G. Corach, A.R. Larotonda: *A stabilization theorem for Banach algebras*, J. Algebra 101 (1986), no. 2, 433–449

[13] A.M. Davie: *Homotopy in Fréchet algebras*, Proc. London Math. Soc. (3) 23 (1971), 31–52

[14] P. de la Harpe: *Groupes hyperboliques, algèbres d'opérateurs et un théorème de Jolissaint*, C. R. Acad. Sci. Paris Sér. I Math. 307 (1988), no. 14, 771–774

[15] O. Forster: *Funktionentheoretischer Hilfsmittel in der Theorie der kommutativen Banach Algebren*, Jber. Deutsch. Math.-Verein. 76 (1974), 1–17

[16] U. Haagerup: *An example of a nonnuclear C*-algebra, which has the metric approximation property*, Invent. Math. 50 (1978/79), no. 3, 279–293

[17] R.H. Herman, L.N. Vaserstein: *The stable range of C^*-algebras*, Invent. Math. 77 (1984), no. 3, 553–555

[18] J.W. Jenkins: *Symmetry and nonsymmetry in the group algebras of discrete groups*, Pacific J. Math. 32 (1970), 131–145

[19] R. Ji: *Smooth dense subalgebras of reduced group C^*-algebras, Schwartz cohomology of groups, and cyclic cohomology*, J. Funct. Anal. 107 (1992), no. 1, 1–33

[20] P. Jolissaint: *Rapidly decreasing functions in reduced C^*-algebras of groups*, Trans. Amer. Math. Soc. 317 (1990), no. 1, 167–196

[21] M. Karoubi: *K-theory. An introduction*, Grundlehren der Mathematischen Wissenschaften, Band 226, Springer 1978

[22] V. Lafforgue: *A proof of property (RD) for cocompact lattices of* $\mathrm{SL}(3,\mathbb{R})$ *and* $\mathrm{SL}(3,\mathbb{C})$, J. Lie Theory 10 (2000), no. 2, 255–267

[23] V. Lafforgue: *K-théorie bivariante pour les algèbres de Banach et conjecture de Baum-Connes*, Invent. Math. 149 (2002), no. 1, 1–95

[24] V.Ya. Lin: *Holomorphic fiberings and multivalued functions of elements of a Banach algebra*, Funkcional. Anal. i Priložen. 7 (1973), no. 2, 43–51

[25] J. Ludwig: *A class of symmetric and a class of Wiener group algebras*, J. Funct. Anal. 31 (1979), no. 2, 187–194

[26] M.A. Naimark: *Normed algebras*, Third edition, Wolters-Noordhoff Series of Monographs and Textbooks on Pure and Applied Mathematics, Wolters-Noordhoff Publishing 1972

[27] B. Nica: *Relatively spectral morphisms and applications to K-theory*, J. Funct. Anal. 255 (2008), 3303–3328

[28] B. Nica: *Homotopical stable ranks for Banach algebras*, J. Funct. Anal. 261 (2011), 803–830

[29] V. Nistor: *Stable range for tensor products of extensions of $\mathcal{K}$ by $C(X)$*, J. Operator Theory 16 (1986), no. 2, 387–396

[30] M.E. Novodvorskii: *Certain homotopic invariants of the space of maximal ideals* (Russian), Mat. Zametki 1 (1967), no. 4, 487–494; *Certain homotopical invariants of spaces of maximal ideals* (English), Math. Notes 1 (1967), no. 4, 321–325

[31] M.A. Rieffel: *Dimension and stable rank in the K-theory of C^*-algebras*, Proc. London Math. Soc. (3) 46 (1983), no. 2, 301–333

[32] M.A. Rieffel: *The homotopy groups of the unitary groups of noncommutative tori*, J. Operator Theory 17 (1987), no. 2, 237–254

[33] H.L. Royden: *Function algebras*, Bull. Amer. Math. Soc. 69 (1963), 281–298

[34] L.B. Schweitzer: *A short proof that $M_n(A)$ is local if A is local and Fréchet*, Internat. J. Math. 3 (1992), no. 4, 581–589

[35] G.E. Shilov: *On decomposition of a commutative normed ring in a direct sums of ideals*, (Russian) Mat. Sbornik N.S. 32(74), (1953), 353–364

[36] A.A. Suslin: *Stability in algebraic K-theory*, Algebraic K-theory, Part I (Oberwolfach, 1980), 304–333, Lecture Notes in Math. 966, Springer 1982

[37] R.G. Swan: *Topological examples of projective modules*, Trans. Amer. Math. Soc. 230 (1977), 201–234

[38] J.L. Taylor: *Banach algebras and topology*, Algebras in analysis (Proc. Instructional Conf. and NATO Advanced Study Inst., Birmingham, 1973), 118–186, Academic Press 1975

[39] J.L. Taylor: *Topological invariants of the maximal ideal space of a Banach algebra*, Advances in Math. 19 (1976), no. 2, 149–206

[40] K. Thomsen: *Nonstable K-theory for operator algebras*, K-Theory 4 (1991), no. 3, 245–267

[41] W. van der Kallen: *Homology stability for linear groups*, Invent. Math. 60 (1980), no. 3, 269–295

[42] L.N. Vaserstein: *On the stabilization of the general linear group over a ring*, Math. USSR-Sb. 8 (1969), 383–400

[43] L.N. Vaserstein: *The stable range of rings and the dimension of topological spaces*, Funkcional. Anal. i Priložen. 5 (1971), no. 2, 17–27

Fields Institute Communications
Volume **61**, 2011

A Survey of Braided Hopf Cyclic Cohomology

Arash Pourkia
Department of Mathematics
The University of Western Ontario
London, ON Canada N6A 5B7
apourkia@uwo.ca

Abstract. We give a brief survey of some aspects of Hopf cyclic cohomology for Hopf algebras in a braided monoidal abelian category.

1 Introduction

Hopf cyclic cohomology was introduced by Connes and Moscovici in [10, 11, 12]. The original motivation was to express the local index formula of a transversally elliptic operator as the image of a certain Hopf cyclic cohomology class, under a characteristic map.

The main motivation for braided Hopf cyclic cohomology is as follows. There are many interesting Hopf algebra-like objects that are close to being a Hopf algebra but are not Hopf algebras in the usual sense. There exists a desire to extend the Hopf cyclic theory to those types of objects. Many of those Hopf algebra-like objects are Hopf algebras in some *braided monoidal categories*. Super Hopf algebras, quasitriangular quasi-Hopf algebras [17, 28], and also multiplier Hopf algebras (based on some new research [15]), are examples of this. Therefore, to achieve the desired goal, it seems natural to extend the Hopf cyclic theory to the context of braided monoidal categories.

This article is a very short survey on the work in [21]. We also refer the reader to the recent works of other authors regarding the subjects of braided Hopf algebras, Hopf cyclic cohomology, and (Hopf) cyclic cohomology in (braided) monoidal categories [1], [2], [4, 5], [15], [18, 19], [23, 24, 25].

A detailed report on all those works does not fit in this very short survey, but will appear elsewhere.

2 Hopf cyclic cohomology

Cyclic (co)homology for (noncommutative) algebras [7, 8], was introduced by Connes, in the early 80's, as the noncommutative analog of the de Rham homology of currents on smooth manifolds. He, in fact, defined more generally the cyclic

2010 *Mathematics Subject Classification.* Primary 19D55; Secondary 16E40, 16T05, 18D10.

(co)homology for *(co)cyclic objects* in any category $\mathcal{C}$ by introducing the *cyclic category*, Λ, as the significant generalization of the *simplicial category*, Δ. The cyclic category Λ [7, 20, 26], has the same set of objects as Δ, namely, the totally ordered sets

$$[n] = \{0 < 1 < \cdots < n\}, \qquad n = 0, 1, 2, \ldots.$$

In fact, Λ contains Δ as a subcategory. Morphisms of Λ are generated by the standard simplicial morphisms

$$\delta_i : [n-1] \to [n], \quad i = 0, 1, \ldots, n$$

$$\sigma_j : [n+1] \to [n], \qquad j = 0, 1, \ldots, n,$$

plus a new set of morphisms $\tau_n : [n] \to [n]$ for $n \geq 0$, called *cyclic morphisms*, defined by

$$\tau_n(i) = \begin{cases} n & i = 0 \\ i-1 & i \neq 0 \end{cases}$$

satisfying the standard simplicial relations:

$$\begin{aligned} \delta_j \delta_i &= \delta_i \delta_{j-1} \quad \text{if} \quad i < j, \\ \sigma_j \sigma_i &= \sigma_i \sigma_{j+1} \quad \text{if} \quad i \leq j, \\ \sigma_j \delta_i &= \begin{cases} \delta_i \sigma_{j-1} \quad \text{if} & i < j \\ \text{id} \quad \text{if} & i = j \text{ or } i = j+1 \\ \delta_{i-1} \sigma_j \quad \text{if} & i > j+1, \end{cases} \end{aligned} \tag{2.1}$$

plus the *cyclic relations*:

$$\begin{aligned} \tau_n \delta_i &= \delta_{i-1} \tau_{n-1} & 1 \leq i \leq n \\ \tau_n \delta_0 &= \delta_n \\ \tau_n \sigma_i &= \sigma_{i-1} \tau_{n+1} & 1 \leq i \leq n \\ \tau_n \sigma_0 &= \sigma_n \tau_{n+1}^2 \\ \tau_n^{n+1} &= \text{id}. \end{aligned} \tag{2.2}$$

A cocyclic object (resp. cyclic object) in any category $\mathcal{C}$ is defined as a functor $\Lambda \to \mathcal{C}$ (resp. $\Lambda^{op} \to \mathcal{C}$). More precisely, a cocyclic object in $\mathcal{C}$ is given by the data

$$X = (X^n, \delta_i, \sigma_i, \tau_n), \quad n \geq 0,$$

where X^n, $n \geq 0$, are objects of $\mathcal{C}$, and faces $\delta_i : X^{n-1} \to X^n$, $0 \leq i \leq n$, degeneracies $\sigma_i : X^{n+1} \to X^n$, $0 \leq i \leq n$, and cyclic maps $\tau_n : X^n \to X^n$ are morphisms of $\mathcal{C}$ satisfying all the above relations in (2.1) and (2.2). In a dual fashion one can define a cyclic object.

In the special case, if $\mathcal{C}$ is the category of k-modules, for a field k, the terminology *(co)cyclic module* is used instead of (co)cyclic object.

Remark 2.1 It is important to mention that among all the relations in (2.2), the last one, $\tau_n^{n+1} = \text{id}$, is a crucial datum in cyclic theory and is called the *cyclic condition*. If all other relations in (2.1) and (2.2) hold except the cyclic condition, we say we have a *para-(co)cyclic* object. It is well known that starting from a para-cocyclic object one can always obtain a cocyclic object by restricting to the subcomplex $\ker(1 - \tau_n^{n+1})$. The same is true for turning a para-cyclic object into a cyclic object, in a dual manner.

Now, if $\mathcal{C}$ is an abelian category, to any (co)cyclic object, $X = (X^n, \delta_i, \sigma_i, \tau_n)$, $n \geq 0$, in $\mathcal{C}$, one can assign a cyclic (co)homology using the so-called Connes complex as follows. The cyclic cohomology of X, denoted by $HC^n(X)$, $n \geq 0$, is the cohomology of the cochain complex

$$X_\lambda^0 \xrightarrow{\ b\ } X_\lambda^1 \xrightarrow{\ b\ } X_\lambda^2 \xrightarrow{\ b\ } X_\lambda^3 \cdots,$$

Here

$$X_\lambda^n = \mathrm{Ker}\,(1 - \lambda_n),$$

where

$$\lambda_n = (-1)^n \tau_n,$$

and $b : X^{n-1} \to X^n$ is defined by

$$b = \sum_{i=0}^{n} (-1)^i \delta_i.$$

Again, in a dual fashion one can define the cyclic homology of a cyclic object in $\mathcal{C}$.

Remark 2.2 One also defines the periodic cyclic cohomology of X, denoted by $HP^n(X)$, $n = 0, 1$, as the direct limit of $HC^n(X)$, $n \geq 0$, under the so called Connes' periodicity map [7, 8]:

$$HC^n(X) \to HC^{n+2}(X).$$

In the late 90's Connes and Moscovici applied the cyclic (co)homology theory to Hopf algebras [10, 11, 12], to solve the problem of the index theory of elliptic operators on foliations in noncommutative geometry [9]. This theory, known as *Hopf cyclic cohomology* theory (or just Hopf cyclic theory), is a cyclic cohomology theory for a Hopf algebra H equipped with a *modular pair in involution* (MPI) (δ, σ), where δ is a character and σ is a co-character of H satisfying some compatibility relations. More precisely, to $(H, (\delta, \sigma))$, they associated a cocyclic module, which is a cocyclic object in the category of $\mathbb{C}$-modules (Theorem 2.1 below). It is important to mention that most of the interesting Hopf algebras are equipped with a MPI. Examples include (co-)quasitriangular Hopf algebras and quantum groups.

Here is the main theorem of Connes and Moscovici in which they associate a cocyclic module to a Hopf algebra H equipped with a modular pair in involution (MPI):

Theorem 2.1 [10, 11, 12] *Let $(H, \Delta, \varepsilon, S)$ be a Hopf algebra endowed with a modular pair in involution (δ, σ). Define the δ-twisted antipode by $\widetilde{S} = \delta * S$. Set $C^n(H) = H^{\otimes n}$, $n \geq 0$ and define faces, degeneracies, and cyclic maps by:*

$$\delta_i(h_1 \otimes \cdots \otimes h_{n-1}) = \begin{cases} (1 \otimes h_1 \otimes \cdots \otimes h_{n-1}) & i = 0 \\ (h_1 \otimes \cdots \otimes \Delta h_i \otimes \cdots \otimes h_{n-1}) & 1 \leq i \leq n-1 \\ (h_1 \otimes \cdots \otimes h_{n-1} \otimes \sigma) & i = n \end{cases}$$

$$\sigma_i(h_1 \otimes \cdots \otimes h_{n+1}) = \varepsilon(h_{i+1})(h_1 \otimes \cdots \otimes h_i \otimes h_{i+1} \otimes \cdots \otimes h_{n+1}),\ 0 \leq i \leq n$$

$$\tau_n(h_1 \otimes \cdots \otimes h_n) = \Delta^{n-1}\widetilde{S}(h_1) \cdot (h_2 \otimes \cdots \otimes h_n \otimes \sigma).$$

Then $(C^\bullet(H),\ \delta_i,\ \sigma_i,\ \tau)$ is a cocyclic module. □

Therefore one can define the (periodic) cyclic cohomology of this cocyclic module, which is called the (periodic) Hopf cyclic cohomology of $(H, (\delta, \sigma))$.

2.1 Hopf cyclic cohomology with coefficients. Ever since the introduction of Hopf cyclic cohomology theory by Connes and Moscovici, [10, 11, 12], much work has been done regarding the different aspects and the generalizations of this theory, including, [1], [18, 19], [23, 24, 25].

Among all the developments in Hopf cyclic theory, our focus is the theory of Hopf cyclic cohomology with coefficients, [13, 14], which we now briefly recall. Hopf cyclic cohomology was generalized to Hopf cyclic cohomology with coefficients [14], by introducing the notion of *Stable Anti-Yetter Drinfeld* (SAYD) modules [13], which is the generalization of the notion of modular pair in involution (MPI). In fact, in this context, SAYD H-modules are the system of coefficients and any MPI corresponds to a one-dimensional SAYD (Example 2.1 below). In [14] a cocyclic module is associated to the triple (H, C, M), where C is an H*-module coalgebra* and M a SAYD H-module. (Theorem 2.2 below).

Definition 2.1 A right-left anti-Yetter-Drinfeld H-module is a right H-module and left H-comodule M such that

$$(mh)_{(-1)} \otimes (mh)_{(0)} = S(h^{(3)})m_{(-1)}h^{(1)} \otimes m_{(0)}h^{(2)},$$

for all $h \in H$ and $m \in M$. We say that M is stable if in addition we have

$$m_{(0)}m_{(-1)} = m,$$

for all $m \in M$.

Remark 2.3 Notice that if one exchanges the role of antipode, S, for its inverse, S^{-1}, in (2.1), one gets the relations for the right-left Yetter-Drinfeld modules. Therefore if $S^2 = \mathrm{id}$ the notions of Yetter-Drinfeld and anti Yetter-Drinfeld modules coincide. Also notice that there are analogous definitions for left-left, left-right and right-right stable anti-Yetter-Drinfeld (SAYD) modules [13].

Example 2.1 [13] There is a one-one correspondence between modular pairs in involution (MPI) (δ, σ) on H and right-left SAYD module structure on $M = k$, defined by

$$rh = \delta(h)r, \quad r \mapsto \sigma \otimes r,$$

for all $h \in H$ and $r \in k$. We denote this module by $M = {}^{\sigma}k_{\delta}$.

Definition 2.2 A left H-module coalgebra is a coalgebra $(C, \Delta_C, \varepsilon_C)$ equipped with a left H-module structure via the left action

$$\phi_C : H \otimes C \to C, \;\; \phi_C(h \otimes c) = hc$$

such that Δ_C and ε_C are H-module maps, i.e.,

$$(hc)^{(1)} \otimes (hc)^{(2)} = h^{(1)}c^{(1)} \otimes h^{(2)}c^{(2)},$$

or equivalently,

$$\Delta_C\phi_C = (\phi_C \otimes \phi_C)(\mathrm{id}_H \otimes \tau \otimes \mathrm{id}_C)(\Delta_H \otimes \Delta_C),$$

and

$$\varepsilon_C(hc) = \varepsilon_H(h)\varepsilon_C(c),$$

or equivalently,

$$\varepsilon_C\phi_C = \varepsilon_H \otimes \varepsilon_C.$$

Example 2.2 The multiplication $m_H : H \otimes H \to H$ turns H into a left H-module coalgebra.

Theorem 2.2 *Let H be a Hopf algebra, C an H-module coalgebra and M a SAYD H-module. Then the complex $\mathcal{C}_H^n(C,M) := M \otimes_H C^{\otimes(n+1)}$ with the following faces, degeneracies and cyclic maps, defines a cocyclic module attached to the triple (H, C, M):*

$$\delta_i(m \otimes_H c_0 \otimes \cdots \otimes c_{n-1}) = \begin{cases} m \otimes_H c_0 \otimes \cdots \otimes c_i^{(1)} \otimes c_i^{(2)} \otimes c_{n-1} & 0 \leq i < n \\ m^{(0)} \otimes_H c_0^{(2)} \otimes c_1 \otimes \cdots \otimes c_{n-1} \otimes m^{(-1)} c_0^{(1)} & i = n \end{cases}$$

$$\sigma_i(m \otimes_H c_0 \otimes \cdots \otimes c_{n+1}) = m \otimes_H c_0 \otimes \cdots \otimes \varepsilon(c_{i+1}) \otimes \cdots \otimes c_{n+1},\ 0 \leq i \leq n$$

$$\tau_n(m \otimes_H c_0 \otimes \cdots \otimes c_n) = m^{(0)} \otimes_H c_1 \otimes \cdots \otimes c_n \otimes m^{(-1)} c_0.$$

□

Example 2.3 In the above theorem, if we let $C = H$ as in Example 2.2, we recover the notion called the Hopf cyclic cohomology of H with coefficients in M. If we let $C = H$ and $M = k$ as in Example 2.1, we recover the cocyclic module of Theorem 2.1, which defines the standard Connes-Moscovici Hopf cyclic theory.

3 Braided monoidal categories and braided Hopf algebras

3.1 Braided monoidal categories. The study of braided monoidal categories and braided Hopf algebras goes back to [16] and [33].

A monoidal category is a unital associative algebra object in the category of categories. More precisely, a monoidal category $(\mathcal{C}, \otimes, I, a, l, r)$ consists of a category $\mathcal{C}$, a functor $\otimes : \mathcal{C} \times \mathcal{C} \to \mathcal{C}$, an object $I \in \mathcal{C}$ (called *unit object*), and natural isomorphisms, defined for all objects A, B, C, of $\mathcal{C}$ by:

$$a = a_{A,B,C} : A \otimes (B \otimes C) \to (A \otimes B) \otimes C,$$
$$l = l_A : I \otimes A \to A, \qquad r = r_A : A \otimes I \to A,$$

(called the *associativity* and *unit constraints*, respectively), such that the so-called *pentagon* and *triangle* diagrams commute [17, 27, 28].

A *braided monoidal* category is a monoidal category $\mathcal{C}$ endowed with a natural family of isomorphisms

$$\psi_{A,B} : A \otimes B \to B \otimes A,$$

called *braiding* such that the so called *hexagon diagrams* commute.

If we show the braiding map $\psi_{A,B}$ by the following standard diagram,

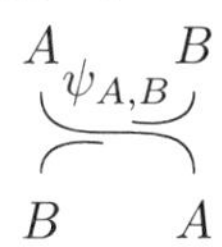

for any A and B in $\mathcal{C}$, then the naturality of ψ can be visualized by the following identity:

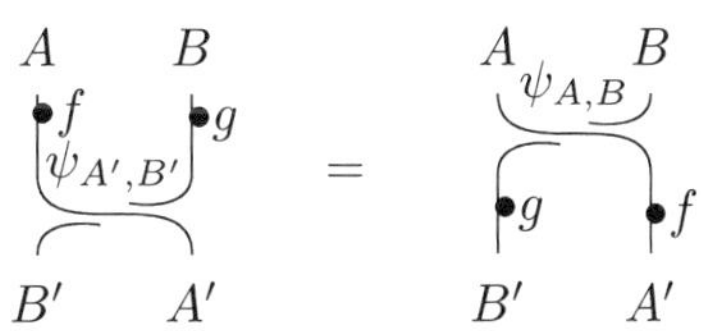

(3.3)

for any two morphisms $f : A \to A'$ and $f : B \to B'$ in $\mathcal{C}$.

A braiding is called a *symmetry* if we have, for any A and B in $\mathcal{C}$,

$$\psi_{B,A} \circ \psi_{A,B} = \mathrm{id}_{A \otimes B},$$

or in terms of braiding diagrams:

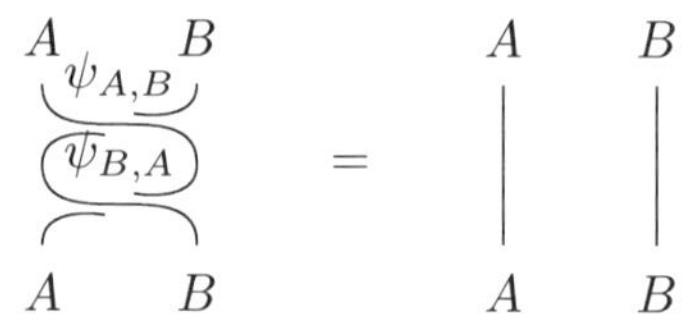

Sometimes we just write $\psi^2 = \mathrm{id}$ to signify the symmetry condition. A *symmetric monoidal category* is a monoidal category endowed with a symmetry.

By a theorem of Mac Lane [27] (cf. also [28]), any (braided) monoidal category is *monoidal equivalent* to a (braided) *strict* monoidal category in which a, l and r are just equalities and the above aforementioned commuting diagrams are reduced to the following equalities:

$$\begin{aligned} (A \otimes B) \otimes C &= A \otimes (B \otimes C), \\ I \otimes A = A &\otimes I = A, \\ \psi_{A, B \otimes C} &= (\mathrm{id}_B \otimes \psi_{A,C})(\psi_{A,B} \otimes \mathrm{id}_C), \\ \psi_{A \otimes B, C} &= (\psi_{A,C} \otimes \mathrm{id}_B)(\mathrm{id}_A \otimes \psi_{B,C}), \end{aligned}$$

for all objects A, B, C of $\mathcal{C}$. In terms of braided diagrams, the last identity, for instance, is as follows:

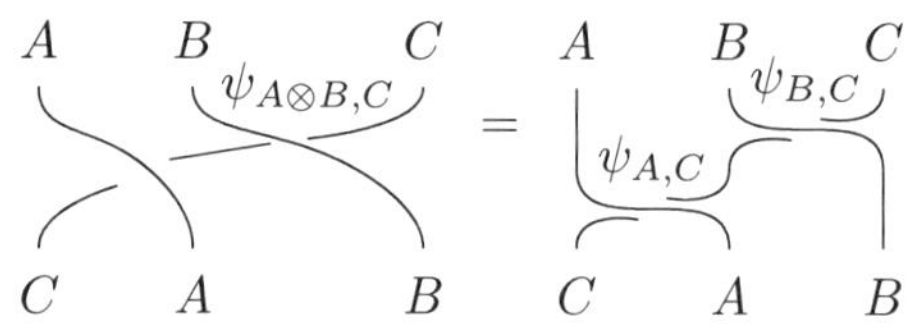

Example 3.1 The category of vector spaces over a field k, Vec_k, with the usual flip as the braiding map, $\psi_{V \otimes W}(v \otimes w) = f(v \otimes w) = (w \otimes v)$, is a symmetric braided monoidal category.

Example 3.2 The category of super vector spaces over a field k, Vec_k^s, also denoted by $\mathbb{Z}_2$-Mod, with even morphisms and with the following braiding map for any two super spaces $V = V_0 \oplus V_1$ and $W = W_0 \oplus W_1$:

$$\psi_{V \otimes W}(v \otimes w) = (-1)^{|v||w|}\,(w \otimes v), \tag{3.4}$$

is another examples of a symmetric braided monoidal category.

There is also a category of differential graded (DG) super vector spaces whose objects are super complexes $V_0 \underset{d}{\overset{d}{\leftrightarrows}} V_1$, and whose morphisms are even chain maps. It is a braided monoidal category with the same braiding map as (3.4).

The above examples are special cases of the following one. In this approach, one turns the category H-Mod of all (left) H-modules into a braided monoidal category, by restricting to certain Hopf algebras, called quasitriangular Hopf algebras [17, 28]. A quasitriangular Hopf algebra, $(H,\ R)$, is a Hopf algebra, $(H, m, \eta, \Delta, \varepsilon, S)$, with an invertible element, $R = R_1 \otimes R_2$ in $H \otimes H$, satisfying some conditions and controlling the cocommutativity of H by $R(h^{(1)} \otimes h^{(2)}) = (h^{(2)} \otimes h^{(1)})R$, where $\Delta(h) = h^{(1)} \otimes h^{(2)}$ for any h in H.

Example 3.3 Let $(H, R = R_1 \otimes R_2)$ be a quasitriangular Hopf algebra and $\mathcal{C}$ be the category of all left H-modules. Then $\mathcal{C}$ is a braided monoidal abelian category. The monoidal structure is defined by

$$h \triangleright (v \otimes w) = h^{(1)} \triangleright v \otimes h^{(2)} \triangleright w,$$

and the braiding map $\psi_{V \otimes W}$ acts by

$$\psi_{V \otimes W}(v \otimes w) := (R_2 \triangleright w \otimes R_1 \triangleright v),$$

for any V and W in $\mathcal{C}$, where $\triangleright$ denotes the action of H. $\mathcal{C}$ is symmetric if and only if $R^{-1} = R_2 \otimes R_1$.

There is also a second approach to obtain a braided monoidal category from H-Mod. Here, one starts with any Hopf algebra $(H, m, \eta, \Delta, \varepsilon, S)$, with a bijective antipode S, and restricts oneself to the center of H-Mod (Example 3.4 bellow). Let us recall that a left-left Yetter-Drinfeld (YD) H-module consists of a vector space V, with a left H-module structure [32, 35]:

$$H \otimes V \to V, \qquad h \otimes v \mapsto hv,$$

and a left H-comodule structure:

$$V \to H \otimes V, \qquad v \mapsto v_{(-1)} \otimes v_{(0)},$$

such that the left action and coaction satisfy the so-called Yetter-Drinfeld (YD) compatibility condition

$$(hv)_{(-1)} \otimes (hv)_{(0)} = h^{(1)} v_{(-1)} S(h^{(3)}) \otimes h^{(2)} v_{(0)},$$

for all $h \in H$ and $v \in V$. The category of all YD H-modules is called the Yetter-Drinfeld category of H, and is usually denoted by ${}_H^H\mathcal{YD}$.

Example 3.4 Let H be a Hopf algebra over a field k with comultiplication $\Delta(h) = h^{(1)} \otimes h^{(2)}$ and a bijective antipode S. ${}_H^H\mathcal{YD}$ is the center of H-Mod. It is a braided monoidal abelian category with the braiding map:

$$\psi_{V \otimes W}(v \otimes w) = v_{(-1)} w \otimes v_{(0)}.$$

This category is in general not symmetric. In fact the inverse of the braiding is given by

$$\psi^{-1}_{V \otimes W}(w \otimes v) = v_{(0)} \otimes S^{-1}(v_{(-1)}) w.$$

3.2 Braided Hopf algebras. Braided Hopf algebras, that is, Hopf algebra objects in braided monoidal categories, are very interesting objects to study, because of their important role, for example, in the study of *pointed* Hopf algebras [3, 31], or their applications in low dimensional topology, for instance, in quantum invariants of knots [34].

Definition 3.1 Let $\mathcal{C}$ be a strict braided monoidal category. A Hopf algebra $(H, m, \eta, \Delta, \varepsilon, S)$ in $\mathcal{C}$ consists of an object $H \in \operatorname{obj} \mathcal{C}$, morphisms $m : H \otimes H \to H$, $\eta : I \to H$, $\Delta : H \to H \otimes H$, $\varepsilon : H \to I$ *and* $S : H \to H$ called multiplication, unit, comultiplication, counit and antipode maps satisfying the relations (only the last three ones are visualized in terms of braiding diagrams):

$$m(\mathrm{id} \otimes m) = m(m \otimes \mathrm{id}), \quad \text{associativity}$$

$$m(\eta \otimes \mathrm{id}) = m(\mathrm{id} \otimes \eta) = \mathrm{id}, \quad \text{unit}$$

$$(\mathrm{id} \otimes \Delta)\Delta = (\Delta \otimes \mathrm{id})\Delta, \quad \text{coassociativity}$$

$$(\varepsilon \otimes \mathrm{id})\Delta = (id \otimes \varepsilon)\Delta = \mathrm{id}, \quad \text{counit}$$

$$\Delta m = (m \otimes m)(\mathrm{id} \otimes \psi \otimes \mathrm{id})(\Delta \otimes \Delta), \quad \text{compatibility}$$

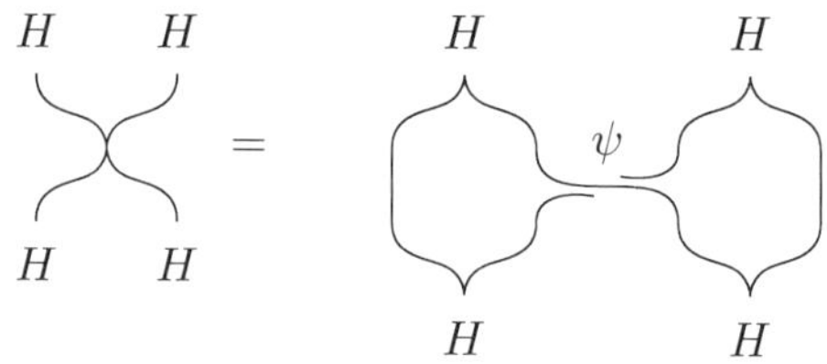

(3.5)

$$\Delta\eta = \eta \otimes \eta, \ \varepsilon m = \varepsilon \otimes \varepsilon, \ \varepsilon\eta = \mathrm{id}_I$$

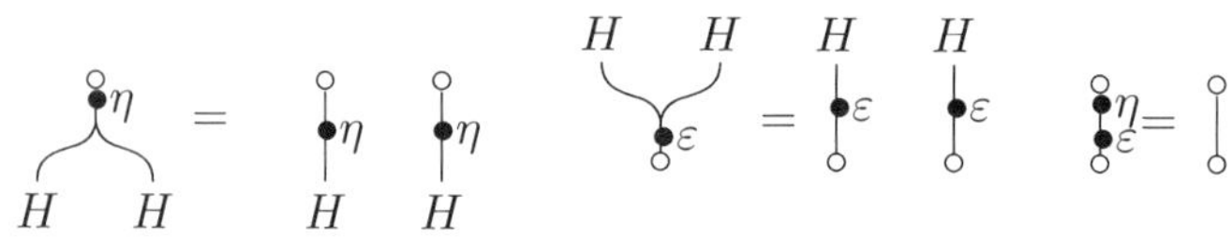

$$m(S \otimes \mathrm{id})\Delta = m(\mathrm{id} \otimes S)\Delta = \eta\varepsilon, \text{antipode}$$

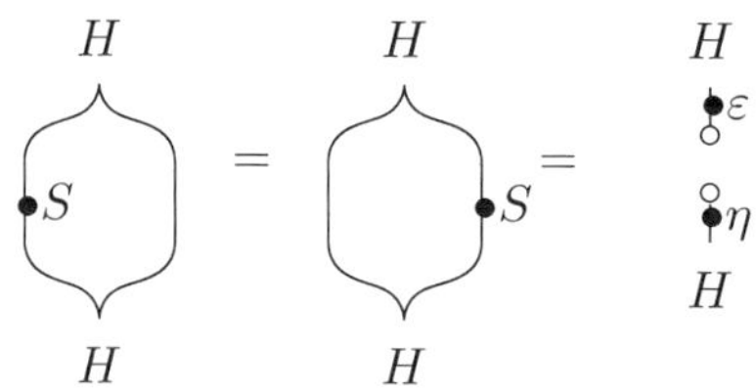

By a braided Hopf algebra we mean a Hopf algebra in a braided monoidal category.

Notice that relation (3.5), which expresses the algebra property of the comultiplication, is the only relation that involves the braiding map ψ.

Example 3.5 A Hopf algebra object in the category of vector spaces, Vect_k (Example 3.1), is a Hopf algebra in the standard sense.

Hopf algebras in ${}^H_H\mathcal{YD}$, of Example 3.4, also called *Yetter-Drinfeld Hopf algebras* [17], form an important classes of braided Hopf algebras. For instance, *Nichols* algebras, Example 3.8 below, play a crucial role in the study of *pointed* Hopf algebras [3, 31].

Example 3.6 Any Hopf algebra H over a field k is a Yetter-Drinfeld Hopf algebra in ${}^k_k\mathcal{YD}$.

Example 3.7 For any V in ${}^H_H\mathcal{YD}$ of Example 3.4, the tensor algebra $T(V)$ is a braided Hopf algebra in ${}^H_H\mathcal{YD}$. Its comultiplication, counit, and antipode are defined by $\Delta(v) = 1 \otimes v + v \otimes 1$, $\varepsilon(v) = 0$, and $S(v) = -v$, for all v in V.

Example 3.8 For any V in ${}^H_H\mathcal{YD}$, the Nichols algebra of V, denoted by $\mathfrak{B}(V)$, is defined as the quotient $T(V)/I$, where I is the largest ideal of $T(V)$ such that, $I \subset \bigoplus_{n=2}^{\infty} T^n(V)$, and $\Delta(I) \subset I \otimes T(V) + T(V) \otimes I$.

Example 3.9 A Hopf algebra object in the category of (DG) super vector spaces, $Vect_k^s$ (Example 3.2), is called a (DG) super Hopf algebra.

Example 3.10 [6, 28] Any quasitriangular quasi-Hopf algebra $(H, R, \Phi, \alpha, \beta)$ can be turned into a Hopf algebra object, $\underline{H}$, in the braided monoidal category of (left) H-modules (Example 3.3), in the following sense. As a vector space $\underline{H} = H$, with H-module structure given by conjugation

$$a \triangleright h = a^{(1)} h S(a^{(2)}).$$

The Hopf algebra structure on $\underline{H}$, which is a deformation of the Hopf algebra structure of H, is given by [6, 28]

$$\underline{m}(a,b)=a\,\underline{\cdot}\,b=X_1aS(x_1X_2)\alpha x_2X_3^{(1)}bS(x_3X_3^{(2)}),$$

with β as the unit, and

$$\underline{\Delta}(h)=h^{(\underline{1})}\otimes h^{(\underline{2})}=$$
$$x_1X_1h^{(1)}g_1S(x_2R_2y_3X_3^{(2)})\otimes x_3R_1\rhd y_1X_2h^{(2)}g_2S(y_2X_3^{(1)}),$$

with counit $\underline{\varepsilon}=\varepsilon$. The antipode is given by

$$\underline{S}(h)=x_1R_2p_2S(q_1(X_2R_1p_1\rhd h)S(q_2)X_3).$$

Here we have used the following notations:

$$\Phi=X_1\otimes X_2\otimes X_3,$$
$$\Phi^{-1}=x_1\otimes x_2\otimes x_3=y_1\otimes y_2\otimes y_3,$$
$$p_1\otimes p_2=x_1\otimes x_2\beta S(x_3),$$
$$q_1\otimes q_2=X_1\otimes S^{-1}(\alpha X_3)X_2,$$

and

$$g_1\otimes g_2=\Delta(S(x_1)\alpha x_2)\xi(S\otimes S)(\Delta^{\mathrm{op}}(x_3)),$$

where,

$$\xi=B_1\beta S(B_4)\otimes B_2\beta S(B_3),$$
$$B_1\otimes B_2\otimes B_3\otimes B_4=(\Delta\otimes\mathrm{id}\otimes\mathrm{id})(\Phi)(\Phi^{-1}\otimes 1).$$

4 Hopf cyclic cohomology in braided monoidal categories

Many interesting objects, like super Hopf algebras, quasi-Hopf algebras, multiplier Hopf algebras, Hopf algebroids, and locally compact quantum groups, are close to being a Hopf algebra but are not a Hopf algebra in the usual sense. The main motivation for our work [21] came from a desire to extend the Hopf cyclic theory [10, 11, 12, 14] to those types of objects.

Fortunately, many of those Hopf algebra-like objects are Hopf algebras in some monoidal categories. Super Hopf algebras and quasi-Hopf algebras are examples of this (Examples 3.9 and 3.10). Also, recently, there has been some work done toward presenting multiplier Hopf algebras as (Hopf algebra) objects in an appropriate braided monoidal category [15].

To achieve the desired goal, we first extend the general Hopf cyclic theory with coefficients in [14] to the context of symmetric braided monoidal abelian categories (cf. Subsection 2.1). This includes, as a very special case, a *braided version* of Connes-Moscovici Hopf cyclic theory in any symmetric braided monoidal abelian category $(\mathcal{C},\otimes,I,\psi)$. It provides us, as an example, with a Hopf cyclic theory for super Hopf algebras. Next we proceed to eliminate the restrictive symmetry condition $\psi^2=\mathrm{id}$. We show that to any Hopf algebra, in a (not necessarily symmetric) braided monoidal abelian category $(\mathcal{C},\otimes,I,\psi)$, endowed with a *braided modular pair in involution*, one can always associate a para-cocyclic object which is an "almost" cocyclic object lacking the cyclic condition $\tau_n^{n+1}=\mathrm{id}$. Of course, as with any para-cocyclic object, by restricting to an appropriate subspace we obtain a cocyclic object (cf. Remark 2.1). Then, as an application, we are able to define a Hopf cyclic cohomology for any (quasitriangular) quasi Hopf algebra by associating a para-cocyclic object to it. We also indicate a beautiful relation between the symmetry condition on the braiding and the cyclic condition. For example, in the case of $n = 2$ we prove the "Kepler-like" formula

$$\tau_2^3 = \psi^2.$$

We should mention that in recent years there have been some other works done by other authors, with different motivations, to study (Hopf) cyclic cohomology in (braided) monoidal categories, [2], [4, 5], [18, 19]. A full report on these works will appear elsewhere.

Here is a very short summary of the work in [21]. We start with the definition of a (right-left) SAYD module of [13], in any braided monoidal category. Let us fix a braided monoidal category $(\mathcal{C}, \otimes, I, \psi)$, and let $(H, \Delta, \varepsilon, S)$ to be a braided Hopf algebra in $\mathcal{C}$.

Definition 4.1 A right-left braided stable anti-Yetter-Drinfeld (SAYD) H-module in $\mathcal{C}$ is an object M in $\mathcal{C}$ such that:
(i) M is a right H-module via an action $\phi_M : M \otimes H \to M$,
(ii) M is a left H-comodule via a coaction $\rho_M : M \to H \otimes M$,
(iii) M satisfies the braided anti-Yetter-Drinfeld condition, i.e.:

$$(\rho)(\phi) = [(m)(S \otimes m) \otimes \phi][(\psi_{H^{\otimes 2},H} \otimes id_M \otimes \mathrm{id}_H)(id_{H^{\otimes 2}} \otimes \psi_{M,H} \otimes \mathrm{id}_H) \\ (\mathrm{id}_{H^{\otimes 2}} \otimes \mathrm{id}_M \otimes \psi_{H,H})(\mathrm{id}_H \otimes \psi_{M,H} \otimes \mathrm{id}_{H^{\otimes 2}})][\rho \otimes \Delta^2].$$

(iv) M is stable, i.e.:

$$(\phi)(\psi_{H,M})(\rho) = \mathrm{id}_M.$$

The above braided SAYD conditions (iii) and (iv) in terms of braided diagrams are as follows:

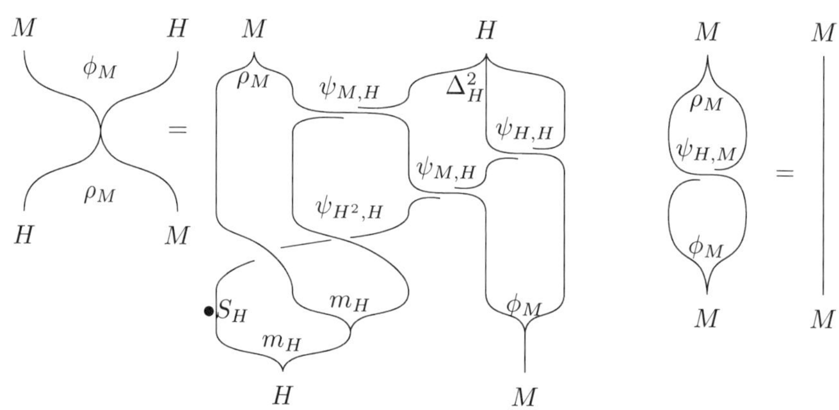

(4.6)

Definition 4.2 A quadruple $(C, \Delta_C, \epsilon_C, \phi_C)$ is called a left (braided) H-module-coalgebra in $\mathcal{C}$ if $(C, \Delta_C, \epsilon_C)$ is a coalgebra in $\mathcal{C}$, and C is a left H-module via an action $\phi_C : H \otimes C \to C$ such that ϕ_C is a coalgebra map in $\mathcal{C}$, i.e., we have:

$$\Delta_C \phi_C = (\phi_C \otimes \phi_C)(\mathrm{id}_H \otimes \psi_{H,C} \otimes \mathrm{id}_C)(\Delta_H \otimes \Delta_C),$$

$$\varepsilon_C \phi_C = \varepsilon_H \otimes \varepsilon_C.$$

4.1 Symmetric case. Now let us assume that $\mathcal{C}$ is symmetric and also abelian. Then we can associate a cocyclic object to any triple (H, C, M), where H is a Hopf algebra, C is a H-module coalgebra and M is a SAYD H-module, all in $\mathcal{C}$, as in the next theorem.

Theorem 4.1 *If $\mathcal{C}$ is a symmetric monoidal abelian category, then the structure $(C_H^{\bullet}, \delta_i, \sigma_i, \tau)$ defined by the following datum attaches a cocyclic object in $\mathcal{C}$ to the triple (H, C, M).*

For $n \geq 0$, we let

$$C_H^n = C_H^n(C, M) := M \otimes_H C^{n+1},$$

and we define faces $\delta_i : C^{n-1} \to C^n$, degeneracies $\sigma_i : C^{n+1} \to C^n$ and cyclic maps $\tau_n : C^n \to C^n$ by:

$$\delta_i = \begin{cases} (1_M \otimes_H 1_{C^i} \otimes \Delta_C \otimes 1_{C^{n-i-1}}) & 0 \leq i < n \\ (1_M \otimes_H \psi_{C,C^n})(1_M \otimes \phi_C \otimes 1_{C^n})(\psi_{H,M} \otimes 1_{C^{n+1}}) & \\ \qquad (\rho_M \otimes \Delta_C \otimes 1_{C^{n-1}}) & i = n \end{cases}$$

$$\sigma_i = (1_M \otimes_H 1_{C^{i+1}} \otimes \varepsilon_C \otimes 1_{C^{n-i}}), \qquad 0 \leq i \leq n$$

$$\tau_n = (1_M \otimes_H \psi_{C,C^n})(1_M \otimes \phi_C \otimes 1_{C^n})(\psi_{H,M} \otimes 1_{C^{n+1}})(\rho_M \otimes 1_{C^{n+1}})$$

□

Defining the notions of braided modular pair in involution (BMPI) and twisted antipode in a categorical framework is straightforward. After doing so [21], we are able to give a braided version of Connes-Moscovici's Hopf cyclic theory, as a corollary of the above theorem.

Corollary 4.1 *Suppose $(H, (\delta, \sigma))$ is a braided Hopf algebra in a symmetric braided monoidal abelian category $\mathcal{C}$, where (δ, σ) is a BMPI. If we put $(C;\ \phi_C,\ \Delta_C) = (H;\ m_H,\ \Delta_H)$, and $M = {}^{\sigma}I_{\delta}$, then the cocyclic module of Theorem (4.1) reduces to the following one which is a braided version of Connes-Moscovici's Hopf cyclic theory:*

$$C^0(H) = I \text{ and } C^n(H) = H^n,\ n \geq 1,$$

with faces, degeneracies, and cyclic maps given by:

$$\delta_i = \begin{cases} (\eta \otimes 1 \otimes 1 \otimes \cdots \otimes 1) & i = 0 \\ (1 \otimes 1 \otimes \cdots \otimes 1 \otimes \underset{i-th}{\Delta} \otimes 1 \otimes 1 \otimes \cdots \otimes 1) & 1 \leq i \leq n-1 \\ (1 \otimes 1 \otimes \cdots \otimes 1 \otimes \sigma) & i = n \end{cases}$$

$$\sigma_i = (1 \otimes 1 \otimes \cdots \otimes \underset{(i+1)-th}{\varepsilon} \otimes 1 \otimes 1 \otimes \cdots \otimes 1),\ 0 \leq i \leq n$$

$$\tau_n = \begin{cases} \mathrm{id}_I & n = 0 \\ (m_n)(\Delta^{n-1}\widetilde{S} \otimes 1_{H^{n-1}} \otimes \sigma) & n \neq 0 \end{cases}$$

Here by m_n we mean, $m_1 = m$, and for $n \geq 2$:

$$m_n = m_{H^n} = \underbrace{(m \otimes m \otimes \cdots \otimes m)}_{n\ times}\mathcal{F}_n(\psi),$$

where

$$\mathcal{F}_n(\psi) := \prod_{j=1}^{n-1} (1_{H^j} \otimes \underbrace{\psi \otimes \psi \otimes \cdots \otimes \psi}_{n-j\ times} \otimes 1_{H^j}).$$

□

Example 4.1 Using the result of Corollary 4.1, one is able to define Hopf cyclic cohomology for super Hopf algebras as follows.

Let $H = H_0 \oplus H_1$ be a super Hopf algebra endowed with a braided modular pair in involution, (δ, σ). Then the complex, faces, degeneracies and cyclic maps of Corollary 4.1 can be written as:

$$C^0(H) = \mathbb{C} \textit{ and } C^n(H) = H^n, n \geq 1,$$

$$\delta_i(h_1, ..., h_{n-1}) = \begin{cases} (1, h_1, h_2, ..., h_{n-1}) & i = 0 \\ (h_1, h_2, ..., h_i^{(1)}, h_i^{(2)}, ..., h_{n-1}) & 1 \leq i \leq n-1 \\ (h_1, h_2, ..., h_{n-1}, \sigma) & i = n \end{cases}$$

$$\sigma_i(h_1, h_2, ..., h_{n+1}) = \varepsilon(h_{i+1})\ (h_1, h_2, ..., h_i, h_{i+2}, ..., h_{n+1}), 0 \leq i \leq n,$$

$$\tau_n(h_1, h_2, ..., h_n) = \alpha\beta\ (S(h_1^{(n)})h_2, S(h_1^{(n-1)})h_3, ..., S(h_1^{(2)})h_n, \widetilde{S}(h_1^{(1)})\sigma),$$

where h_i's are homogeneous elements and:

$$\begin{aligned} \alpha &= \prod_{i=1}^{n-1} (-1)^{(|h_1^{(1)}|+\cdots+|h_1^{(i)}|)(|h_1^{(i+1)}|)}, \\ \beta &= \prod_{j=1}^{n-1} (-1)^{|h_1^{(j)}|(|h_2|+|h_3|+\cdots+|h_{n-j+1}|)}. \end{aligned}$$

Remark 4.1 In [22] we have defined a super version of the Connes-Moscovici Hopf algebra $\mathcal{H}_1$ of [10]. We denote that super Hopf algebra by $\mathcal{H}_1^s$. The Hopf cyclic theory for super Hopf algebras can be applied to compute the Hopf cyclic theory of $\mathcal{H}_1^s$. More on this will appear elsewhere.

4.2 Non-symmetric case. By eliminating the restrictive symmetry condition $\psi^2 = \text{id}$, we have the analogue of Corollary 4.1 in a *non-symmetric* braided monoidal abelian category as follows:

Theorem 4.2 *Let $(H, (\delta, \sigma))$ be a braided Hopf algebra in a braided abelian monoidal category $\mathcal{C}$ (not necessarily symmetric), with (δ, σ) a BMPI. Then the structure provided in Corollary 4.1, defines a para-cocyclic object in $\mathcal{C}$ attached to $(H, (\delta, \sigma))$. The relation between the symmetry condition on the braiding and the cyclic condition is as follows. For all $n \neq 0$:*

$$\tau_n^{n+1} = (\psi_{H^{(n-1)},H})^n,$$

which is equal to id*, for all $n \neq 0$, iff $\psi^2 = \text{id}$. This means that the attached para-cocyclic object is a cocyclic object if and only if the ambient category is symmetric.*

The formula (4.2) in the case of $n = 2$ becomes

$$\tau_2^3 = \psi^2,$$

which can be visualized in terms of braiding diagrams by

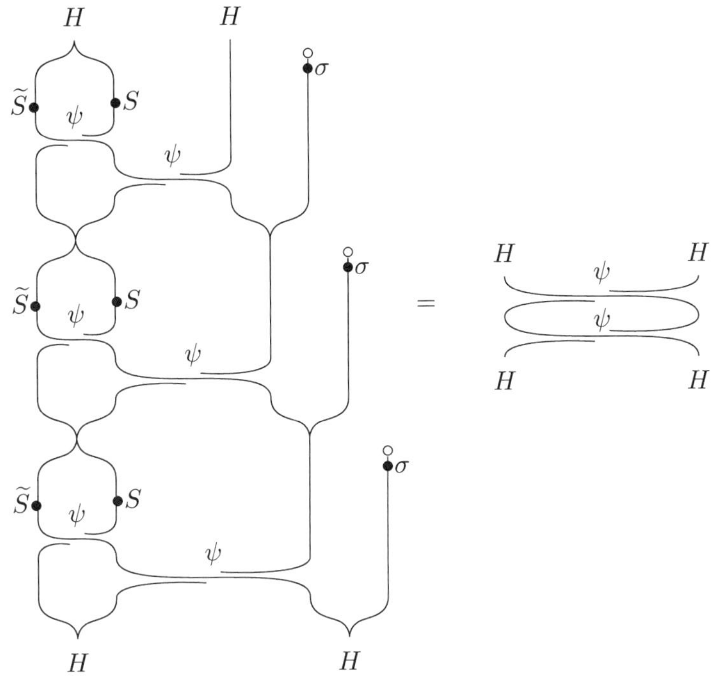

4.3 Hopf cyclic cohomology for quasitriangular quasi-Hopf algebras. Using the result of Theorem 4.2, we can define the desired Hopf cyclic cohomology for quasi triangular quasi-Hopf algebras as we shortly explain below [21].

As was explained in Example 3.10, a quasitriangular quasi-Hopf algebra $(H, R, \Phi, \alpha, \beta)$ (which is not, necessarily, a Hopf algebra in the standard sense) can be turned into a braided Hopf algebra $\underline{H}$ living in the braided monoidal category, H-Mod, of (left) H-modules.

Now, let (δ, σ) be a braided modular pair in involution for $\underline{H}$. Then using Theorem (4.2) we can associate a para-cocyclic module to $\underline{H}$. By passing to subspaces $\ker(1 - \tau_n^{n+1})$ we obtain a cocyclic module. Therefore we have the following corollary.

Corollary 4.2 *For any quasitriangular quasi-Hopf algebra $(H, R, \Phi, \alpha, \beta)$ endowed with a braided modular pair in involution (δ, σ), the complex obtained in Theorem (4.2) defines a Hopf cyclic cohomology for H, denoted by*

$$HC^*_{(\delta,\sigma)}(H, R, \Phi, \alpha, \beta).$$

References

[1] R. Akbarpour and M. Khalkhali, Hopf algebra equivariant cyclic homology and cyclic homology of crossed product algebras. J. Reine Angew. Math. 559 (2003), 137–152.

[2] S. E. Akrami and S. Majid, Braided cyclic cocycles and nonassociative geometry. J. Math. Phys. 45 (2004), no. 10, 3883–3911.

[3] N. Andruskiewitsch and H-J. Schneider, Pointed Hopf algebras. New directions in Hopf algebras, 1–68, Math. Sci. Res. Inst. Publ., 43, Cambridge Univ. Press, Cambridge, 20.

[4] G. Böhm and D. Stefan, A categorical approach to cyclic duality. preprint *arXiv:0910.4622*, (2009), to appear in J. Noncommutative Geometry.

[5] G. Böhm and D. Stefan, (Co)cyclic (co)homology of bialgebroids: an approach via (co)monads. Comm. Math. Phys. 282 (2008), no. 1, 239286.

[6] D. Bulacu and E. Nauwelaerts, Radford's biproduct for quasi-Hopf algebras and bosonization. *J. Pure Appl. Algebra* **174** (2002), no. 1, 1–42.
[7] A. Connes, Cohomologie cyclique et foncteurs Ext^n . (French) (Cyclic cohomology and functors Ext^n) C. R. Acad. Sci. Paris Sr. I Math. **296** (1983), no. 23, 953–958.
[8] A. Connes, Noncommutative differential geometry. Inst. Hautes Études Sci. Publ. Math. No. **62** (1985), 257–360.
[9] A. Connes, Noncommutative Geometry. Academic Press, San Diego, CA, 1994, 661 p., available online at http://alainconnes.org/.
[10] A. Connes and H. Moscovici, Hopf algebras, cyclic cohomology and the transverse index theorem. Comm. Math. Phys. **198** (1998), no. 1, 199–246.
[11] A. Connes and H. Moscovici, Cyclic cohomology and Hopf algebras. Lett. Math. Phys. **48** (1999), no. 1, 97–108.
[12] A. Connes, and H. Moscovici, Cyclic cohomology and Hopf algebra symmetry. Conference Moshé Flato 1999 (Dijon). Lett. Math. Phys. **52** (2000), no. 1, 1–28.
[13] P. M. Hajac, M. Khalkhali, B. Rangipour, and Y. Sommerhäuser, Stable anti-Yetter-Drinfeld modules. C. R. Math. Acad. Sci. Paris **338** (2004), no. 8, 587–590.
[14] P. M. Hajac, M. Khalkhali, B. Rangipour, and Y. Sommerhäuser, Hopf-cyclic homology and cohomology with coefficients. C. R. Math. Acad. Sci. Paris **338** (2004), no. 9, 667–672.
[15] K. Janssen and J. Vercruysse, Multiplier bi- and Hopf algebras. J. Algebra Appl. **9** (2010), no. 2, 275–303.
[16] A. Joyal and R Street, Braided tensor categories. Adv. Math. 102 (1993), no. 1, 20-78.
[17] Ch. Kassel, *Quantum groups*. Graduate Texts in Mathematics 155, Springer-Verlag, New York, 1995.
[18] A. Kaygun, Bialgebra cyclic homology with coefcients. K -Theory 34 (2005), no. 2, 151194.
[19] A. Kaygun, The universal Hopf-cyclic theory. J. Noncommut. Geom. 2 (2008), no. 3, 333351.
[20] M. Khalkhali, Basic noncommutative geometry. EMS Series of Lectures in Mathematics. European Mathematical Society (EMS), Z urich, 2009.
[21] M. Khalkhali and A. Pourkia, Hopf cyclic cohomology in braided monoidal categories. Homology, Homotopy and Applications (HHA), Vol. 12 (2010), No. 1, pp.111-155. http://intlpress.com/HHA/v12/n1/a9/
[22] M. Khalkhali and A. Pourkia, a super version of the Connes-Moscovici Hopf algebra. To appear in Contemporary Mathematics, AMS publications.
[23] M. Khalkhali, and B. Rangipour, A new cyclic module for Hopf algebras. K-Theory **27** (2) (2002), 111-131.
[24] M. Khalkhali, and B. Rangipour, Invariant cyclic homology. K-Theory **28** (2) (2003), 183-205.
[25] M. Khalkhali, and B. Rangipour, Introduction to Hopf-cyclic cohomology. Noncommutative geometry and number theory, 155–178, Aspects Math., E37, Vieweg, Wiesbaden, 2006.
[26] J. L. Loday, Cyclic homology. Second edition, Grundlehren der Mathematischen Wissenschaften [Fundamental Principles of Mathematical Sciences], 301, Springer-Verlag, Berlin, 1998.
[27] S. Mac Lane, Categories for the working mathematician. Second edition. Graduate Texts in Mathematics, 5. Springer-Verlag, New York, 1998.
[28] S. Majid, Foundations of quantum group theory. Cambridge University Press, Cambridge, 1995.
[29] H. Moscovici and B. Rangipour, Cyclic cohomology of Hopf algebras of transverse symmetries in codimension 1. Adv. Math. **210** (2007), no. 1, 323–374.
[30] H. Moscovici and B. Rangipour, Hopf algebras of primitive Lie pseudogroups and Hopf cyclic cohomology. Adv. Math. **220** (2009), no. 3, 706–790.
[31] D. Radford, The structure of Hopf algebras with a projection. J. Algebra 92 (1985), no. 2, 322347.
[32] D. Radford, and J. Towber, Yetter-Drinfeld categories associated to an arbitrary bialgebra. J. Pure Appl. Algebra **87** (1993), 259-279.
[33] N. Saavedra Rivano, Catégories Tannakiennes. *Lecture Notes in Mathematics* **265**, Springer-Verlag, Berlin, 1972.
[34] V. G. Turaev, Quantum invariants of knots and 3-manifolds. de Gruyter studies in Mathematics, no. 18, W de Gruyter, Berlin, 1994.

[35] D. Yetter, Quantum groups and representations of monoidal categories. Math. Proc. Cambridge Philos. Soc. **108** (1990), no. 2, 261–290.

Fields Institute Communications
Volume **61**, 2011

A Survey on Rankin-Cohen Deformations

Richard Rochberg
Department of Mathematics
Washington University
St. Louis, MO 63130 USA
rr@math.wustl.edu

Xiang Tang
Department of Mathematics
Washington University
St. Louis, MO 63130 USA
xtang@math.wustl.edu

Yi-jun Yao
School of Mathematical Sciences
Fudan University
Shanghai, 200433, P.R. China
yaoyijun@fudan.edu.cn

Abstract. This is a survey about recent progress in Rankin-Cohen deformations. We explain a link between Rankin-Cohen brackets and higher order Hankel forms.

1 Introduction

The famous *Erlanger Programm* of Klein says that geometry is about the study of the transformation groups of various spaces, or more precisely the properties invariant under the actions of such groups, *i.e.*, the symmetries.

Noncommutative geometry (NCG), which originated from Connes' study in operator algebras in 1970's, brought the landscape of geometry many new objects and some astonishing phenomena.

Back in the early 1990s, Connes and Moscovici pointed out that in noncommutative geometry (NCG) while noncommutative spaces are represented by the algebras (usually noncommutative C^*-algebras) of "continuous functions" over noncommutative spaces, the *local* symmetries are reflected in some Hopf algebras. One of the first noncommutative spaces studied in NCG is the C^*-algebra of a foliated space. In the case of codimension-n oriented foliations, Connes and Moscovici discovered a Hopf algebra $\mathcal{H}_n$, which governs the local symmetry of leaf spaces of foliations of codimension n. The Hopf algebra $\mathcal{H}_n$ is universal in the sense that

2010 *Mathematics Subject Classification.* Primary 11F37; Secondary 16T05, 46L87, 53D55.

it depends only on the codimension of a foliation. This family of Hopf algebras $\{\mathcal{H}_n\}$ is very useful in the study of transverse index theory, and later was found to have connections with various different areas of mathematics, c.f. [10], [13]. In this paper, we review the application of the Hopf algebra $\mathcal{H}_1$ in Rankin-Cohen deformations, which was initiated by Connes and Moscovici [14].

We start by recalling the general setting of transverse geometry. Let M be a smooth manifold and $\mathcal{F}$ be a foliation on M of codimension n. Let X be a complete flat transversal of $\mathcal{F}$, and F^+X be the oriented frame bundle of X. The holonomy pseudogroup Γ acts on X and therefore F^+X by parallel transforming X along paths in leaves of $\mathcal{F}$. The "transverse geometry" is the study of the transversal X along with the action by the holonomy pseudogroup Γ.

In what follows we focus on the case when $n = 1$, and define Connes-Moscovici's Hopf algebra $\mathcal{H}_1$. Now the complete transversal X is a flat 1-dimensional manifold; and the oriented frame bundle F^+X is diffeomorphic to $X \times \mathbb{R}^+$, and Γ is a discrete holonomy pseudogroup acting on X as local diffeomorphisms. We introduce coordinates x on X and y on $\mathbb{R}^+$. The lifted action of Γ on F^+X is

$$(x, y) \mapsto (\phi(x), \phi'(x)y), \quad \phi \in \Gamma. \tag{1.1}$$

On F^+X, there is a Γ-invariant volume form $\omega = \dfrac{\mathrm{d}x \wedge \mathrm{d}y}{y^2}$, which is also a symplectic form. This allows one to consider the Hilbert space $L^2\left(F^+X, \dfrac{\mathrm{d}x \wedge \mathrm{d}y}{y^2}\right)$ of square-integrable functions on F^+X. We are interested in two types of linear operators acting on this Hilbert space $L^2\left(F^+X, \dfrac{\mathrm{d}x \wedge \mathrm{d}y}{y^2}\right)$:

1. for $f \in C_c^\infty(F^+X)$, we define $M_f : \xi \mapsto f\xi$ for all $\xi \in L^2\left(F^+X, \dfrac{\mathrm{d}x \wedge \mathrm{d}y}{y^2}\right)$;
2. for $\phi \in \Gamma$, we define $U_\phi : \xi \mapsto \phi^*(\xi) = \xi \circ \phi$ for all $\xi \in L^2\left(F^+X, \dfrac{\mathrm{d}x \wedge \mathrm{d}y}{y^2}\right)$.

The extended smooth foliation algebra $\mathcal{A}_\Gamma$ is the algebra generated by M_f's and U_ϕ's with the relation

$$U_\phi M_f = M_{\phi^*(f)} U_\phi.$$

From this relation, we can say that $\mathcal{A}_\Gamma$ is the crossed product algebra $C_c^\infty(F^+X) \rtimes \Gamma$. In noncommutative geometry, this algebra $\mathcal{A}_\Gamma$ is viewed as the algebra of smooth functions on the leaf space[1] associated to the foliation $\mathcal{F}$ on M. The Hopf algebra $\mathcal{H}_1$ acts on the extended smooth foliation algebra by linear operators.

By choosing a flat connection on F^+X, we consider two vector fields, i) $X = y\partial_x$ as a lifting of the vector filed ∂_x on X, and ii) $Y = y\partial_y$ the Euler vector field along the fiber direction. X and Y act on the extended smooth foliation algebra $\mathcal{A}_\Gamma$ by

$$X(fU_\phi) = X(f)U_\phi, \; Y(fU_\phi) = Y(f)U_\phi.$$

It is easy to check that Y is invariant under the action of Γ, but X is not:

$$U_\phi X U_\phi^{-1} = X - y\frac{\phi^{-1\prime\prime}(x)}{\phi^{-1\prime}(x)}Y.$$

[1]More precisely, let N be the normal bundle of the foliation $\mathcal{F}$ on M, and F^+N be the oriented frame bundle of N. The foliation $\mathcal{F}$ lifts to define a foliation $\widetilde{\mathcal{F}}$ on F^+N. $\mathcal{A}_\Gamma$ can be viewed as the algebra of smooth functions on the leaf space of the lifted foliation $\widetilde{\mathcal{F}}$.

The failure of X being Γ-invariant inspires higher operations. Define a linear operator δ_1 on $\mathcal{A}_\Gamma$ in the following way:

$$\delta_1(fU_\phi) = \mu_{\phi^{-1}} fU_\phi,$$

where $\mu_{\phi^{-1}}(x,y) = y\dfrac{\phi^{-1\prime\prime}(x)}{\phi^{-1\prime}(x)}$.

We compute $[Y, \delta_1]$, which turns out to be δ_1 itself. But $[X, \delta_1]$ leads to a new operator, which we call δ_2:

$$\delta_2(fU_\phi) = X(\mu_{\phi^{-1}}) fU_\phi.$$

Iterating the procedure of computing the commutator with X, we are led to a sequence of operators δ_n acting on $\mathcal{A}_\Gamma$ by

$$\delta_n(fU_\phi) = X^{n-1}(\mu_{\phi^{-1}}) fU_\phi.$$

With the above preparation, we are ready to present the Hopf algebra $\mathcal{H}_1$. As an algebra, it is the universal enveloping algebra of an infinite dimensional Lie algebra H_1 whose generators are labeled by $\{X, Y, \delta_n,\ n \in \mathbb{N}\}$ with the following commutation relations:

$$\begin{aligned} [Y, X] &= X, & [X, \delta_n] &= \delta_{n+1}, \\ [Y, \delta_n] &= n\delta_n, & [\delta_n, \delta_m] &= 0. \end{aligned}$$

We define the following structures on $\mathcal{H}_1$:

1. product $\cdot : \mathcal{H}_1 \otimes \mathcal{H}_1 \to \mathcal{H}_1$ is defined as the product on the universal enveloping algebra of H_1.
2. coproduct $\Delta : \mathcal{H}_1 \to \mathcal{H}_1 \otimes \mathcal{H}_1$ is an algebra homomorphism generated by

$$\begin{aligned} \Delta Y &= Y \otimes 1 + 1 \otimes Y, \\ \Delta\delta_1 &= \delta_1 \otimes 1 + 1 \otimes \delta_1, \\ \Delta X &= X \otimes 1 + 1 \otimes X + \delta_1 \otimes Y, \\ \Delta\delta_n &= [\Delta X, \Delta\delta_{n-1}]. \end{aligned}$$

3. counit $\epsilon : \mathcal{H}_1 \to \mathbb{C}$ is defined by taking the constant component in $\mathcal{H}_1$.
4. antipode $S : \mathcal{H}_1 \to \mathcal{H}_1$ is an algebra anti-homomorphism generated by

$$S(X) = -X + \delta_1 Y, \quad S(Y) = -Y, \quad S(\delta_1) = -\delta_1.$$

It is straightforward to check that $(\mathcal{H}_1, \cdot, \Delta, S, \epsilon, \mathrm{id})$ defines a Hopf algebra, which acts naturally on the extended smooth foliation algebra $\mathcal{A}_\Gamma$.

The Hopf algebra $\mathcal{H}_1$ and its higher dimensional generalizations were discovered by Connes and Moscovici [11] as local symmetries on $\mathcal{A}_\Gamma$. In the following, we will survey some recent developments about the applications of $\mathcal{H}_1$ in the study of modular forms and in particular Rankin-Cohen deformations.

The results reviewed in this paper are interactions between transverse geometry and modular form theory. These two "distant" subjects are mysteriously connected due to the fact that the Hopf algebra $\mathcal{H}_1$ appears in both theories as the local symmetry of the corresponding systems. What we will develop in the last section of this paper is to add one more subject to this story, namely the Hankel forms and transvectant theory in harmonic analysis. There, we will introduce an algebra $\mathcal{B}_\Gamma$ associated to a pseudogroup Γ acting on a 1-dimensional complex domain by holomorphic transformations. We will show that the Hopf algebra $\mathcal{H}_1$ acts on $\mathcal{B}_\Gamma$. And interestingly, through the Hopf algebra action the Rankin-Cohen brackets on modular forms are translated to Hankel forms of higher weights [21].

Acknowledgments: We would like to express our gratitude to Connes and Moscovici for insightful suggestions and constant supports in our work. We would also like to thank Pevzner, Sündall, and Zagier for helpful discussion. Rochberg's work was partially supported by NSF grant 0700238; Tang's work was partially supported by NSF grant 0703775 and 0900985; Yao's work was partially supported by NSF grant 0903985.

2 Rankin-Cohen brackets and deformations

In number theory, modular forms are very important because the coefficients of their Fourier expansions encode a great amount of number theoretical information. We recall the definition of a modular form. Let Γ be a congruence subgroup of $SL_2(\mathbb{Z})$ (a subgroup of $SL_2(\mathbb{R})$ such that the entries of a matrix are all integers), a modular form of weight $2k$ is a function f which satisfies:

- (holomorphicity) f is holomorphic on the upper-half plane $\mathbb{H}$,
- (modularity) for $\gamma = \begin{pmatrix} a & b \\ c & d \end{pmatrix} \in \Gamma$ and $z \in \mathbb{H}$, $f\Big|_{2k}\gamma = f$, where
$$\left(f\Big|_{2k}\gamma\right)(z) = (cz+d)^{-2k} f\left(\frac{az+b}{cz+d}\right),$$
(this means the invariance of the form $f\,dz^k$);
- (growth condition at the boundary) $|f(z)|$ is assumed to be controlled by a polynomial in $\max\{1, \mathrm{Im}(z)^{-1}\}$.

We denote by $\mathcal{M}(\Gamma)$ the (weight) graded algebra of modular forms with respect to the group Γ.

The Rankin-Cohen brackets are a family of universal formulas describing all bilinear operations (up to a scalar) between spaces of modular forms (of even weight) that can be defined in terms of derivatives. In the 50's Rankin started the research on bi-differential operators on $\mathcal{M}(\Gamma)$ which produce new modular forms, and twenty years later Henri Cohen[2] gave a complete answer (cf. [5]) to this question by showing that all these operators are linear combinations of the bracket

$$[f,g]_n = \sum_{r=0}^{n}(-1)^r \binom{n+2k-1}{n-r}\binom{n+2l-1}{r} f^{(r)} g^{(n-r)} \in \mathcal{M}_{2k+2l+2n}(\Gamma), \quad (2.2)$$

where $f \in \mathcal{M}_{2k}$ and $g \in \mathcal{M}_{2l}$ are two modular forms, and $f^{(r)} = \left(\frac{1}{2\pi i}\frac{\partial}{\partial z}\right)^r f$. The above bilinear operation $[\ ,\ ]_n$ is called the n-th Rankin-Cohen bracket.

The original and obvious importance of the Rankin-Cohen brackets in number theory is that these brackets often give rise to non-trivial identities between the Fourier coefficients of modular forms, e.g. H. Cohen's foundational paper [5]. Zagier [33] has found several other "raisons d'être" for the Rankin-Cohen brackets. He

[2] We have given the first name of the author because in the literature of Rankin-Cohen deformations, there are two different and important authors with the same last name, "Cohen".

discovered the following procedure to obtain the Rankin-Cohen brackets: we denote Ramanujan's derivation on modular forms by $X : \mathcal{M}_{2k}(\Gamma) \to \mathcal{M}_{2k+2}(\Gamma)$:

$$\begin{aligned} Xf &= \frac{1}{2\pi i}\frac{df}{dz} - \frac{1}{2\pi i}\frac{\partial}{\partial z}(\log \eta^4)\cdot kf, \\ \eta(z) &= q^{1/24}\prod_{n=1}^{\infty}(1-q^n), \quad q = e^{2\pi i z}, \end{aligned}$$

and define two sequences of modular forms by induction: for some $\Phi \in \mathcal{M}_4(\Gamma)$, $f_0 = f, g_0 = g$,

$$\begin{aligned} f_{r+1} &= Xf_r + r(r+2k-1)\Phi f_{r-1}, \\ g_{s+1} &= Xg_s + s(s+2l-1)\Phi g_{s-1}. \end{aligned} \tag{2.3}$$

Zagier showed [33] that the Rankin-Cohen brackets can be written as

$$\sum_{r=0}^{n}(-1)^r\binom{n+2k-1}{n-r}\binom{n+2l-1}{r} f_r g_{n-r} = [f,g]_n.$$

These forms of the brackets have the advantage that one can easily see the modularity of $[f,g]_n$ without any extra effort from the definition of f_r and g_s, while in the original presentation the modularity of the brackets is far from being obvious as the derivative of a modular form is in general not a modular form any longer.

Zagier and his collaborators [6] showed that the collection of all Rankin-Cohen brackets together gives rise to (non-commutative) associative deformations of the algebra $\mathcal{M}(\Gamma)$ of modular forms. We review their constructions in more detail.

Paula Cohen[3], Manin, and Zagier [6] established a bijection between formal series with modular form coefficients and formal invariant pseudodifferential operators. For a modular form f of weight $2k$ we define

$$\mathcal{D}_{-k}(f,\partial) = \sum_{n=0}^{\infty}\frac{\binom{-k}{n}\binom{-k-1}{n}}{\binom{-2k}{n}} f^{(n)}\partial^{-k-n}, \tag{2.4}$$

and the composition of two such formal pseudodifferential operators (plus the bilinearity) gives us the following product:

$$f \star g = \sum_{n=0}^{\infty} c_n(k,l)[f,g]_n, \tag{2.5}$$

$$c_n(k,l) = t_n(k,l) := \frac{1}{\binom{-2l}{n}}\sum_{r+s=n}\frac{\binom{-k}{r}\binom{-k-1}{r}}{\binom{-2k}{r}}\frac{\binom{n+k+l}{s}\binom{n+k+l-1}{s}}{\binom{2n+2k+2l-2}{s}}. \tag{2.6}$$

Meanwhile, by checking the first several terms, Eholzer [33] conjectured that

$$f \star g := \sum_{n=0}^{\infty}[f,g]_n \tag{2.7}$$

is also an associative product.

In order to include this phenomenon in their framework, P. Cohen, Manin, and Zagier [6]modified their definition of invariant formal pseudodifferential operators, and they finally obtained a whole family of such deformations, parametrized by the

---

[3]Here is the other author with the same last name "Cohen".

Riemann sphere([6],[33]). In this most general case, we consider the $\gamma \in SL_2(\mathbb{C})$ action on functions $\xi \in C^\infty(\mathbb{H})$:

$$(W_\gamma^\kappa \xi)(z) = \xi\left(\frac{az+b}{cz+d}\right)(cz+d)^\kappa.$$

Then we can define

$$\mathcal{D}^\kappa_{-k}(f,\partial) = \sum_{n=0}^\infty \frac{\binom{-k}{n}\binom{-k+\kappa-1}{n}}{\binom{-2k}{n}} f^{(n)}\partial^{-k-n}, \tag{2.8}$$

which has the following invariance property:

$$\mathcal{D}^\kappa_{-k}(f|_{2k}\gamma,\partial) = W_\gamma^{\kappa *}\mathcal{D}^\kappa_{-k}(f,\partial)W_\gamma^\kappa.$$

By composing two such formal invariant pseudodifferential operators, we obtain, over the algebra of formal series with coefficients in modular forms $\mathcal{M}[[\hbar]]$, that the natural linear extension plus the following formula gives an associative product: for $f \in \mathcal{M}_{2k}$, $g \in \mathcal{M}_{2l}$,

$$\mu^\kappa(f,g) = \sum_{n=0}^\infty t_n^\kappa(k,l)[f,g]_n \hbar^n, \tag{2.9}$$

where the coefficients are

$$t_n^\kappa(k,l) = \frac{1}{\binom{-2l}{n}} \sum_{r+s=n} \frac{\binom{-k}{r}\binom{-k+\kappa-1}{r}}{\binom{-2k}{r}} \frac{\binom{n+k+l-\kappa}{s}\binom{n+k+l-1}{s}}{\binom{2n+2k+2l-2}{s}}. \tag{2.10}$$

Moreover, the coefficients $t_n^\kappa(k,l)$ are conjectured to be equal to

$$t_n^\kappa(k,l) = \left(-\frac{1}{4}\right)^n \sum_{j\geq 0} \binom{n}{2j} \frac{\binom{-\frac{1}{2}}{j}\binom{\kappa-\frac{3}{2}}{j}\binom{\frac{1}{2}-\kappa}{j}}{\binom{-k-\frac{1}{2}}{j}\binom{-l-\frac{1}{2}}{j}\binom{n+k+l-\frac{3}{2}}{j}}. \tag{2.11}$$

By taking $\kappa = \frac{1}{2}$ or $\frac{3}{2}$ under this form, (1.3) turns to be Eholzer's product Eq. (2.5), and until 2004 this has the only possible way to prove the associativity of Eholzer's product. (Unfortunately, Zagier's original proof [35] of the identity Eq. (2.11) is not published, and one can find an elementary but rather long proof in [31]).

3 Modular Hecke algebras and Connes-Moscovici's deformation

The interaction between the theory of modular forms and noncommutative geometry goes back to December 2001, when Zagier gave a course at Collège de France([34]) and Connes was in the audience. One year later, Connes and Moscovici discovered that the Hopf algebra $\mathcal{H}_1$ that controls the local symmetry of the transverse geometry of codimension one foliations does act on some big algebra constructed from the algebra of modular forms, and named it the "modular Hecke algebra", which we will briefly recall now.

We first define

$$\mathcal{M}(\Gamma(N)) := \Sigma^{\oplus}\mathcal{M}_{2k}(\Gamma(N))\,,\ \mathcal{M}^0(\Gamma(N)) := \Sigma^{\oplus}\mathcal{M}^0_{2k}(\Gamma(N)),$$

These algebras form a projective system with respect to the divisibility of the integer N. Define

$$\mathcal{M} := \varinjlim_{N\to\infty} \mathcal{M}(\Gamma(N))\,, \quad \text{resp.} \quad \mathcal{M}^0 := \varinjlim_{N\to\infty} \mathcal{M}^0(\Gamma(N))\,.$$

An *operator Hecke form of level* Γ [13] is a map

$$\begin{aligned} F : \Gamma\backslash GL_2^+(\mathbb{Q}) &\to \mathcal{M}\,, \\ \Gamma_\alpha &\mapsto F_\alpha \in \mathcal{M}\,, \end{aligned}$$

which has a finite support, and satisfies the *covariance condition*:

$$F_{\alpha\gamma}(z) = F_\alpha|\gamma(z), \quad \forall \alpha \in GL_2^+(\mathbb{Q}), \gamma \in \Gamma, z \in \mathbb{H}\,.$$

The modular Hecke algebra $\mathcal{A}(\Gamma)$ is an associative algebra consisting of operator Hecke forms of level Γ with the product

$$(F^1 * F^2)_\alpha := \sum_{\beta\in\Gamma\backslash GL_2^+(\mathbb{Q})} F^1_\beta \cdot F^2_{\alpha\beta^{-1}}\Big|\beta.$$

An important discovery of Connes and Moscovici in [13] is that the Hopf algebra $\mathcal{H}_1$ acts on $\mathcal{A}(\Gamma)$.

Before we give the details of this action, we briefly recall the general definition of a Hopf algebra action on an algebra. Let H be a Hopf algebra and M be an algebra. We say that $\alpha : H \otimes M \to M$ defines an action of H on M if the following two conditions hold;

1. M is an H-module with respect to the algebra structure on H;
2. the following property holds with respect to the coalgebra structure on H;

$$\alpha(h, a_1a_2) = m\big((\alpha\otimes\alpha)(\Delta(h), a_1\otimes a_2)\big),$$

   where a_1, a_2 are elements of M, Δ is the coproduct of H, and $m : M\otimes M \to M$ is the multiplication operator.

With this definition, one can easily check that Connes-Moscovici's Hopf algebra $\mathcal{H}_1$ acts on the extended smooth foliation algebra $\mathcal{A}_\Gamma$. In the following, we review the action of $\mathcal{H}_1$ on the modular Hecke algebra $\mathcal{A}(\Gamma)$ as was introduced by Connes-Moscovici [13].

For an $f \in \mathcal{M}_{2k}$, we define

$$Xf = \frac{1}{2\pi i}\frac{\partial}{\partial z} - \frac{1}{2\pi i}\frac{\partial}{\partial z}(\log\eta^4)\cdot kf\,,\ Yf = k\cdot f.$$

For $\gamma = \begin{pmatrix} a & b \\ c & d \end{pmatrix} \in GL^+(2,\mathbb{Q})$, we define

$$\mu_\gamma(z) = \frac{1}{2\,\pi^2}\left(G_2^*|\gamma(z) - G_2^*(z) + \frac{2\pi i\,c}{cz+d}\right),$$

$$G_2^*(z) = 2\zeta(z) + 2\sum_{m\geq 1}\sum_{n\in\mathbb{Z}}\frac{1}{(mz+n)^2} = \frac{\pi^2}{3} - 8\pi^2\sum_{m,n\geq 1} me^{2\pi imnz}.$$

We notice that here $\mu_\gamma \equiv 0$ if $\gamma \in SL_2(\mathbb{Z})$.

We define the action of X, Y, δ_n on $\mathcal{A}(\Gamma)$ as follows, for $F \in \mathcal{A}(\Gamma)\,, \alpha \in G^+(\mathbb{Q})$,

$$\begin{array}{rcl} X(F)_\alpha & := & X(F_\alpha)\,, \\ Y(F)_\alpha & := & Y(F_\alpha)\,, \\ \delta_n(F)_\alpha & := & \mu_{n,\alpha}\cdot F_\alpha\,, \text{ where } \mu_{n,\,\alpha} := X^{n-1}(\mu_\alpha)\,, \quad n\in\mathbb{N}. \end{array}$$

With the above preparation, it is not difficult to check the following theorem.

Theorem 3.1 *([13]) Let Γ be a congruence subgroup of $SL_2(\mathbb{Z})$.*

1. *The Hopf algebra $\mathcal{H}_1$ acts on the algebra $\mathcal{A}(\Gamma)$.*
2. *The Schwarzian derivative $\delta_2' = \delta_2 - \frac{1}{2}\delta_1^2$ is inner and is implemented by $\omega_4 = -\frac{1}{72}E_4 \in \mathcal{A}(\Gamma)$, where*

$$E_4 = 1 + 240\sum_{n=1}^{\infty} n^3 \frac{e^{2\pi i n z}}{1-e^{2\pi i n z}}$$

is the normalized Eisenstein series of degree 4.

In fact, Connes and Moscovici [13] pointed out that $\mathcal{A}(\Gamma)$ can actually be obtained as the crossed product of the algebra of modular forms by the Hecke ring, and then reduced by a projection which is determined by the congruence subgroup Γ. Based on this fact, and assuming the associativity of the Eholzer product, Connes and Moscovici [14] subsequently proved that this associative formal deformation can be canonically extended from the algebra of modular forms to the modular Hecke algebra associated to a congruence subgroup.

Remark 3.2 Inspired by Theorem 3.1 (2), Connes and Moscovici introduced a concept of projectivity of an $\mathcal{H}_1$ action on an algebra $\mathcal{A}$ as follows: there is an element $\Omega \in \mathcal{A}$ such that,

1. $$\delta_2'(a) := \left(\delta_2 - \frac{1}{2}\delta_1^2\right)(a) = \Omega\, a - a\,\Omega\,,\ \forall a \in \mathcal{A}; \qquad (3.12)$$
2. due to the commutativity of the δ_k's,

$$\delta_k(\Omega) = 0\,,\ \forall k \in \mathbb{N}. \qquad (3.13)$$

We point out that this $\mathcal{H}_1$ projectivity structure is a generalization of the projective structure on elliptic curves. On an elliptic curve, a projective structure means a choice of atlas such that the transition function is between different charts can be chosen to be in $SL_2(\mathbb{R})$. The Ramanujan differential appears as a connection associated to such a projective structure.

A crucial observation of Connes and Moscovici [14] is that the extended Rankin-Cohen brackets on the modular Hecke algebra $\mathcal{A}(\Gamma)$ can be represented using elements in $\mathcal{H}_1$ and the element $\omega_4 \in \mathcal{A}(\Gamma)$. For example, the first Rankin-Cohen bracket can be realized by

$$RC_1(a\otimes b) = m((X\otimes Y - Y\otimes X - \delta_1 Y\otimes Y)(a\otimes b)),$$

where $m : \mathcal{A}(\Gamma)\otimes\mathcal{A}(\Gamma) \to \mathcal{A}(\Gamma)$ is the multiplication on $\mathcal{A}(\Gamma)$. Generalizing the above formula of RC_1, we define

$$A_{n+1} \;:=\; S(X)\,A_n - n\,\Omega^o\left(Y - \frac{n-1}{2}\right)A_{n-1},$$

$$B_{n+1} \quad := \quad X\,B_n - n\,\Omega\left(Y - \frac{n-1}{2}\right)B_{n-1},$$

where $A_{-1} := 0$, $A_0 := 1$ and $B_0 := 1$, $B_1 := X$, and Ω^o is the multiplication operator from the right by Ω. We notice that this is very similar to the sequences Eq. (2.3) of modular forms that Zagier used to reformulate the classical Rankin-Cohen brackets with $\Omega = \omega_4$. In general, the n-th Rankin-Cohen bracket can be written as

$$RC_n(a,b) := \sum_{k=0}^{n} \frac{A_k}{k!}\,(2Y+k)_{n-k}(a)\,\frac{B_{n-k}}{(n-k)!}\,(2Y+n-k)_k(b).$$

We remark that the above RC_n is a generalization of the classical n-th Rankin-Cohen bracket in the sense that when we take the modular Hecke algebra and we restrict the n-th Rankin-Cohen bracket on modular forms, we get the classical n-th Rankin-Cohen bracket. The advantage of this generalization is that now we can apply RC_n to an arbitrary algebra A on which $\mathcal{H}_1$ acts with a projective structure.

By a technique called *full injectivity*, Connes and Moscovici [14] proved their main theorem, which states that the associative product can ultimately be lifted to a universal deformation formula for projective actions of the Hopf algebra $\mathcal{H}_1$:

Theorem 3.3 *([14]) The functor $RC_* := \sum RC_n$ applied to any algebra $\mathcal{A}$ endowed with a projective structure yields a family of formal associative deformations of $\mathcal{A}$, whose products are given by*

$$f \star g = \sum RC_n(f,g)\hbar^n.$$

The full injectivity method used in the proof of the above theorem essentially says that there are enough different actions of $\mathcal{H}_1$ with projective structures so that any cocycle properties which lead to an associative deformation (on the algebras on which $\mathcal{H}_1$ acts) can always be lifted to the Hopf algebra level.

4 Rankin-Cohen deformation via Fedosov

In the previous section, we have seen a beautiful result of Connes and Moscovici extending structures in modular form theory to study the Hopf algebra $\mathcal{H}_1$. In this section, we look at the universal deformation formula obtained in Theorem 3.3 from the viewpoint of transverse geometry of codimension one foliation.

In geometry, deformation of the algebra of smooth functions on a manifold has been studied for a long time. In particular, it is not difficult to see that the first order limit of an associative deformation of the commutative algebra of smooth functions on a manifold defines a Poisson bracket on the manifold. Here, by a Poisson bracket on a manifold P we mean a bilinear map $\{\ ,\ \} : C^\infty(P) \otimes C^\infty(P) \to C^\infty(P)$ such that for any $f, g, h \in C^\infty(P)$,

1. $\{f,g\} = -\{g,f\}$,
2. $\{f,gh\} = g\{f,h\} + \{f,g\}h$,
3. $\{\{f,g\},h\} + \{\{h,f\},g\} + \{\{g,h\},f\} = 0$.

When P has a symplectic structure, namely, a non-degenerate closed 2-form ω on P, then the inverse of ω defines a Poisson bracket on $C^\infty(P)$ by $\{f,g\} := \omega^{-1}(df,dg)$. ($\omega$ is viewed as a skew-symmetric bilinear form on TM, and its inverse defines a skew-symmetric bilinear form on T^*M.) In mathematical physics, the phase space of a physical system is usually described by a symplectic manifold, and observables of a physical system are smooth functions on the symplectic manifold. Bayer-Flato-Fronsdal-Lichnerowicz-Sternheimer [1] pointed out that we can use deformations of the algebra of smooth functions on a symplectic (Poisson) manifold, the phase space of a physical system, to study the corresponding quantum system. They call a deformation of the algebra $C^\infty(P)$ with the first order term equal to the Poisson structure a *deformation quantization* of the symplectic (Poisson) manifold. An easy and beautiful example of such a theory is that in quantizing a free particle on $\mathbb{R}$, we have the algebra of quantum observables are generated by the position operator $\hat{q}$ (the multiplication operator by the function q) and the momentum operator $i\hbar\frac{d}{dx}$ on $L^2(\mathbb{R})$. Such an algebra is a deformation of the algebra of smooth functions on $\mathbb{R}^2$ (the corresponding phase space) with the standard symplectic structure $dp\wedge dq$. The product of this deformation quantization of $C^\infty(\mathbb{R}^2)$ can be written as

$$f\star g(x)=\exp\left(-\frac{i\hbar}{2}\omega^{ij}\frac{\partial}{\partial y^i}\frac{\partial}{\partial z^j}\right)f(y)g(z)|_{x=y=z},\qquad f,g\in C^\infty(\mathbb{R}^2),\tag{4.14}$$

which is called the Moyal product [22]. With this geometry and physics in mind, we ask ourselves whether there is a geometric (maybe physical) interpretation of the Rankin-Cohen deformation.

We start by looking for a better understanding of a simplified version of the Rankin-Cohen deformation obtained in Theorem 3.3. We notice that if we set all δ_i ($i=1,\ldots,\infty$) and Ω (the projective structure) to be zero in the Rankin-Cohen bracket RC_n, we obtain a Universal Deformation Formula (UDF) of the 2-dimensional solvable Lie algebra h_1 associated to the $ax+b$ group, i.e., $h_1=\operatorname{span}\{X,Y\}$ with $[Y,X]=X$. We call the simplified deformation the *reduced Rankin-Cohen product*:

$$RC_{red}=\sum_{n=0}^{\infty}\frac{\hbar^n}{n!}\sum_{k=0}^{n}\left[(-1)^k\binom{n}{k}X^k(2Y+k)_{n-k}\otimes X^{n-k}(2Y+n-k)_k\right],\tag{4.15}$$

with $Y_k=Y(Y+1)\cdots(Y+k-1)$.

The $ax+b$ group acts on the upper half-plane by translation, $(x,y)\mapsto(ax+b,ay)$. This induces an action of h_1 on $C^\infty(\mathbb{R}\times\mathbb{R}^+)$. Now applying the reduced Rankin-Cohen bracket (4.15) on $C^\infty(\mathbb{R}\times\mathbb{R}^+)$, we obtain a deformation of the algebra $C^\infty(\mathbb{R}\times\mathbb{R}^+)$. By using an argument via the orbit method (more precisely by a theorem of Gutt [19]), we together with Bieliavsky [2] proved that the reduced Rankin-Cohen deformation on $C^\infty(\mathbb{R}\times\mathbb{R}^+)$ is isomorphic to the Moyal product (4.14) over the half-plane.

In the mid 90's, Giaquinto and Zhang proposed another UDF for h_1 (cf. [20]):

$$F=\sum_{n=0}^{\infty}\frac{\hbar^n}{n!}\sum_{r=0}^{n}(-1)^r\binom{n}{r}X^{n-r}Y_r\otimes X^rY_{n-r}.$$

With the same idea, we [2] were able to prove that the Giaquinto-Zhang UDF applied to $C^\infty(\mathbb{R}\times\mathbb{R}^+)$ is also isomorphic to the Moyal product as well. We remark that the action of the universal enveloping algebra of h_1 on $C^\infty(\mathbb{R}\times\mathbb{R}^+)$ is fully injective. Therefore, the associativity of the Moyal product can be used to prove that the realization of the reduced Rankin-Cohen deformation on $C^\infty(\mathbb{R}\times\mathbb{R}^+)$ is associative. By the full injectivity property, we can give a new proof that the reduced Rankin-Cohen deformation and also the Giaquinto-Zhang deformation are associative.

The above study provides a connection between the reduced Rankin-Cohen brackets and the well-known Moyal product. This also inspired us to pursue further whether we can give a geometric reconstruction of Connes-Moscovici's Rankin-Cohen deformation of the Hopf algebra $\mathcal{H}_1$.

When asking ourselves this question, we realized that there is already a potential direction towards the answer. Recall that $\mathcal{H}_1$ was introduced in the study of the transverse geometry of codimension one foliation. In particular, $\mathcal{H}_1$ acts on the extended smooth foliation algebra $\mathcal{A}_\Gamma$. In noncommutative geometry, $\mathcal{A}_\Gamma$ is viewed as the algebra of smooth functions on the quotient space F^+X/Γ, which is usually a non-Hausdorff topological space. A key observation here is that the volume form $dx \wedge dy/y^2$ defines a symplectic form on F^+X, which is invariant under the action of Γ. This leads us to the consideration of deformation quantization of F^+X with the symmetry of the pseudogroup Γ.

As a first step, we need to understand the first order term of the Rankin-Cohen deformation. As we have mentioned, when we consider the deformation of $C^\infty(P)$, the first order term is a Poisson structure. Now, for a noncommutative algebra $\mathcal{A}_\Gamma$ what is the first order term of a deformation? The answer to this question is that, it is a noncommutative Poisson structure.

A noncommutative Poisson structure on an algebra A is a degree 2 Hochschild cocycle Π such that the Gerstenhaber bracket $[\Pi, \Pi]_G$ is a coboundary.

We are able to prove the following result by a long but direct computation.

Theorem 4.1 *([2]) Let A be an algebra equipped with an $\mathcal{H}_1$ action. Then $\Pi(a,b) = X(a)Y(b) - Y(a)X(b) + \delta_1(Y(a))Y(b) = m(RC_1(a \otimes b))$ defines a noncommutative Poisson structure on A.*

In the second step, we aim to understand the geometric meaning of a projective structure. As we mentioned, classically a projective atlas on a Riemann surface X assign a principal $SL_2(\mathbb{R})$ bundle on X. In the case of a codimension one foliation, we obtained the following theorem giving a geometric interpretation of a projective structure on $\mathcal{A}_\Gamma$ analogous to this classical picture.

Theorem 4.2 *([2]) In the standard action of $\mathcal{H}_1$ on a smooth foliation algebra $\mathcal{A}_\Gamma$, the projective structure is equivalent to the existence of an invariant connection on F^+X of the form*

$$\nabla_{\partial_x}\partial_x = \mu(x,y)\partial_y, \qquad \nabla_{\partial_x}\partial_y = \frac{1}{2y}\partial_x,$$
$$\nabla_{\partial_y}\partial_x = \frac{1}{2y}\partial_x, \qquad \nabla_{\partial_y}\partial_y = -\frac{1}{2y}\partial_y.$$

We point out that the connection introduced in Theorem 4.2 is a symplectic torsion-free connection on F^+X. This reminds us of Fedosov's theory [18] of deformation quantization of a symplectic manifold.

We briefly explain Fedosov's construction of a deformation quantization of the manifold (F^+X, ω). We consider a Weyl algebra bundle W whose fiber at every point p of F^+X is the Weyl algebra W_p consisting of formal power series

$$a(u,\hbar) = \sum_{k,|\alpha|\geq 0} \hbar^k a_{k,\alpha} u^\alpha.$$

Here $\hbar$ is the formal parameter, $y = (u^1, u^2) \in T_x F^+X$, and $\alpha = (\alpha_1, \alpha_2) \in \mathbb{Z}_{\geq 0} \times \mathbb{Z}_{\geq 0}$, and $y^\alpha = (u^1)^{\alpha_1}(u^2)^{\alpha_2}$.

The product on W_p is the Moyal product, for $a, b \in W_p$,

$$\begin{aligned} a \circ b(u) &= \exp\left(-\frac{i\hbar}{2}\omega^{ij}\frac{\partial}{\partial v^i}\frac{\partial}{\partial w^j}\right) a(v,\hbar)b(w,\hbar)|_{v=w=u} \\ &= \sum_{k=0}^{\infty}\left(-\frac{i\hbar}{2}\right)^k \frac{1}{k!}\omega^{i_1 j_1}\cdots\omega^{i_k j_k}\frac{\partial^k a}{\partial u^{i_1}\cdots\partial u^{i_k}}\frac{\partial^k b}{\partial u^{j_1}\cdots\partial u^{j_k}}. \end{aligned}$$

Let $\bigwedge$ be the bundle $\wedge^\bullet T^* F^+ X$. An abelian connection $D : \Gamma^\infty(W \otimes \bigwedge) \to \Gamma^\infty(W \otimes \bigwedge)$ is a connection on W with $D^2 a = 0$, for any $a \in \Gamma^\infty(W \otimes \bigwedge)$. When considering $W_D := \{a, Da = 0\}$, Fedosov proved that

Theorem. ([18]) *For all $a_0 \in C^\infty(F^+X)[[\hbar]]$, there exists a unique section $a \in W_D$, noted denoted by $\sigma^{-1}(a_0)$, such that $\sigma(a) := a(x, 0, \hbar) = a_0$. Hence, σ is a bijection between W_D and $C^\infty(F^+X)[[\hbar]]$. And we can define on $C^\infty(F^+X)[[\hbar]]$ an associative product*

$$a \star b = \sigma(\sigma^{-1}(a) \circ \sigma^{-1}(b)).$$

In the case that F^+X has a Γ-invariant symplectic connection, the second named author proved [28] that the above deformation quantization of $C_c^\infty(F^+X)$ defines a deformation of the corresponding crossed product algebra $\mathcal{A}_\Gamma = C_c^\infty(F^+X) \rtimes \Gamma$.

When we are given the Γ-invariant symplectic torsion-free connection ∇ (Thm. 4.2) on F^+X, if we assume furthermore that $\mu(x, y)$ is of the form $y\nu(x)$ with $\nu(x)$ an arbitrary smooth function of variable x, the connection is actually flat. This allows to find an explicit formula for the abelian connection D. The equation $Da = 0$ gives us a system of differential equations. And starting from $a_{0,0} = f$ and solving the system by induction, we obtain two sequences of elements:

$$A_{m+1} = -XA_m - m\frac{\mu}{y^3}\left(Y - \frac{m-1}{2}\right)A_{m-1}, \tag{4.16}$$

$$B_{m+1} = XB_m - m\frac{\mu}{y^3}\left(Y - \frac{m-1}{2}\right)B_{m-1}. \tag{4.17}$$

where $X = \dfrac{1}{y}\dfrac{\partial}{\partial x}$, $Y = -y\dfrac{\partial}{\partial y}$. Extending the above formulas onto the extended smooth foliation algebra $\mathcal{A}_\Gamma$, we obtained the same recurrence relation that Connes and Moscovici used in their definition of generalized Rankin-Cohen brackets, so one can consider what we get as another realization of the Rankin-Cohen deformation. The advantage of this argument is that one gets the associativity without any extra effort. We show [2] that in this context we can also prove the "full injectivity", which allows us to conclude the associativity of the Rankin-Cohen deformation at the Hopf algebra level.

We notice that in Theorem 4.1, the first Rankin-Cohen bracket always defines a Poisson structure no matter whether there is an $\mathcal{H}_1$ projective structure or not. This inspires us to ask whether $\mathcal{H}_1$ has a Universal Deformation Formula without assuming the existence of a projective structure. In the language of deformation quantization, this question is whether there is a deformation of the algebra $\mathcal{A}_\Gamma$ without the existence of an invariant symplectic connection. We learned from Fedosov [18] and also Gorokhovsky-Bressler-Nest-Tsygan [1] that the existence of an invariant symplectic torsion-free connection is not a necessary condition for a smooth foliation algebra to have a deformation quantization. Actually, we can always construct a deformation quantization of an extended smooth foliation algebra $\mathcal{A}_\Gamma$ using the idea of algebroid stacks. With this in mind, we proved the following theorem, where we can drop the assumption of the existence of a projective structure.

Theorem 4.3 *([29]) The Hopf algebra $\mathcal{H}_1$ has a universal deformation formula, i.e. there is an element $R \in \mathcal{H}_1[[\hbar]] \otimes_{\mathbb{C}[[\hbar]]} \mathcal{H}_1[[\hbar]]$ satisfying*

$$((\Delta \otimes 1)R)(R \otimes 1) = ((1 \otimes \Delta)R)(1 \otimes R), \tag{4.18}$$

$$(\epsilon \otimes 1)(R) = 1 \otimes 1 = (1 \otimes \epsilon)(R). \tag{4.19}$$

As a side remark, we also obtained [29] a proof of the associativity of the Eholzer product of a reasonable length using some elementary methods.

5 Rankin-Cohen via Representations

Rankin-Cohen brackets and related deformation questions can also be studied using the theory of infinite dimensional representations of $SL_2(\mathbb{R})$. First of all, one can explicitly give an interpretation of these brackets using unitary representation theory of $SL_2(\mathbb{R})$. This fact has been known by experts for a long time[4] but does not seem to be clearly written down anywhere. The main result is the following theorem [32]:

Theorem. *Let $f \in \mathcal{M}_{2k}, g \in \mathcal{M}_{2l}$ be two modular forms. Let $\pi_f \cong \pi_{\deg f}, \pi_g \cong \pi_{\deg g}$ be associate representations which are discrete series of the group $SL_2(\mathbb{R})$. The tensor product of these two representations is decomposed into a direct sum of discrete series,*

$$\pi_f \otimes \pi_g = \bigoplus_{n=0} \pi_{\deg f + \deg g + 2n}.$$

The Rankin-Cohen bracket $[f,g]_n$ gives (up to a scalar) the minimal K-weight vectors in the representation space of the component $\pi_{\deg f + \deg g + 2n}$.

We remark that the result is about the tensor product between $SL_2(\mathbb{R})$ representations are well-known (c.f. [27] by Repka). The new part in the above theorem is the relation between Rankin-Cohen brackets and representation theory.

The representation π_f is constructed in the following way: let $f \in \mathcal{M}_{2k}(\Gamma)$ be a modular form; one can associate to it a function on $\Gamma \backslash SL_2(\mathbb{R})$ by using the following mapping:

$$(\sigma_{2k} f)(\gamma) = f_{|_k}|\gamma(i) = (ci+d)^{-2k} f\left(\frac{ai+b}{ci+d}\right), \quad \text{for } \gamma = \begin{pmatrix} a & b \\ c & d \end{pmatrix} \in SL_2(\mathbb{R}).$$

This function belongs to

$$C^\infty(\Gamma \backslash SL_2(\mathbb{R}), 2k) = \{F \in C^\infty(\Gamma \backslash SL_2(\mathbb{R})), F(\gamma \cdot r_\theta) = \exp(i2k\theta) F(\gamma)\},$$

where

$$r_\theta = \begin{pmatrix} \cos(\theta) & -\sin(\theta) \\ \sin(\theta) & \cos(\theta) \end{pmatrix}.$$

[4]For example, Deligne wrote in 1973 (cf. [15]): Remarque 2.1.4. L'espace $F(G, GL_2(\mathbb{Z}))$ ci-dessus est stable par produit. D'autre part, $D_{k-1} \otimes D_{l-1}$ contient les $D_{k+l+2m-1} (m \geq 0)$. Pour $m = 0$, ceci correspond au fait que le produit fg d'une forme modulaire holomorphe de poids k par une de poids l, en est une de poids $k+l$. Pour $m = 1$, en coordonnées (1.1.5.2), on trouve que $l\frac{\partial f}{\partial z} \cdot g - kf \cdot \frac{\partial g}{\partial z}$ est modulaire holomorphe de poids $k+l+2$, et ainsi de suite. De même dans le cadre adélique.

By taking into account the natural action of $SL_2(\mathbb{R})$ on $C^\infty(\Gamma\backslash SL_2(\mathbb{R}))$:

$$(\pi(h)F)(g) = F(gh),$$

one gets a representation of $SL_2(\mathbb{R})$ and also a representation of the complexified Lie algebra $\mathfrak{sl}_2(\mathbb{C})$ by taking the smallest invariant subspace which contains the orbit of $\sigma_{2k}f$. One can show that this representation is a discrete series representation of weight $2k$. In this step, we take all the vectors in a basis of the representation space to a subspace of $C^\infty(\mathbb{H})$, by using the inverse of $\sigma_{2(k+n)}, n \geq 0$.

With this interpretation, we studied the deformed products in a more general setting. From now on we drop the holomorphicity condition on a modular form, and consider all functions which satisfy the modularity condition with respect to Γ. We denote the algebra under consideration by $\widetilde{\mathcal{M}}(\Gamma)$. Enlarging $\widetilde{\mathcal{M}}(\Gamma)$, we consider the tensor algebra

$$\widetilde{\mathcal{M}}(\Gamma)^{\otimes} = \sum_n \widetilde{\mathcal{M}}(\Gamma)^{\otimes n},$$

and define

$$\begin{aligned} \mathrm{M} : \widetilde{\mathcal{M}}(\Gamma)^{\otimes} &\to \widetilde{\mathcal{M}}(\Gamma), \\ f_1 \otimes f_2 \otimes \cdots \otimes f_n &\mapsto f_1 f_2 \cdots f_n, \end{aligned} \tag{5.20}$$

and we extend the degree operator by

$$\deg(f_1 \otimes f_2 \otimes \cdots \otimes f_n) = \sum_{i=1}^n \deg f_i.$$

We define a derivation on $\widetilde{\mathcal{M}}(\Gamma)$, for an element $f \in \widetilde{\mathcal{M}}(\Gamma)$ of degree $2k$,

$$\tilde{X} f = \frac{1}{2\pi i}\frac{df}{dz} - \frac{2kf}{4\pi Im(z)}. \tag{5.21}$$

We extend $\tilde{X}$ to $\widetilde{\mathcal{M}}(\Gamma)^{\otimes}$ by the Leibniz rule

$$\tilde{X}(f_1 \otimes f_2 \otimes \cdots \otimes f_n) = \sum_{i=1}^n f_1 \otimes f_2 \otimes \cdots \otimes \tilde{X} f_i \otimes \cdots \otimes f_n.$$

Inspired by the Rankin-Cohen brackets (2.2), we study the following two families of formal products. The first family is $\star : \widetilde{\mathcal{M}}(\Gamma)^{\otimes}[[\hbar]] \otimes \widetilde{\mathcal{M}}(\Gamma)^{\otimes}[[\hbar]] \to \widetilde{\mathcal{M}}(\Gamma)^{\otimes}[[\hbar]]$ defined by linear extension and the following formula:

$$\begin{aligned} f \star g &= \sum \frac{A_n(\deg f, \deg g)}{(\deg f)_n (\deg g)_n} \\ &\times \left(\sum_{r=0}^n (-1)^r \tilde{X}^r \binom{\deg f + n - 1}{n - r} f \otimes \tilde{X}^{n-r} \binom{\deg g + n - 1}{r} g \right) \hbar^n, \end{aligned} \tag{5.22}$$

where $f, g \in \widetilde{\mathcal{M}}(\Gamma)^{\otimes}$. The second family of products $*$ is the restriction of $\star$ defined by Eq. (5.22) to $\widetilde{\mathcal{M}}(\Gamma) \subset \widetilde{\mathcal{M}}(\Gamma)^{\otimes}$ composed with the application M defined in (5.20). More concretely, $* : \widetilde{\mathcal{M}}(\Gamma)[[\hbar]] \times \widetilde{\mathcal{M}}(\Gamma)[[\hbar]] \to \widetilde{\mathcal{M}}(\Gamma)[[\hbar]]$ is defined as follows:

$$\begin{aligned} f * g &= \mathrm{M}(f \star g) \\ &= \sum \frac{A_n(\deg f, \deg g)}{(\deg f)_n (\deg g)_n} \left(\sum_{r=0}^{n} (-1)^r \tilde{X}^r \binom{2k+n-1}{n-r} f \tilde{X}^{n-r} \binom{2l+n-1}{r} g \right) \hbar^n \\ &= \sum \frac{A_n(\deg f, \deg g)}{(\deg f)_n (\deg g)_n} [f, g]_n \hbar^n, \end{aligned} \tag{5.23}$$

where $f, g \in \widetilde{\mathcal{M}}$, and the notation $(\alpha)_n := \alpha(\alpha+1)\cdots(\alpha+n-1)$. We will always make the (natural) assumption $A_0 = 1$ and $A_1(x,y) = xy$, and A_i $(i \geq 2)$ is a polynomial of two variables.

Our main goal is to study the associativity of the above two family of products. We name the associativity of the product $\star$ *strong associativity*, and the associativity of $*$ *weak associativity*.

The third named author proved [32] that, while it is obvious that a family of coefficient functions $A_n(\deg f, \deg g)$ which defines a strongly associative product also defines a weakly associative product, the converse, which is not clear *a priori*, is also true. This equivalence between the two types of associativity is used to show the following results.

Theorem 5.1 *([32]) Cohen-Manin-Zagier [6] have found all associative formal deformations* $* : \widetilde{\mathcal{M}}[[\hbar]] \times \widetilde{\mathcal{M}}[[\hbar]] \to \widetilde{\mathcal{M}}[[\hbar]]$ *defined by linearity and the formula*

$$f * g \quad = \quad \sum \frac{A_n(\deg f, \deg g)}{(\deg f)_n (\deg g)_n} [f, g]_n \hbar^n, \tag{5.24}$$

where $\widetilde{\mathcal{M}}(\Gamma)$ *is the space of functions which satisfy the modularity condition, and we use the notation* $(\alpha)_n := \alpha(\alpha+1)\cdots(\alpha+n-1)$. *One again assumes* $A_0 = 1$ *and* $A_1(x,y) = xy$.

The main reason for this claim to be valid is that the determinant of some 2×2 linear system is non-zero, and the drop of holomorphicity allows us to have the total freedom to modify a function in the interior of a fundamental domain. Furthermore, with the help of some computations of certain multivariable polynomials done by Mathematica, the third named author was able to prove the following proposition.

Proposition 5.2 *([32]) Let* Γ *be a congruence subgroup of* $SL_2(\mathbb{Z})$ *such that* $\mathcal{M}(\Gamma)$ *admits the unique factorization property (for example* $SL_2(\mathbb{Z})$ *itself), let* $F_1, F_2, G_1, G_2 \in \mathcal{M}(\Gamma)$ *such that*

$$RC(F_1, G_1) = RC(F_2, G_2),$$

as formal series in $\mathcal{M}(\Gamma)[[\hbar]]$*; then there exists a constant* C *such that*

$$F_1 = CF_2, G_2 = CG_1.$$

6 Hankel forms and a new $\mathcal{H}_1$ action

In this section, we aim to set up a connection between the Hopf algebra $\mathcal{H}_1$ and the theory of higher order Hankel forms introduced by Janson and Peetre[5] [21]. We hope that such a connection will inspire more interactions among transverse geometry, number theory, and harmonic analysis. Similar to what we did in Sec. 5, we will consider a set of functions more general than modular forms. Instead

[5]In the literature, there is a huge amount of study of Hankel forms and transvectants. Due to our limited knowledge, we have only cited those references related to our work.

of dropping the holomorphicity property of a modular form, in this section we will drop the modularity property. For example, we will consider holomorphic functions on the unit disk of $\mathbb{C}$ that are square integrable with respect to some measure.

Let D be the unit disk in the complex plane $\mathbb{C}$. Let $\eta(z) = 1 - |z|^2$, and $d\sigma(z) = (1/\pi)(1-|z|)^{-2}dxdy$. The weighted Bergman space $A^{2\alpha}(D)$ $(\alpha > 0)$ is defined to be

$$\{f : \bar{\partial} f = 0, \int_D |f(z)|^2 \eta^\alpha d\sigma(z) < \infty\}.$$

The group $SU(1,1) \cong SL_2(\mathbb{R})$ acts on the weighted Bergman space $A^{2\alpha}(D)$ as follows, for $\gamma = \begin{pmatrix} a & b \\ \bar{b} & \bar{a} \end{pmatrix}$ with $|a|^2 - |b|^2 = 1$,

$$\pi_\alpha(\gamma)(f)(z) = f\left(\frac{az - \bar{b}}{-bz + \bar{a}}\right)(-bz + \bar{a})^{-\alpha}. \tag{6.25}$$

It is easy to check that the above action π_k defines a unitary representation of $SL_2(\mathbb{R})$, which is actually irreducible. Such a representation is called a discrete series representation of $SL_2(\mathbb{R})$. Applying the result of Repka [27], a tensor product of representations in the discrete series is a direct sum of irreducible $SL_2(\mathbb{R})$ representations in the discrete series. More explicitly, as $SL_2(\mathbb{R})$ representations, we have the following decomposition:

$$\pi_k \otimes \pi_l = \bigoplus_{i \geq 0} \pi_{k+l+2i}.$$

We denote the projection from $A^{2k}(D) \otimes A^{2l}(D)$ to $A^{2(k+l+2i)}$ by Π_i, which is an $SL_2(\mathbb{R})$ equivariant bilinear map from $A^{2k}(D) \times A^{2l}(D)$ to $A^{2(k+l+2i)}$. Such an equivariant bilinear form is called by Janson and Peetre [21] a *Hankel form of weight* $2i$. Similar to the Rankin-Cohen brackets, P_{2i} can be expressed using the derivatives and the weights of the two components, i.e. up to a constant,

$$\Pi_{2i}(f,g)(z) = \sum_{s=0}^{i} (-1)^{i-s} \begin{pmatrix} i \\ s \end{pmatrix} \frac{1}{(w(f))_s (w(g))_{i-s}} \partial^s f \partial^{i-s} g,$$

where $w(f)$ and $w(g)$ are the weights of f and g.

We remark that the above discussion of Hankel forms of higher weights is closely related to the theory of transvectant. Olver-Sanders [23], Choi-Mourrain-Solé [4], and El Gradechi [17] provided an explicit relation between transvectants, Rankin-Cohen brackets, and the Moyal product [23]. One can even use these interesting relations to give a proof of the associativity of the Rankin-Cohen deformation, which was explained by Pevzner [24]. Recently, higher-order Hankel operators were discussed by Pittman-Polletta [25]. We refer readers to the cited references for more details.

In the remainder of this section, we aim to introduce an algebra $\mathcal{B}_\Gamma$ associated to a pseudogroup Γ of holomorphic transformations on a complex domain Σ of dimension one. And we will show that there is a natural action of $\mathcal{H}_1$ on $\mathcal{B}_\Gamma$. In the special case when Γ is trivial and Σ is D, the unit disk in $\mathbb{C}$, the Rankin-Cohen deformation gives rise to Hankel forms of higher weights.

Let Σ be a complex domain of dimension 1 (maybe with several components). Γ is a pseudogroup acting on Σ by holomorphic transformations. Let $T^{1,0}\Sigma$ be the holomorphic tangent bundle of Σ, which is equipped with a Γ action

$$\gamma(x,y)=(\gamma(x),\partial_x\gamma(x)y),$$

where (x,y) is a point on σ. We assume that the holomorphic tangent bundle is trivialized. Namely, we have chosen global coordinates (x,y) on $T^{1,0}\Sigma$.

Remark 6.1 We have required the bundle $T^{1,0}\Sigma$ to be trivialized in order to work with holomorphic functions and holomorphic vector fields on Σ. If $T^{1,0}\Sigma$ is not a trivial bundle over Σ, then we will have to work with sheaves of holomorphic functions and holomorphic vector fields. Accordingly, we will have to generalize the notion of Hopf algebra to Hopf algebroid like in [12].

We consider the space of holomorphic functions on $T^{1,0}\Sigma$, and denote it by $A_{T^{1,0}\Sigma}$. The action of Γ on $T^{1,0}\Sigma$ induces an action of Γ on $A_{T^{1,0}\Sigma}$. Therefore, we define $\mathcal{B}_\Gamma$ to be the crossed product algebra $A_{T^{1,0}\Sigma}\rtimes\Gamma$.

We remark that when Σ is the unit disk in $\mathbb{C}$, the space of holomorphic functions on $T^{1,0}D$ has a natural decomposition,

$$A_{T^{1,0}D}=\bigoplus_{i\geq 0}A_D y^i,$$

where y is the coordinate along the fiber direction of $T^{1,0}D$, and A_D is the space of holomorphic functions on D. For $i\geq 1$, $SU(1,1)$, the group of holomorphic automorphisms of D, acts on each component $A_D y^i$ exactly like the $SU(1,1)$ action on the weighted Bergman space $A^{4i}(D)$. Actually both A_D and $A^{4i}(D)$ share a dense subalgebra, the algebra of polynomials in the variable x.

To introduce the Hopf algebra $\mathcal{H}_1$ action on B_Γ, we first introduce two holomorphic vector fields on $T^{1,0}\Sigma$. Define

$$X=y\partial_x,\qquad Y=y\partial_y.$$

It is easy to check that for an element $\gamma\in\Gamma$,

$$U_\gamma XU_{\gamma^{-1}}=X-y\frac{\partial_x^2\gamma^{-1}}{\partial_x\gamma^{-1}}Y,\qquad U_\gamma YU_{\gamma^{-1}}=Y,$$

where U_γ is the action of γ on $A_{T^{1,0}\Sigma}$.

The vector fields X and Y lift to act on the algebra $\mathcal{B}_\Gamma$ by acting on the component of $A_{T^{1,0}\Sigma}$. We define $\delta_1:\mathcal{B}_\Gamma\to\mathcal{B}_\Gamma$ by

$$\delta_1(fU_\gamma)=y\partial_x\big(\log(\partial_x\gamma^{-1})\big)fU_\gamma.$$

It is not difficult to check that the operators X, Y, δ_1 satisfy

$$[Y,X]=X,\quad [Y,\delta_1]=\delta_1,$$

and $[X,\delta_1]=\delta_2$, where $\delta_2:\mathcal{B}_\Gamma\to\mathcal{B}_\Gamma$ is the operator defined by

$$\delta_2(fU_\gamma)=y^2\partial_x^2\big(\log(\partial_x(\gamma^{-1}))\big).$$

Continuing this procedure, we have $\delta_n:\mathcal{B}_\Gamma\to\mathcal{B}_\Gamma$ by

$$\delta_n(fU_\gamma)=X\big(\delta_{n-1}(fU_\gamma)\big).$$

Proposition 6.2 *Connes-Moscovici's Hopf algebra $\mathcal{H}_1$ acts naturally on the algebra $\mathcal{B}_\Gamma$.*

Proof The proof that $\mathcal{H}_1$ acts on $\mathcal{B}_\Gamma$ is a repetition of Connes-Moscovici's proof that $\mathcal{H}_1$ acts on the extended smooth foliation algebra $\mathcal{A}_\Gamma$. Here we are replacing smooth functions by holomorphic functions and differentiations by holomorphic differentiations. But the Hopf algebraic structure involved does not change at all. □

As an application of the Rankin-Cohen deformation, assuming Γ is trivial, we apply the reduced Rankin-Cohen deformation (4.15) to the algebra $A_{T^{1,0}D}$; we have for $f(x)y^k, g(x)y^l$,

$$RC_{red}(f(x)y^k, g(x)y^l) = \sum_{n=0}^{\infty} \frac{\hbar^n}{n!} \cdot$$
$$\sum_{s=0}^{n}(-1)^s \left[\begin{pmatrix} n \\ s \end{pmatrix} \frac{(2k+n-1)!}{(2k+s-1)!}\partial_x^s(f)(x)\frac{(2l+n-1)!}{(2l+n-s-1)!}\partial_x^{n-s}(g)(x)\right] y^{k+l+n}.$$

When we take the component of $\hbar^n$, we find

$$R_n(fy^k, gy^l) = (-1)^n \frac{(2k+n-1)!(2l+n-1)}{((2k-1)!(2l-1)!)} \cdot$$
$$\sum_{s=0}^{n}(-1)^{n-s} \left[\begin{pmatrix} n \\ s \end{pmatrix} \frac{1}{(2k)_s(2l)_{n-s}}\partial_x^s(f)(x)\partial_x^{n-s}(g)(x)\right] y^{k+l+n}.$$

Recalling the relation between $A_{T^{1,0}D}$ and the weighted Bergman space $A^{4n}(D)$, we can extend the action of h_1 onto $\oplus_{n\geq 1}A^{4n}(D)$ and conclude with the following proposition.

Proposition 6.3 *The reduced i-th Rankin-Cohen bracket on $\oplus_{n\geq 1}A^{4n}(D)$ is a Hankel form of weight $2i$. The associativity of the reduced Rankin-Cohen deformation implies that this family of Hankel forms defines an associative deformation of $\oplus_{n\geq 1}A^{4n}(D)$.*

We remark that in [30] (and also [16]), Unterberger-Unterberger introduced an operator symbol calculus associated to weighted Bergman spaces. And they are also able to construct an associative deformation on $\oplus_{n\geq 1}A^{4n}(D)$. It is an interesting question to compare our deformation in Proposition 6.3 and the deformation obtained in [30], while our coefficients are much simpler than the ones in [30].

References

[1] Bayen, F., Flato, M., Fronsdal, C., Lichnerowicz, A., and Sternheimer, D., Deformation theory and quantization. I. Deformations of symplectic structures, *Ann. Physics*, **111** (1978), no. 1, 61–110.

[2] Bieliavsky, P., Tang, X., and Yao, Y., Rankin-Cohen brackets and formal deformations, *Adv. Math.* **212** (2007), no. 1, 293–314.

[3] Bressler, P., Gorokhovsky, A., Nest, R., and Tsygan, B., Deformation quantization of gerbes, *Adv. Math.*, **214** (2007), no. 1, 230–266.

[4] Choie, Y., Mourrain, B., and Solé, P., Rankin-Cohen brackets and invariant theory, *J. Algebraic Combin.*, **13**, (2001), no. 1, 5–13.

[5] Cohen, H., Sums involving the values at negative integers of L-functions of quadratic characters, *Math. Ann.*, **217** (1975), 271-285.

[6] Cohen, P., Manin, Y., and Zagier, D., Automorphic pseudodifferential operators, *Algebraic aspects of integrable systems*, 17–47, Progr. Nonlinear Differential Equations Appl., **26**, Birkaüser Boston, Boston, MA, 1997.

[7] Connes, A., Noncommutative differential geometry, *Inst. Hautes Etudes Sci. Publ. Math.*, No. **62** (1985), 257–360.

[8] Connes, A., Cyclic cohomology and the transverse fundamental class of a foliation. In *Geometric methods in operator algebras*, pp. 52–144, *Pitman Res. Notes in Math.* **123**, Longman, Harlow, 1986.

[9] Connes, A., *Noncommutative geometry*, Academic Press, Inc., San Diego, CA, 1994.

[10] Connes, A. and Kreimer, D., Hopf algebras, renormalization and noncommutative geometry, *Comm. Math. Phys.*, **199** (1998), no. 1, 203–242.

[11] Connes, A. and Moscovici, H., Hopf algebras, cyclic cohomology and the transverse index theorem, *Commun. Math. Phys.*, **198** (1998), 199-246.

[12] Connes, A., Moscovici, H., Differentiable cyclic cohomology and Hopf algebraic structures in transverse geometry, *Essays on geometry and related topics*, Vol. **1, 2**, 217–255, Monogr. Enseign. Math., 38, Enseignement Math., Geneva, 2001.

[13] Connes, A., Moscovici, H., Modular Hecke algebras and their Hopf symmetry, *Mosc. Math. J.*, **4** (2004), no. 1, 67–109, 310.

[14] Connes, A., and Moscovici, H., Rankin-Cohen brackets and the Hopf algebra of transverse geometry, *Mosc. Math. J.* **4** (2004), no. 1, 111–130, 311.

[15] Deligne, P., Formes modulaires et representations de GL(2),(French) *Modular functions of one variable, II,* (Proc. Internat. Summer School, Univ. Antwerp, Antwerp, 1972), pp. 55–105., Lecture Notes in Math., Vol. **349**, Springer, Berlin, 1973.

[16] van Dijk, G., Pevzner, M., Ring structures for holomorphic discrete series and Rankin-Cohen brackets, *J. Lie Theory*, 1 **7** (2007), no. 2, 283–305.

[17] El Gradechi, A., The Lie theory of the Rankin-Cohen brackets and allied bi-differential operators, *Adv. Math.* **207** (2006), no. 2, 484–531.

[18] Fedosov, B., *Deformation quantization and index theory*, Mathematical Topics, 9. Akademie Verlag, Berlin, 1996. 3

[19] Gutt, S., *Déformations formelles de l'algèbre des fonctions différentiables sur une variété sympletique*, thesis, Universite Libre de Bruxelles, 1983.

[20] Giaquinto, A., and Zhang, J., Bialgebra actions, twists, and universal deformation formulas, *J. Pure Appl. Algebra*, **128** (1998), no. 2, 133–151.

[21] Janson, S., Peetre, J., A new generalization of Hankel operators (the case of higher weights), *Math. Nachr.*, **132** (1987), 313–328.

[22] Moyal, J., Quantum mechanics as a statistical theory, *Proc. Cambridge Philos. Soc.*, **45**, (1949). 99–124.

[23] Olver, P., Sanders, J., Transvectants, modular forms, and the Heisenberg algebra, *Adv. in Appl. Math.*, **25** (2000), no. 3, 252–283.

[24] Pevzner, M., Rankin-Cohen brackets and associativity, *Lett. Math. Phys.* **85** (2008), no. 2-3, 195–202.

[25] Pittman-Polletta, B., A geometric interpretation and explicit form for higher-order Hankel operators, *arXiv:0901.2953*, (2009).

[26] Rankin, R. A., The construction of automorphic forms from the derivatives of a given form. *J. Indian Math. Soc.*, (N.S.) **20** (1956), 103–116.

[27] Repka, J., Tensor products of unitary representations of $SL_2(R)$, *Amer. J. Math.* **100** (1978), no. 4, 747–774.

[28] Tang, X., Deformation quantization of pseudo-symplectic (Poisson) groupoids, *Geom. Funct. Anal.*, **16** (2006), no. 3, 731–766.

[29] Tang, X., and Yao, Y., A universal deformation formula for $\mathcal{H}_1$ without projectivity assumption, *J. Noncommut. Geom.*, **3** (2009), no. 2, 151–179.

[30] Unterberger, A., and Unterberger, J., Algebras of symbols and modular forms, *J. Anal. Math.*, **68** (1996), 121–143.

[31] Yao, Y., Autour des Déformations de Rankin-Cohen, thesis, Ecole Polytechnique, (2007).

[32] Yao, Y., Rankin-Cohen deformations and representation theory, *arXiv:0708.1528*, (2007).

[33] Zagier, D., Modular forms and differential operators. K. G. Ramanathan memorial issue, *Proc. Indian Acad. Sci. Math. Sci.* **104** (1994), no. 1, 57–75.

[34] Zagier, D., *Formes modulaires et Opérateurs différentiels*, Cours 2001-2002 au Collège de France.

[35] Zagier, D., Some combinatorial identities occurring in the theory of modular forms, in preparation.

Fields Institute Communications
Volume **61**, 2011

Pseudo-Differential Operators and Regularity of Spectral Triples

Otgonbayar Uuye
Department of Mathematical Sciences
University of Copenhagen
Universitetsparken 5
DK-2100 Copenhagen E, Denmark
otogo@math.ku.dk
http://www.math.ku.dk/~otogo

Abstract. We introduce the notion of an algebra of generalized pseudo-differential operators and prove that a spectral triple is regular if and only if it admits an algebra of generalized pseudo-differential operators. We also provide a self-contained proof of the fact that the product of regular spectral triples is regular.

Introduction

In their work on the local index theorem [CM95], Connes and Moscovici introduced the notion of a regular spectral triple. Later, Higson showed that a spectral triple is regular if and only if a certain algebra constructed from the spectral triple is what he calls an algebra of generalized differential operators [Hig04, Hig06]. See Theorem 2.4 for the precise statement.

Motivated by their work we introduce a notion of an algebra of generalized pseudo-differential operators; it clarifies the role of the regularity condition and serves as a convenient framework to study index theory via complex powers. I should note that most of the main ideas in this paper can be traced back to [CM95, Appendix B] or [Hig06, Subsection 4.5], but for the convenience of the reader we tried to remain self-contained.

Now we describe the contents of the paper. After some preliminaries in Section 1, we recall Higson's notion of an algebra of generalized differential operators in Section 2. In Section 3, we develop the notion of an algebra of generalized pseudo-differential operators. As usual, there are two versions – operators of all orders and operators of order at most zero. Extending Higson's result, we show that a spectral triple is regular if and only if it admits an algebra of generalized pseudo-differential operators (see Theorem 4.1).

2010 *Mathematics Subject Classification.* Primary 58J42; Secondary 58B34.
Key words and phrases. spectral triple, pseudo-differential operators, regularity.

Finally, we show that the product of regular spectral triples is again regular (see Theorem 4.4). This is a folklore result and it is implicitly contained in or follows from results in [CM95, Hig04, Hig06]. But no direct reference seems to exist in the literature.

Acknowledgments This paper is essentially contained in my thesis. I would like to thank my advisor Nigel Higson for his constant encouragement and guidance. I would also like to thank the NCG group at the University of Copenhagen for support.

1 Preliminaries

1.1 Spectral triples. The following definition of a regular spectral triple is due to Connes and Moscovici [CM95].

Definition 1.1 A *spectral triple* $(\mathcal{A}, \mathcal{H}, \mathscr{D})$ consists of an associative algebra $\mathcal{A}$, a graded Hilbert space $\mathcal{H}$ equipped with an even representation of $\mathcal{A}$ and a densely defined, self-adjoint, odd operator $\mathscr{D}$ such that

1. $\mathcal{A}$ is contained in the domain of the derivation $[\mathscr{D}, -]$, that is, if $\mathrm{dom}(\mathscr{D})$ is the domain of $\mathscr{D}$, then any $a \in \mathcal{A}$ satisfies $a \cdot \mathrm{dom}(\mathscr{D}) \subseteq \mathrm{dom}(\mathscr{D})$ and the commutator $[\mathscr{D}, a] : \mathrm{dom}(\mathscr{D}) \to \mathcal{H}$ extends by continuity to a bounded operator on $\mathcal{H}$ and
2. the operators $a \cdot (\mathscr{D} \pm i)^{-1}$ and $(\mathscr{D} \pm i)^{-1} \cdot a$ are compact for any $a \in \mathcal{A}$.

We say that a spectral triple $(\mathcal{A}, \mathcal{H}, \mathscr{D})$ is *regular* if

3. the space $\mathcal{A} + [\mathscr{D}, \mathcal{A}]$ is contained in the *smooth domain* of the derivation $\delta(T) := [|\mathscr{D}|, T]$ on $\mathcal{L}(\mathcal{H})$.

We remind the reader that the smooth domain of δ is defined as follows. The *domain* $\mathrm{dom}(\delta)$ of $\delta = [|\mathscr{D}|, -]$ is the set of $T \in \mathcal{L}(\mathcal{H})$ such that

(i) $T \cdot \mathrm{dom}(|\mathscr{D}|) \subseteq \mathrm{dom}(|\mathscr{D}|)$ and
(ii) $[|\mathscr{D}|, T]$ extends to a bounded operator on $\mathcal{H}$.

For $k \geq 2$, we define $\mathrm{dom}(\delta^k)$ inductively as

$$\mathrm{dom}(\delta^k) := \{b \in \mathrm{dom}(\delta) \mid \delta(b) \in \mathrm{dom}(\delta^{k-1})\}. \tag{1.1}$$

The smooth domain of δ is $\mathrm{dom}^\infty(\delta) := \bigcap_{k=1}^\infty \mathrm{dom}(\delta^k)$.

For simplicity, we do not consider any additional structure on $\mathcal{A}$; even though in most natural examples $\mathcal{A}$ will have a topology or a norm, it will not play any role in our analysis.

Obviously, the following is the basic example that should be mentioned in any paper on spectral triples.

Example 1.2 Let M be a complete Riemannian manifold of even dimension and let $S \to M$ be a complex spinor bundle. Let $\mathcal{H} := L^2(M, S)$ denote the graded Hilbert space of L^2-sections of S. Let $\mathscr{D}$ be a Dirac-type operator acting on $C_c^\infty(M, S)$. Then standard ΨDO theory implies that $\mathscr{D}$ with domain $C_c^\infty(M, S) \subset L^2(M, S)$ is essentially self-adjoint and

$$(C_c^\infty(M), L^2(M, S), \bar{\mathscr{D}}), \tag{1.2}$$

is a *regular* spectral triple, where $\bar{\mathscr{D}}$ is the closure of $\mathscr{D}$. See [HR00].

The material in the following two subsections is elementary and well-known. We include it for reference and the convenience of the reader.

1.2 Sobolev spaces. Let Δ be an *invertible, positive, self-adjoint operator* on a Hilbert space $\mathcal{H}$.

Definition 1.3 The Δ-*Sobolev space* of order $s \in \mathbb{R}$, denoted

$$W^s = W^s(\Delta) = W^s(\mathcal{H}, \Delta), \tag{1.3}$$

is the Hilbert space completion of $\operatorname{dom}(\Delta^{\frac{s}{2}})$ with respect to the inner product given by

$$\langle \xi, \eta \rangle_{W^s} := \langle \Delta^{\frac{s}{2}} \xi, \Delta^{\frac{s}{2}} \eta \rangle_{\mathcal{H}} \tag{1.4}$$

for ξ, $\eta \in \operatorname{dom}(\Delta^{\frac{s}{2}})$.

Note that the invertibility hypothesis guaranties that $|| \cdot ||_{W^s}$ is nondegenerate. The following is well-known.

Lemma 1.4 *For $s \geq t$, we have a continuous inclusion*

$$W^s \subseteq W^t. \tag{1.5}$$

Moreover, for $s \geq 0$, $\operatorname{dom}(\Delta^{\frac{s}{2}})$ is complete and thus $W^s = \operatorname{dom}(\Delta^{\frac{s}{2}})$. □

Example 1.5 If Δ is *bounded*, then nothing interesting happens: for any $s \in \mathbb{R}$, $W^s = \mathcal{H}$ and the inner products are different but equivalent.

Example 1.6 Suppose that Δ has compact resolvents. Then there exists a complete orthonormal basis $\{\xi_n\}$ of $\mathcal{H}$ consisting of eigenvectors: $\Delta \xi_n = \lambda_n \xi_n$ with $0 < \lambda_1 \leq \lambda_2 \leq \cdots \to \infty$. Hence the Sobolev spaces can be identified as the weighted l^2-spaces

$$W^s = \{(a_n)_{n=1}^{\infty} \mid \sum_n \lambda_n^s |a_n|^2 < \infty\}. \tag{1.6}$$

Example 1.7 (Classical Sobolev Spaces) Let M be a closed manifold and let Δ_0 be a *strictly positive* order-two elliptic partial differential operator acting on the smooth functions on M. Let $\Delta = \bar{\Delta}_0$ be its closure. Then Δ is invertible and self-adjoint and has compact resolvents (cf. [HR00]). It follows from the basic estimate that as a Hilbert space W^s is equivalent to the "standard" s-Sobolev space (cf. [Shu01, Proposition I.7.3]).

Definition 1.8 Let Δ be an invertible, positive, self-adjoint operator. The space of Δ-*smooth vectors* is

$$W^\infty := \bigcap_{s \in \mathbb{R}} W^s = \bigcap_{n=0}^{\infty} W^{2n} = \bigcap_{n=0}^{\infty} \operatorname{dom}(\Delta^n) \subseteq \mathcal{H}. \tag{1.7}$$

The space W^∞ contains $1_{[0,M]}(\Delta)\mathcal{H}$ for any $M > 0$, and thus W^∞ is dense in $\operatorname{dom}(\Delta^z)$ for any $z \in \mathbb{C}$. It follows that W^∞ is dense in W^s for any $s \in \mathbb{R}$.

Lemma 1.9 *Let $z \in \mathbb{C}$. For any $s \in \mathbb{R}$, the operator $\Delta^z_{|W^\infty} : W^\infty \to \mathcal{H}$ extends*[1] *to an* isometry

$$\Delta^z : W^{s+2\operatorname{Re}(z)} \to W^s. \tag{1.8}$$

In particular, $\Delta^z(W^\infty) = W^\infty$.

Proof First note that $\Delta^{i\operatorname{Im}(z)}$ is a unitary operator on $\mathcal{H}$. Therefore, for any $\xi \in W^\infty$,

$$||\Delta^z \xi||_{W^s} = ||\Delta^{\operatorname{Re}(z)+\frac{s}{2}} \xi||_{\mathcal{H}} = ||\xi||_{W^{s+2\operatorname{Re}(z)}}. \tag{1.9}$$

□

[1] Note that if $s > 0$ the range actually shrinks to $W^s \subset \mathcal{H}$.

Lemma 1.10 *The space $W^\infty \subset \mathcal{H}$ is a common core for the operators Δ^z, $z \in \mathbb{C}$, i.e. Δ^z is essentially self-adjoint on $W^\infty \subset \mathcal{H}$.*

Proof Any vector $\xi \in 1_{[0,M]}(\Delta)\mathcal{H} \subset W^\infty$ is an analytic vector for Δ^z. Indeed, for any $\xi \in 1_{[0,M]}(\Delta)\mathcal{H}$,

$$||(\Delta^z)^n\xi|| \leq M^{|\operatorname{Re}(z)n|}||\xi|| \tag{1.10}$$

and consequently

$$\sum_{n=0}^{\infty} \frac{||(\Delta^z)^n\xi||}{n!}t^n < \infty \tag{1.11}$$

for any $t > 0$. Since Δ^z preserves W^∞, all the analytic vectors for Δ^z are also analytic for the restriction $\Delta^z_{|W^\infty}$ and applying Nelson's theorem [RS75, Theorem X.39] to $\Delta^z_{|W^\infty}$, we see that Δ^z is essentially self-adjoint on W^∞. □

Another convenient way to express the complex powers Δ^z is using the Cauchy integral formula: for $\operatorname{Re}(z) < 0$,

$$\Delta^z = \frac{1}{2\pi i}\int \lambda^z(\lambda - \Delta)^{-1}d\lambda, \tag{1.12}$$

where the integral is a contour integral along a downwards pointing vertical line in $\mathbb{C}$ which separates 0 from $\operatorname{Spec}(\Delta)$. Let $\Delta \geq c > 0$ and let $s \in \mathbb{R}$. Then it follows from the spectral theorem that for any $\lambda \notin \operatorname{Spec}(\Delta)$, the resolvent $(\lambda-\Delta)^{-1}$ is a bounded operator on the Sobolev spaces W^s with norm at most $\left((\operatorname{Re}(\lambda) - c)^2 + \operatorname{Im}(\lambda)^2\right)^{-\frac{1}{2}}$. Hence, for any $\operatorname{Re}(z) < 0$, the integral converges to a bounded operator on W^s.

More generally, we have the following.

Lemma 1.11 *For $k \in \mathbb{N}$ and $\operatorname{Re}(z) < k$,*

$$\binom{z}{k}\Delta^{z-k} = \frac{1}{2\pi i}\int \lambda^z(\lambda - \Delta)^{-k-1}d\lambda \quad in \quad \mathcal{L}(W^s),\ s \in \mathbb{R}, \tag{1.13}$$

where the integral is a contour integral along a downwards pointing vertical line in $\mathbb{C}$ which separates 0 from $\operatorname{Spec}(\Delta)$. □

1.3 Operators of finite analytic order. We consider various class of linear operators on W^∞. The algebra of all linear operators $W^\infty \to W^\infty$ is denoted by $\operatorname{End}(W^\infty)$.

Example 1.12 If an (unbounded) operator P has domain $\operatorname{dom}(P) \supseteq W^\infty$ and preserves W^∞, *i.e.* $P(W^\infty) \subseteq W^\infty$, then the restriction $P_{|W^\infty}$ gives an element of $\operatorname{End}(W^\infty)$. We often write, simply, P for $P_{|W^\infty}$.

Definition 1.13 We say that a linear operator $W^\infty \to W^\infty$ has *analytic order at most* $t \in \mathbb{R}$ if it extends by continuity to a *bounded* linear operator $W^{s+t} \to W^s$ for every $s \in \mathbb{R}$. We write

$$\operatorname{Op}^t = \operatorname{Op}^t(\Delta) = \operatorname{Op}^t(\mathcal{H}, \Delta) \tag{1.14}$$

for the class of operators of analytic order at most t and define

$$\operatorname{Op} = \operatorname{Op}^\infty := \bigcup_t \operatorname{Op}^t \text{ and } \operatorname{Op}^{-\infty} := \bigcap_t \operatorname{Op}^t. \tag{1.15}$$

Lemma 1.14 *Operators with finite analytic order form a filtered algebra:*
(a) $\operatorname{Op}^s \subseteq \operatorname{Op}^t$ *for* $s \leq t$ *and*
(b) $\operatorname{Op}^s \cdot \operatorname{Op}^t \subseteq \operatorname{Op}^{s+t}$.

In particular, $\mathrm{Op}^0 \subset \mathrm{Op}$ *is a subalgebra and* $\mathrm{Op}^{-\infty} \subset \mathrm{Op}$ *and* $\mathrm{Op}^t \subset \mathrm{Op}^0$, $t \in [-\infty, 0)$ *are two-sided ideals.* □

Notice that operators with analytic order at most 0 extend, in particular, to *bounded* linear operators on $\mathcal{H} = W^0$, allowing us to identify Op^0 with a subalgebra of $\mathcal{L}(\mathcal{H})$, the algebra of bounded linear operators on $\mathcal{H}$.

Example 1.15 We see from Lemma 1.9 that the operator Δ^z belongs to $\mathrm{Op}^{2\,\mathrm{Re}(z)}$ for any $z \in \mathbb{C}$. It follows from the spectral theorem that for $z, w \in \mathbb{C}$

$$\Delta^z \cdot \Delta^w = \Delta^{z+w} \quad \text{in Op}. \tag{1.16}$$

Lemma 1.16 *If* Ψ *is a filtered subalgebra of* Op *such that* $\Delta^{\frac{t}{2}}$ *belongs to* Ψ^t *for all* $t \in \mathbb{R}$, *then* Ψ^0 *is unital and*

$$\Psi^t = \Delta^{\frac{t}{2}}\Psi^0 = \Psi^0\Delta^{\frac{t}{2}}. \tag{1.17}$$

Conversely, if a unital subalgebra $\Psi^0 \subseteq \mathrm{Op}^0$ *satisfies* $\Delta^{\frac{t}{2}}\Psi^0 = \Psi^0\Delta^{\frac{t}{2}}$ *for all* $t \in \mathbb{R}$, *then* $\Psi^t := \Delta^{\frac{t}{2}}\Psi^0$ *defines a filtered subalgebra of* Op *such that* $\Delta^{\frac{t}{2}}$ *belongs to* Ψ^t.

Proof For the first statement: since $\Delta^0 = 1$, Ψ^0 is unital. Moreover, for any $t \in \mathbb{R}$,

$$\Psi^t = \Delta^{\frac{t}{2}}\Delta^{-\frac{t}{2}}\Psi^t \subseteq \Delta^{\frac{t}{2}}\Psi^{-t}\Psi^t \subseteq \Delta^{\frac{t}{2}}\Psi^0 \subseteq \Psi^t\Psi^0 \subseteq \Psi^t. \tag{1.18}$$

Similarly for the other side.

For the second statement: since Ψ^0 is unital,

$$\Psi^t\Psi^s = \Delta^{\frac{t}{2}}\Psi^0\Delta^{\frac{s}{2}}\Psi^0 = \Delta^{\frac{t}{2}}\Delta^{\frac{s}{2}}\Psi^0\Psi^0 = \Psi^{t+s} \quad \text{and} \tag{1.19}$$

$$\Delta^{\frac{t}{2}} \text{ belongs to } \Psi^t. \tag{1.20}$$

□

Corollary 1.17 *For any* $t \in \mathbb{R}$,

$$\mathrm{Op}^t = \Delta^{\frac{t}{2}}\mathrm{Op}^0 = \mathrm{Op}^0\Delta^{\frac{t}{2}}. \tag{1.21}$$

□

Corollary 1.18 *Let* Δ *be an invertible positive self-adjoint operator. Let* $\mathcal{K}$, $\mathcal{L}^p$ *and* $\mathcal{L}^{(p,\infty)}$ *denote the ideals of compact, p-Schatten and p-Dixmier operators.*

(a) If $\Delta^{-\frac{1}{2}} \in \mathcal{K}$ *then* $\mathrm{Op}^{-t} \subseteq \mathcal{K}$ *for* $t > 0$.
(b) If $\Delta^{-\frac{1}{2}} \in \mathcal{L}^p$, $p \geq 1$ *then* $\mathrm{Op}^{-t} \subseteq \mathcal{L}^{p/t}$ *for* $0 < t \leq p$.
(c) If $\Delta^{-\frac{1}{2}} \in \mathcal{L}^{(p,\infty)}$, $p \geq 1$ *then* $\mathrm{Op}^{-t} \subseteq \mathcal{L}^{(p/t,\infty)}$ *for* $0 < t \leq p$.

Proof Using the fact that $\mathrm{Op}^{-t} \subseteq \Delta^{-\frac{t}{2}}\mathcal{L}$, these follow immediately from well-known properties of the respective ideals. □

2 Algebra of generalized differential operators

In this section, we recall Higson's notion of an algebra of generalized differential operators [Hig04, Hig06]. Let $\mathbb{N} = \mathbb{Z}_{\geq 0}$.

Definition 2.1 (Higson) An $\mathbb{N}$-*filtered* subalgebra $\mathcal{D} \subseteq \mathrm{Op}(\Delta)$ is called an *algebra of generalized differential operators* if $\mathcal{D}$ is closed under the derivation $[\Delta, -]$ and satisfies

$$[\Delta, \mathcal{D}^k] \subseteq \mathcal{D}^{k+1}, \quad k \in \mathbb{N}. \tag{2.1}$$

Example 2.2 Let M be a complete Riemannian manifold and let $S \to M$ be a Hermitian vector bundle. Let Δ be the closure of the scalar Laplacian $+1$ (to ensure invertibility). Then the algebra $\mathcal{D} = \mathcal{D}_c(M, S)$ of compactly supported differential operators acting on the sections of S is an example of an algebra of generalized differential operators. Note that in this example Δ is not an element of $\mathcal{D}$. See [HR00].

Example 2.3 (Polynomial Weyl Algebra) Consider the usual Lebesgue measure on $\mathbb{R}^n$ and let $\mathcal{H} := L^2(\mathbb{R}^n)$. The operator

$$1 + \sum_{i=1}^{n} \left(x_i^2 - \frac{\partial^2}{\partial x_i^2} \right) \tag{2.2}$$

with domain $C_c^\infty(\mathbb{R}^n)$ of compactly supported smooth functions is called the *harmonic oscillator*. It is essentially self-adjoint and strictly positive.

Let Δ denote the closure. Then Δ is an invertible positive self-adjoint operator with compact resolvent. Let $\mathcal{W} = \mathcal{W}_n$ be the algebra of *polynomial* differential operators on $\mathbb{R}^n$ acting on $L^2(\mathbb{R}^n)$. As an algebra, it is generated by $x_1, \ldots, x_n$ and $\frac{\partial}{\partial x_1}, \ldots, \frac{\partial}{\partial x_n}$. We filter $\mathcal{W}$ by requiring that

$$\operatorname{order}(x_i) := 1 \quad \text{and} \quad \operatorname{order}\left(\frac{\partial}{\partial x_i}\right) := 1, \quad 1 \leq i \leq n. \tag{2.3}$$

(The nonzero degree of x_i compensates for the noncompactness of $\mathbb{R}^n$.) Then $\mathcal{W}$ is an algebra of generalized differential operators and Δ is an element of $\mathcal{W}^2$. See [Shu01].

Now we relate generalized differential operators to regularity. Let $(\mathcal{A}, \mathcal{H}, \mathscr{D})$ be a spectral triple and let $\Delta = \mathscr{D}^2 + 1$. Then it is clear that $\mathscr{D} \in \operatorname{Op}^1(\Delta) \subset \operatorname{End}(W^\infty)$. Suppose that $\mathcal{A} \cdot W^\infty \subseteq W^\infty$. Define an $\mathbb{N}$-filtered algebra $\mathcal{D} \subseteq \operatorname{End}(W^\infty)$ inductively by

1. $\mathcal{D}^0 :=$ the subalgebra generated by $\mathcal{A} + [\mathscr{D}, \mathcal{A}] \subseteq \operatorname{End}(W^\infty)$ and
2. $\mathcal{D}^1 := \mathcal{D}^0 + [\Delta, \mathcal{D}^0] + \mathcal{D}^0[\Delta, \mathcal{D}^0] \subseteq \operatorname{End}(W^\infty)$ and
3. $\mathcal{D}^k := \mathcal{D}^{k-1} + \sum_{j=1}^{k-1} \mathcal{D}^j \cdot \mathcal{D}^{k-j} + [\Delta, \mathcal{D}^{k-1}] + \mathcal{D}^0[\Delta, \mathcal{D}^{k-1}] \subseteq \operatorname{End}(W^\infty)$ for $k \geq 2$.

Theorem 2.4 (Higson [Hig06, Theorem 4.26]) *Let $(\mathcal{A}, \mathcal{H}, \mathscr{D})$ be a spectral triple and let $\Delta = \mathscr{D}^2 + 1$. Suppose that $\mathcal{A} \cdot W^\infty \subseteq W^\infty$. Then $(\mathcal{A}, \mathcal{H}, \mathscr{D})$ is regular if and only if $\mathcal{D}^k \subseteq \operatorname{Op}^k$, $k \geq 0$.*

We will give a proof in Section 4, essentially rephrasing the arguments in [CM95, Hig06] in the language of algebra is of generalized pseudo-differential operators, which we describe next.

3 Algebra of generalized pseudo-differential operators

In this section, we define and study algebras of generalized pseudo-differential operators. Just as in the classical case, these provide a convenient framework to study index theoretic problems, see [Hig06].

3.1 Operators of all orders.

Definition 3.1 An $\mathbb{R}$-*filtered* subalgebra $\Psi \subseteq \mathrm{Op}(\Delta)$ is called an *algebra of generalized pseudo-differential operators* if Ψ satisfies for $z \in \mathbb{C}$, $t \in \mathbb{R}$,

$$\Delta^{\frac{z}{2}}\Psi^t \subseteq \Psi^{\mathrm{Re}(z)+t} \quad \text{and} \quad \Psi^t\Delta^{\frac{z}{2}} \subseteq \Psi^{\mathrm{Re}(z)+t} \tag{3.1}$$

and

$$[\Delta^{\frac{z}{2}}, \Psi^t] \subseteq \Psi^{\mathrm{Re}(z)+t-1}. \tag{3.2}$$

Remark 3.2 If $\Delta^{\frac{z}{2}}$ belongs to $\Psi^{\mathrm{Re}(z)}$ for $z \in \mathbb{C}$, then (3.1) is trivially satisfied and

$$\Psi^t = \Delta^{\frac{t}{2}}\Psi^0 = \Psi^0\Delta^{\frac{t}{2}}, \quad t \in \mathbb{R}, \tag{3.3}$$

by Lemma 1.16.

The following is the classical example.

Example 3.3 (Pseudo-differential Operators) Let M be a closed manifold. Let $\Psi = \bigcup \Psi^t$ denote the $\mathbb{R}$-filtered algebra of *pseudo-differential operators* on M with scalar symbols and let $\Psi_{\mathrm{cl}} \subset \Psi$ denote the filtered subalgebra of *classical pseudo-differential operators* on M (see for instance [Shu01]).

Let Δ be (the closure of) an invertible, positive, order two, partial differential operator on M. Then we have inclusions of $\mathbb{R}$-*filtered* algebras

$$\Psi_{\mathrm{cl}} \subset \Psi \subset \mathrm{Op}(\Delta). \tag{3.4}$$

Moreover $\Delta^{\frac{z}{2}}$ belongs to $\Psi_{\mathrm{cl}}^{\mathrm{Re}(z)} \subset \Psi^{\mathrm{Re}(z)}$, $z \in \mathbb{C}$, and since

$$[\Psi^s, \Psi^t] \subseteq \Psi^{s+t-1}, \quad s, t \in \mathbb{R}, \tag{3.5}$$

we have

$$[\Delta^{\frac{z}{2}}, \Psi_{\mathrm{cl}}^t] \subseteq \Psi_{\mathrm{cl}}^{\mathrm{Re}(z)+t-1} \quad \text{and} \quad [\Delta^{\frac{z}{2}}, \Psi^t] \subseteq \Psi^{\mathrm{Re}(z)+t-1}. \tag{3.6}$$

Hence Ψ_{cl} and Ψ are algebras of generalized pseudo-differential operators (in view of Remark 3.2).

More generally, we could consider pseudo-differential operators acting on sections of a vector bundle. Then we need to assume that Δ has scalar principal symbol: the property (3.5) would not hold in general, but (3.6) would still hold.

For a calculus of pseudo-differential and classical pseudo-differential operators on $\mathbb{R}^n$, see [Shu01, Chapter IV] or [Fed96, Chapter 3].

Remark 3.4 Let $\Psi \subseteq \mathrm{Op}$ be an algebra of generalized pseudo-differential operators. Then $\mathcal{D}^k := \Psi^k$, $k \in \mathbb{N}$, is clearly an algebra of generalized *differential* operators.

In the other direction, we have the following lemma.

Lemma 3.5 (cf. [Hig06, Proposition 4.31]) *Let $\mathcal{D}$ be an algebra of generalized differential operators. Let Ψ^t denote the space of linear combinations of operators $P \in \mathrm{Op}$ such that for any $l \in \mathbb{R}$, P may be decomposed as*

$$P = X\Delta^{\frac{z-m}{2}} + Q \tag{3.7}$$

with $\mathrm{Re}(z) \leq t$ and $X \in \mathcal{D}^m$, $m \in \mathbb{N}$ and $Q \in \mathrm{Op}^l$. Then Ψ is an algebra of generalized pseudo-differential operators.

First we prove an auxiliary lemma (cf. [CM95, Theorem B1], [Hig06, Lemma 4.20]).

Notation 3.6 In a filtered space, we write

$$P \sim \sum_{k \in \Lambda} P_k \tag{3.8}$$

if for any $l \in \mathbb{R}$, there exists a finite subset $F \subset \Lambda$ such that $P - \sum_{k \in F} P_k$ has order at most l.

Lemma 3.7 (cf. [CM95, Theorem B1], [Hig06, Lemma 4.20]) *Let $z \in \mathbb{C}$ and let $Y \in \mathcal{D}$. Let $Y^{(0)} := Y$ and $Y^{(k)} := [\Delta, Y^{(k-1)}]$, $k \geq 1$. Then*

$$\Delta^z Y \sim \sum_{k=0}^{\infty} \binom{z}{k} Y^{(k)} \Delta^{z-k}. \tag{3.9}$$

Note that $\mathrm{order}(Y^{(k)}) \leq \mathrm{order}(Y) + k$, so $\mathrm{order}(Y^{(k)}\Delta^{z-k}) \leq \mathrm{order}(Y) + 2\,\mathrm{Re}(z) - k$.

Proof Assume $\mathrm{Re}(z) < 0$. Let R denote the resolvent $(\lambda - \Delta)^{-1} \in \mathrm{Op}^{-2}$. Then it is easy to see that

$$RY = YR + RY^{(1)}R. \tag{3.10}$$

Hence, for any $n \in \mathbb{N}$,

$$RY = YR + Y^{(1)}R^2 + \cdots + Y^{(n)}R^{n+1} + RY^{(n+1)}R^{n+1}. \tag{3.11}$$

Applying the Cauchy integral formula (Lemma 1.11), we see that

$$\Delta^z Y = \sum_{k=0}^{n} \binom{z}{k} Y^{(k)} \Delta^{z-k} + \frac{1}{2\pi i} \int \lambda^z (\lambda - \Delta)^{-1} Y^{(n+1)} (\lambda - \Delta)^{-n-1} d\lambda. \tag{3.12}$$

But the last integral converges absolutely in $\mathcal{L}(W^{s+l}, W^s)$ for $l = \mathrm{order}(Y) + n + 1 - 2(n+1)$, and hence has order at most $\mathrm{order}(Y) - n - 1$.

For general $z \in \mathbb{C}$, the identity

$$\Delta^{z+1} Y = \Delta^z Y \Delta + \Delta^z Y^{(1)}$$

allows one to reduce to the case $\mathrm{Re}(z) < 0$. □

Proof of Lemma 3.5 Clearly Ψ is a filtered subspace of Op. It follows from Lemma 3.7 that Ψ is a subalgebra satisfying (3.1). For $P = X\Delta^{\frac{w-m}{2}} + Q$ in Ψ^t with $\mathrm{Re}(w) = t$ and $X \in \mathcal{D}^m$ and $\mathrm{order}(Q)$ small, the expansion

$$[\Delta^{\frac{z}{2}}, P] \sim \sum_{k=1}^{\infty} \binom{\frac{z}{2}}{k} X^{(k)} \Delta^{\frac{z+w-m-2k}{2}} + [\Delta^{\frac{z}{2}}, Q] \tag{3.13}$$

shows that $[\Delta^{\frac{z}{2}}, P] \in \Psi^{\mathrm{Re}(z)+t-1}$, proving (3.2). □

3.2 Operators of order at most zero.

Definition 3.8 A subalgebra $\mathcal{B} \subseteq \mathrm{Op}^0$ is called an *algebra of generalized pseudo-differential operators of order at most zero* if $\mathcal{B}$ is closed under the derivation $\delta := [\Delta^{\frac{1}{2}}, -]$.

Lemma 3.9 (cf. [CM95, Lemma B1]) *Let $\mathcal{B} = \mathrm{dom}^{\infty}(\delta)$ denote the smooth domain of δ. Then $\mathcal{B}$ is an algebra of generalized pseudo-differential operators of order at most zero.*

Proof We just need to show that $\mathcal{B}$ is a subset of Op^0, since by construction $\mathcal{B}$ is an algebra closed under δ.

Let $s = \mathrm{Re}(z) < 0$. Then for any $b \in \mathcal{B}$ and $n \in \mathbb{N}$,

$$\Delta^{\frac{z}{2}} b = \sum_{k=0}^{n} \binom{z}{k} \delta^k(b) \Delta^{\frac{z-k}{2}} \\ + \frac{1}{2\pi i} \int \lambda^z (\lambda - \Delta^{\frac{1}{2}})^{-1} \delta^{n+1}(b) (\lambda - \Delta^{\frac{1}{2}})^{-n-1} d\lambda, \tag{3.14}$$

in $\mathcal{L}(\mathcal{H})$, where the integral is a contour integral along a downwards pointing vertical line in $\mathbb{C}$ which separates 0 from $\mathrm{Spec}(\Delta^{\frac{1}{2}})$. The proof of Lemma 3.7 applied to $\Delta^{\frac{1}{2}}$ goes through verbatim.

The last integral gives a bounded operator in $\mathcal{L}(\mathcal{H}, W^{n+1})$, so taking n large enough we see that $||\Delta^{\frac{s}{2}} b\xi|| \leq C \cdot ||\Delta^{\frac{s}{2}} \xi||$, $\xi \in \mathrm{dom}(\Delta^{\frac{s}{2}})$, for some $C > 0$. Hence $b \cdot W^s \subseteq W^s$ for $s < 0$. For $s > 0$, the identity

$$\Delta^{\frac{s+1}{2}} b = \Delta^{\frac{s}{2}} b \Delta^{\frac{1}{2}} + \Delta^{\frac{s}{2}} \delta(b) \tag{3.15}$$

allows one to reduce to the case $s < 0$. Hence $\mathcal{B} \subseteq \mathrm{Op}^0$. □

Remark 3.10 It is clear that if $\Psi \subseteq \mathrm{Op}$ is an algebra of generalized pseudo-differential operators then $\mathcal{B} := \Psi^0$ is an algebra of generalized pseudo-differential operators of order at most zero.

Conversely, we have the following.

Lemma 3.11 (cf. [Hig06, Lemma 4.27]) *Let $\mathcal{B} \subseteq \mathrm{Op}^0$ be an algebra of generalized pseudo-differential operators of order at most zero. Then*

$$\mathcal{D}^k := \sum_{j=0}^{k} \Delta^{\frac{j}{2}} \mathcal{B} \subseteq \mathrm{Op}^k, \quad k \in \mathbb{N}, \tag{3.16}$$

is an algebra of generalized differential operators.

Proof First note that since

$$\Delta^{\frac{k}{2}} b = \sum_{j=0}^{k} \binom{k}{j} \delta^j(b) \Delta^{\frac{k-j}{2}}, \quad k \in \mathbb{N}, \tag{3.17}$$

we have

$$\Delta^{\frac{k}{2}} \mathcal{B} = \mathcal{B} \Delta^{\frac{k}{2}}, \quad k \in \mathbb{N}. \tag{3.18}$$

Hence $\mathcal{D}$ is an $\mathbb{N}$-filtered subalgebra of Op (see the proof of Lemma 1.16). Then using the facts

$$\delta(\Delta^{\frac{k}{2}} \mathcal{B}) = \Delta^{\frac{k}{2}} \delta(\mathcal{B}) \subseteq \Delta^{\frac{k}{2}} \mathcal{B} \quad \text{and} \tag{3.19}$$

$$[\Delta, P] = 2\Delta^{\frac{1}{2}} \delta(P) - \delta(\delta(P)), \tag{3.20}$$

we see that $\mathcal{D}$ is indeed an algebra of generalized differential operators. □

Remark 3.12 Suppose that $\mathcal{B}$ is an algebra of generalized pseudo-differential operators of order at most zero satisfying, for all $t \in \mathbb{R}$,

$$\Delta^{it} \in \mathcal{B} \quad \text{and} \quad \Delta^{\frac{t}{2}} \mathcal{B} \Delta^{-\frac{t}{2}} \subseteq \mathcal{B}. \tag{3.21}$$

Then one can show that $\Psi^t := \Delta^{\frac{t}{2}}\mathcal{B}$, $t \in \mathbb{R}$, is an algebra of generalized pseudo-differential operators, using Lemma 3.7. This is applicable to $\mathcal{B} = \mathrm{dom}^\infty(\delta)$ of Lemma 3.9.

4 Regularity of spectral triples

See [CM95, Appendix B], [Hig04, Theorem 3.25], [Hig06, Theorem 4.26].

Theorem 4.1 *Let $(\mathcal{A}, \mathcal{H}, \mathscr{D})$ be a spectral triple. Let $\Delta := \mathscr{D}^2 + 1$ and let $\delta := [\Delta^{\frac{1}{2}}, -]$. Then the following conditions are equivalent:*

(1) The spectral triple $(\mathcal{A}, \mathcal{H}, \mathscr{D})$ is regular, that is, $\mathcal{A} + [\mathscr{D}, \mathcal{A}]$ is contained in $\mathrm{dom}^\infty([|\mathscr{D}|, -])$.

(2) The set $\mathcal{A} + [\mathscr{D}, \mathcal{A}]$ is contained in $\mathrm{dom}^\infty(\delta)$.

(3) There exists an algebra of generalized pseudo-differential operators of order at most zero containing $\mathcal{A} + [\mathscr{D}, \mathcal{A}]$.

(4) There exists an algebra of generalized differential operators containing $\mathcal{A} + [\mathscr{D}, \mathcal{A}]$ in degree 0.

(5) There exists an algebra of generalized pseudo-differential operators containing $\mathcal{A} + [\mathscr{D}, \mathcal{A}]$ in degree 0.

Proof (1) $\Leftrightarrow$ (2) is clear, since $\Delta^{\frac{1}{2}}$ is a bounded perturbation of $|\mathscr{D}|$. The implication (2) $\Rightarrow$ (3) follows from Lemma 3.9.

To prove (3) $\Rightarrow$ (2), suppose that $\mathcal{B} \subseteq \mathrm{Op}^0$ is an algebra of generalized pseudo-differential operators of order at most zero. Then $\mathcal{B}$ is contained in $\mathrm{dom}^\infty(\delta)$. Indeed, if $b \in \mathcal{B} \subseteq \mathrm{Op}^0$, then $b \cdot \mathrm{dom}(\Delta^{\frac{1}{2}}) \subseteq \mathrm{dom}(\Delta^{\frac{1}{2}})$ and, being an operator of order 0, the commutator $[\Delta^{\frac{1}{2}}, b] \in \mathcal{B} \subseteq \mathrm{Op}^0$ extends to a bounded operator on $\mathcal{H}$. In other words, b belongs to $\mathrm{dom}(\delta)$. But $\mathcal{B}$ is closed under δ, therefore $\mathcal{B} \subseteq \mathrm{dom}^\infty(\delta)$.

The implication (3) $\Rightarrow$ (4) follows from Lemma 3.11, (4) $\Rightarrow$ (5) follows from Lemma 3.5 and (5) $\Rightarrow$ (3) follows from Remark 3.10. □

As a corollary we obtain a proof of Theorem 2.4.

Proof of Theorem 2.4 If $(\mathcal{A}, \mathcal{H}, \mathscr{D})$ is regular, then by Theorem 4.1, there exists an algebra of generalized differential operators $\mathcal{E} \subseteq \mathrm{Op}$. It is clear, by induction, that $\mathcal{D}^k \subseteq \mathcal{E}^k$, $k \in \mathbb{N}$. Thus $\mathcal{D}^k \subseteq \mathrm{Op}^k$. Conversely, if $\mathcal{D}^k \subseteq \mathrm{Op}^k$, $k \in \mathbb{N}$, then $\mathcal{D}$ is an algebra of generalized differential operators and by Theorem 4.1, $(\mathcal{A}, \mathcal{H}, \mathscr{D})$ is regular. □

Example 4.2 1. The commutative spectral triple of Example 1.2 is regular, because of Example 2.2 or Example 3.3.

2. The spectral triple of [CM95] associated to a triangular structure is regular because the algebra of $\Psi\mathrm{DO}'$-operators is an example of an algebra of generalized pseudo-differential operators.

Lemma 4.3 *Let $(\mathcal{A}_1, \mathcal{H}_1, \mathscr{D}_1)$ and $(\mathcal{A}_2, \mathcal{H}_2, \mathscr{D}_2)$ be spectral triples. Let $\mathcal{A}_1 \otimes_{\mathrm{alg}} \mathcal{A}_2$ denote the algebraic tensor product and let $\mathcal{H}_1 \widehat{\otimes} \mathcal{H}_2$ denote the graded Hilbert space tensor product. Then the operator*

$$\mathscr{D} := \mathscr{D}_1 \widehat{\otimes} 1 + 1 \widehat{\otimes} \mathscr{D}_2 \tag{4.1}$$

with domain

$$\mathrm{dom}(\mathscr{D}) := \mathrm{dom}(\mathscr{D}_1) \widehat{\otimes}_{\mathrm{alg}} \mathrm{dom}(\mathscr{D}_2) \subseteq \mathcal{H}, \tag{4.2}$$

is essentially self-adjoint and $(\mathcal{A}_1 \otimes \mathcal{A}_2, \mathcal{H}_1 \widehat{\otimes} \mathcal{H}_2, \bar{\mathscr{D}})$ *is a spectral triple.*

Proof See [Uuy09]. □

We write $\mathscr{D}_1 \times \mathscr{D}_2$ for the closure $\bar{\mathscr{D}}$ of $\mathscr{D}$ and call

$$(\mathcal{A}_1 \otimes \mathcal{A}_2, \mathcal{H}_1 \widehat{\otimes} \mathcal{H}_2, \mathscr{D}_1 \times \mathscr{D}_2). \tag{4.3}$$

the product of $(\mathcal{A}_1, \mathcal{H}_1, \mathscr{D}_1)$ and $(\mathcal{A}_2, \mathcal{H}_2, \mathscr{D}_2)$.

The following theorem is implicitly contained in or follows from results in [CM95, Hig04, Hig06]. But no direct reference seems to exist in the literature.

Theorem 4.4 *The product of regular spectral triples is again regular.*

Proof Let $(\mathcal{A}_1, \mathcal{H}_1, \mathscr{D}_1)$ and $(\mathcal{A}_2, \mathcal{H}_2, \mathscr{D}_2)$ be regular spectral triples and let $\mathcal{D}_i \subseteq \mathrm{Op}(\mathcal{H}_i, \mathscr{D}_i^2 + 1)$ be algebras of generalized differential operators for $(\mathcal{A}_i, \mathcal{H}_i, \mathscr{D}_i)$ respectively, i.e. $\mathcal{A}_i + [\mathscr{D}_i, \mathcal{A}_i] \subseteq \mathcal{D}_i^0$ for $i = 1, 2$.

We claim that $\mathcal{D}_1 \widehat{\otimes} \mathcal{D}_2$, with the product filtering, is an algebra of generalized differential operators for the product $(\mathcal{A}_1 \otimes \mathcal{A}_2, \mathcal{H}_1 \widehat{\otimes} \mathcal{H}_2, \mathscr{D}_1 \times \mathscr{D}_2)$. Then by Theorem 4.1, we see that the product is regular.

Let $\Delta_i = \mathscr{D}_i^2 + 1$ and $\Delta = (\mathscr{D}_1 \times \mathscr{D}_2)^2 + 1$. Then it is easy to check that

$$\mathrm{Op}^s(\mathcal{H}_1, \Delta_1) \widehat{\otimes} \mathrm{Op}^t(\mathcal{H}_2, \Delta_2) \subseteq \mathrm{Op}^{s+t}(\mathcal{H}_1 \widehat{\otimes} \mathcal{H}_2, \Delta). \tag{4.4}$$

Hence $\mathcal{D}_1 \widehat{\otimes} \mathcal{D}_2$ is a filtered subalgebra of $\mathrm{Op}(\mathcal{H}_1 \widehat{\otimes} \mathcal{H}_2, (\mathscr{D}_1 \times \mathscr{D}_2)^2 + 1)$. For $P_1 \in \mathcal{D}_1^k$, $P_2 \in \mathcal{D}_2^l$, we see that

$$[\Delta, P_1 \widehat{\otimes} P_2] = [\Delta_1, P_1] \widehat{\otimes} P_2 + P_1 \widehat{\otimes} [\Delta_2, P_2], \tag{4.5}$$

belongs to $(\mathcal{D}_1 \widehat{\otimes} \mathcal{D}_2)^{k+l-1}$ and thus $\mathcal{D}_1 \widehat{\otimes} \mathcal{D}_2$ is an algebra of generalized differential operators. Finally, for $a_1 \in \mathcal{A}_1$ and $a_2 \in \mathcal{A}_2$, clearly $a_1 \otimes a_2$ belongs to $(\mathcal{D}_1 \widehat{\otimes} \mathcal{D}_2)^0$ and so does

$$[\mathscr{D}_1 \times \mathscr{D}_2, a_1 \otimes a_2] = [\mathscr{D}_1, a_1] \widehat{\otimes} a_2 + a_1 \widehat{\otimes} [D_2, a_2]. \tag{4.6}$$

This completes the proof. □

References

[CM95] A. Connes and H. Moscovici, *The local index formula in noncommutative geometry*, Geom. Funct. Anal. **5** (1995), no. 2, 174–243. MR MR1334867 (96e:58149)

[Fed96] Boris Fedosov, *Deformation quantization and index theory*, Mathematical Topics, vol. 9, Akademie Verlag, Berlin, 1996. MR MR1376365 (97a:58179)

[Hig04] Nigel Higson, *The local index formula in noncommutative geometry*, Contemporary developments in algebraic K-theory, ICTP Lect. Notes, XV, Abdus Salam Int. Cent. Theoret. Phys., Trieste, 2004, pp. 443–536 (electronic). MR MR2175637

[Hig06] ———, *The residue index theorem of Connes and Moscovici*, Surveys in noncommutative geometry, Clay Math. Proc., vol. 6, Amer. Math. Soc., Providence, RI, 2006, pp. 71–126. MR MR2277669

[HR00] Nigel Higson and John Roe, *Analytic K-homology*, Oxford Mathematical Monographs, Oxford University Press, Oxford, 2000, Oxford Science Publications. MR MR1817560 (2002c:58036)

[RS75] Michael Reed and Barry Simon, *Methods of modern mathematical physics. II. Fourier analysis, self-adjointness analysis, self-adjointness*, Academic Press [Harcourt Brace Jovanovich Publishers], New York, 1975. MR MR0493420 (58 #12429b)

[Shu01] M. A. Shubin, *Pseudodifferential operators and spectral theory*, second ed., Springer-Verlag, Berlin, 2001, Translated from the 1978 Russian original by Stig I. Andersson. MR MR1852334 (2002d:47073)

[Uuy09] Otgonbayar Uuye, *Multiplicativity of the JLO-character*, to appear in the "Journal of Noncommutative Geometry" (2009).

Titles in This Series

61 **Masoud Khalkhali and Guoliang Yu, Editors,** Perspectives on noncommutative geometry, 2011

60 **Alina-Carmen Cojocaru, Kristin Lauter, Rachel Pries, and Renate Scheidler, Editors,** WIN—Women in numbers: Research directions in number theory, 2011

59 **Erhard Neher, Alistair Savage, and Weiqiang Wang, Editors,** Geometric representation theory and extended affine Lie algebras, 2011

58 **V. Kumar Murty, Editor,** Algebraic curves and cryptography, 2010

57 **Siv Sivaloganathan, Editor,** New perspectives in mathematical biology, 2010

56 **Rob de Jeu and James D. Lewis, Editors,** Motives and algebraic cycles: A celebration in honour of Spencer J. Bloch, 2009

55 **Panos M. Pardalos and Thomas F. Coleman, Editors,** Lectures on global optimization, 2009

54 **Noriko Yui, Helena Verrill, and Charles F. Doran, Editors,** Modular forms and string duality, 2008

53 **Mikhail Lyubich and Michael Yampolsky, Editors,** Holomorphic dynamics and renormalization: A volume in honour of John Milnor's 75th birthday, 2008

52 **Luigi Rodino, Bert-Wolfgang Schulze, and M. W. Wong, Editors,** Pseudo-differential operators: Partial differential equations and time-frequency analysis, 2007

51 **Giovanni Forni, Mikhail Lyubich, Charles Pugh, and Michael Shub, Editors,** Partial hyperbolic dynamics, laminations, and Teichmüller flow, 2007

50 **Ilia Binder and Dirk Kreimer, Editors,** Universality and renormalization, 2007

49 **Wayne Nagata and N. Sri Namachchivaya, Editors,** Bifurcation theory and spatio-temporal pattern formation, 2006

48 **Hermann Brunner, Xiao-Qiang Zhao, and Xingfu Zou, Editors,** Nonlinear dynamics and evolution equations, 2006

47 **Hans U. Boden, Ian Hambleton, Andrew J. Nicas, and B. Doug Park, Editors,** Geometry and topology of manifolds, 2005

46 **Thierry Passot, Catherine Sulem, and Pierre-Louis Sulem, Editors,** Topics in kinetic theory, 2005

45 **Ragnar-Olaf Buchweitz and Helmut Lenzing, Editors,** Representations of algebras and related topics, 2005

44 **Lajos Horváth and Barbara Szyszkowicz, Editors,** Asymptotic methods in stochastics, 2004

43 **George Janelidze, Bodo Pareigis, and Walter Tholen, Editors,** Galois theory, Hopf algebras, and semiabelian categories, 2004

42 **Saber Elaydi, Gerry Ladas, Jianhong Wu, and Xingfu Zou, Editors,** Difference and differential equations, 2004

41 **Alf van der Poorten and Andreas Stein, Editors,** High primes and misdemeanours: Lectures in honour of the 60th birthday of Hugh Cowie Williams, 2004

40 **Vlastimil Dlab and Claus Michael Ringel, Editors,** Representations of finite dimensional algebras and related topics in Lie theory and geometry, 2004

39 **Stephen Berman, Yuly Billig, Yi-Zhi Huang, and James Lepowsky, Editors,** Vector operator algebras in mathematics and physics, 2003

38 **Noriko Yui and James D. Lewis, Editors,** Calabi-Yau varieties and mirror symmetry, 2003

37 **Panos Pardalos and Henry Wolkowicz, Editors,** Novel approaches to hard discrete optimization, 2003

36 **Shigui Ruan, Gail S. K. Wolkowicz, and Jianhong Wu, Editors,** Dynamical systems and their applications in biology, 2003

35 **Yakov Eliashberg, Boris Khesin, and François Lalonde, Editors,** Symplectic and contact topology: Interactions and perspectives, 2003

34 **T. J. Lyons and T. S. Salisbury, Editors,** Numerical methods and stochastics, 2002

For a complete list of titles in this series, visit the
AMS Bookstore at **www.ams.org/bookstore/**.